Time, Tense, and Quantifiers

Proceeding of the Stuttgart Conference on the Logic of Tense and Quantification

Edited by Christian Rohrer

Max Niemeyer Verlag
Tübingen 1980

CIP-Kurztitelaufnahme der Deutschen Bibliothek

Time, tense, and quantifiers : proceedings of the Stuttgart Conference on the Logic of Tense and Quantification / ed. by Christian Rohrer. – Tübingen : Niemeyer, 1980.
(Linguistische Arbeiten ; 83)
ISBN 3-484-10374-4

NE: Rohrer, Christian [Hrsg.]; Conference on the Logic of Tense and Quantification ⟨1979, Stuttgart⟩

ISBN 3-484-10374-4 / ISSN 0344-6727

INHALTSVERZEICHNIS

Introduction

The present volume contains the proceedings of the March 21st-24^{1th}, 1979, colloquium on Formal Semantics: the two main topics were the analysis of temporal constructions and quantifiers in natural languages.

This colloquium was organized during the final phase of a research project on tense, aspect and verbclassification[1] at the Institute of Linguistics of the University of Stuttgart. Its purpose was to provide an overview of the research in progress and to prepare the ground for similar meetings.[2]

Although it is too early to draw any definite conclusions one can nevertheless discern certain general trends. Classical tense logic was developed in order to solve certain philosophical problems. It was not intended as a tool for describing tense forms of natural languages. Nevertheless tense logic is still the most precise and adequate means for the description of temporal phenomena expressed in natural languages. Nearly all contributions are based on classical tense logic. There are of course differences, e.g. evaluation at points or at intervals, multiple indexing, two-valued versus many-valued logics, sentence semantics versus discourse semantics, etc., but the framework is basically tense logic and its model-theoretic semantics. However, if one wants to account for the intricate interaction between verb meanings, temporal adverbs, and tense forms, one obtains extremely complicated reconstructions. This degree of complication seems to suggest that in the long run one might need a completely new approach. An indication of how such a new approach might look like can be found in several contributions.

During the last few years pragmatics has become more and more important in linguistics. At the colloquium it was argued quite convincingly that an adequate description of tense forms requires a pragmatic component. Unfortunately, at the conference, nobody presented a formal framework within which the pragmatics of tense forms could be accounted for. Since all the contributions to the colloquium appear in this volume, the reader can draw his own conclusions. I would like to thank the DFG and the University of Stuttgart for supporting the colloquium financially; I also extend my gratitude to the members of my research project for their help in organizing the meeting, to the participants for their papers and the stimulating discussion and above all to Mrs. Zettl for her invaluable administrative help and her excellent typing.

C.Rohrer, University of Stuttgart / Institut Linguistik, September 1979.

---

1 This project (Ro 245/8 and 10) is financed by the DFG.

2 The next meeting will be held in Bar-Ilan (Israel).

ADVERBS OF FREQUENCY

Lennart Åqvist, Jaap Hoepelman and Christian Rohrer* (Universität Stuttgart)

1. *Introductory remarks*

In this paper we propose to give a logical treatment of the following nine adverbs, at the very least:

Always	Very often	Very seldom
Never	Often	Seldom
Sometimes	Fairly often	Fairly seldom

More precisely, we wish to arrive at a plausible analysis of certain sentences involving these adverbs. By "analysis" we then have in mind both a *syntactic* analysis purporting to clarify the logical form of those sentences and a *semantical* analysis intended to account for their meaning. In sec. 3 below we summarize the technique of analysis which emerges from the treatment, given in sec. 2, of our first example sentence. Further examples are dealt with in sec. 4. Our choice of examples to be handled is much inspired by David Lewis's significant article "Adverbs of Quantification" /Lewis (1975) in the reference list below/. The formal theory resulting from our discussion is presented in full detail in the Appendix, secs. 6 - 9.

2. *A pattern of analysis*

Consider the sentence

(1) very often the fog lifts before noon here.

As for the meaning of (1) one might suggest à la Lewis that it asserts that the fog lifts before noon here *on very many days* or, perhaps rather, *on very many days when the fog lifts here* (at all).

---

* The present contribution reports research done under the auspices of the Deutsche Forschungsgemeinschaft (DFG) project "Die Beschreibung mit Hilfe der Zeitlogik von Zeitformen und Verbalperiphrasen im Französischen, Portugiesischen und Spanischen", led by Chr. Rohrer. We are grateful to our colleagues D.M.Gabbay, F.Guenthner and H.Kamp for stimulating discussions.

Assuming this idea to be reasonably viable, how are we then to cast it into an appropriate logical form? Well, trying first-order predicate calculus as our framework, we suggest to begin with that the formalization of (1) should somehow involve the following two *conditions* (sentential functions, *open* sentences containing one *free* variable):

(1a) x is a day & the fog lifts before noon here on x

(1b) x is a day & the fog lifts here on x

Call (1a) the *target* condition and (1b) the *reference* condition with respect to the example sentence (1). Again, from a semantical viewpoint, we claim that these conditions (1a) and (1b) determine or specify certain *sets*, viz. the set of days on which the fog lifts before noon here *and* the set of days on which the fog lifts here (at all), respectively. We then bring out the force of "very often" in (1) by taking (1) to assert that the *conditional probability* of the first set (determined by the target condition (1a)) *given* the second set (determined by the reference condition (1b)) is very high, say, ≥ 0.9. And by this conditional probability we mean, following Laplace, nothing but the *cardinality* of the intersection of the two sets *divided by* the *cardinality* of the second set (which we assume to exceed 0 in order to avoid division by 0). Or, if you prefer, the *ratio* of the number of "favorable" cases in the first set *to* that of all the "admissible" or "possible" cases in the second set.

So far, so good. Several technical aspects of the analysis just indicated remain to be straightened out, though. For one thing, what is the status of the open, free-variable-containing, sentences (1a) and (1b) and exactly how do they "determine" or "specify" the respective sets mentioned? We suggest the following mechanism: first, we associate with the conditions (1a) and (1b) the matching one-place *predicates* using the variable-binding operator λ for *property* (and, in general, *relation*)*abstraction*, viz.

(1c) λx: x is a day & the fog lifts before noon here on x

(1d) λx: x is a day & the fog lifts here on x

Call (1c) the *target* λ-predicate and (1d) the *reference* λ-predicate with respect to the *analysandum* sentence (1).

Furthermore, in the semantics of first-order predicate logic supplemented with λ, there will be *models* assigning to each predicate in the formal language a suitable set as its *extension;* e.g. the extension in a given model of the predicates (1c) and (1d), respectively, may be the two sets of days that were characterized above in an intuitive way. This settles the question how the open sentences (1a) and (1b), involving free variables, "determine" those sets of days; they do so *via* the associated λ-predicates (1c) and (1d), which are in turn interpreted relatively to set-theoretical models in a familiar way.

The next technical problem concerns the exact syntactic status of the frequency adverb "very often". It is an operator, all right. But what are the arguments for this operator? and how many are there? just one, or maybe more than one? In answer to these questions we suggest that the operator, formally represented by VeryOften-When, is a *binary* or *two-place* one which forms sentences out of *pairs* of predicates, like the pair (1c), (1d). The logical structure of (1) would then be captured by the following formalization:

(1*) VeryOften-When(λxFx, λxGx)

where "Fx" and "Gx" are short for the conditions (1a) and (1b), respectively, and the reading of (1*) as a whole is as follows: "it is *very often* the case that the property λxFx is realized *when* the property λxGx is realized". The property λxFx / λxGx/ is then the property of being a day on which the fog lifts before noon here /...the fog lifts here at all/.

Finally, we claimed above that the meaning of (1) could be characterized in terms of a certain conditional probability, interpreted in a Laplacean sense. In order to bring out this idea semantically we must enrich our aforementioned *models* with appropriate cardinality functions and use these to define conditional probabilities. We then lay down a *truth condition* for the formalization (1*) of (1) as follows: (1*) is true in a model M iff

$$P(ext^{M}(\lambda xFx) \; / \; ext^{M}(\lambda xGx)) \geq 0.9,$$

provided that the cardinality of $ext^{M}(\lambda xGx) > 0$; where P(/) is our conditional probability function, and where $ext^{M}($ $)$ means the *extension in* the model M *of*.

For a more rigorous and neat presentation of the formal machinery here suggested the reader should consult the Appendix below, where the theory of our nine frequency adverbs is set forth in full detail.

3. *Summary of analysis technique*

We now make an attempt to describe systematically the procedure applied in the analysis of the sentence (1). Given an *analysandum* sentence involving an adverb of frequency, go through the following steps:

Syntactic level

Step 1. Determine *target* and *reference conditions* to be associated with the *analysandum*, possibly introducing abbreviatory shorthands!
Step 2. Form the corresponding *target* and *reference λ-predicates!*
Step 3. Apply to these λ-predicates the formal counterpart of frequency adverb in order to obtain a *formalization* of the *analysandum!* This formal counterpart will usually be a two-place operator forming sentences out of pairs of predicates. Check the *reading* of the formalization thus obtained!

Semantic level

Step 4. Specify a model built on a finite Laplacean frame (see sec. 8 - 9 in the Appendix below) and determine the extensions of the λ-predicates in the model!
Step 5. Determine the cardinality of these extensions and of their intersection as well as the conditional probability of the extension of the target λ-predicate given that of the reference λ-predicate in accordance with the definition given in sec. 8 below!
Step 6. Use the appropriate truth condition (sec. 9 in the Appendix) to determine the meaning of the *analysandum* sentence and to check its truth-value!

Let us quickly repeat the results of going through Steps 1 - 3 on the syntactic level in the case of our sentence (1), leaving the semantical steps aside for the time being. As applied to (1) Step 1 gave us the target condition (1a) and the reference condition (1b), for which we introduced "Fx" and "Gx" as shorthands. Step 2 then gave us "λxFx" and "λxGx" as target and reference λ-predicates, respectively. Finally, we obtained the formalization (1*) of (1) as a result of taking Step 3; we also indicated how (1*) was to be read.

4. *Further examples*

Consider

(2) Caesar seldom awoke before dawn.

discussed by Lewis (1975) in support of the claim "that the range of quantification is often restricted". We suggest the following analysis of (2):

Step 1.
Target condition:

(2a) x is a day & Caesar awoke before dawn on x; formally: Hx

Reference condition:

(2b) x is a day & Caesar awoke on x (at all); formally: Jx

Step 2.
Target λ-predicate:

(2c) λxHx

Reference λ-predicate:

(2d) λxJx

Step 3.
As a formalization of (2) we propose

(2*) Seldom-When(λxHx, λxJx)

the reading of which is: "it is *seldom* the case that the property λxHx of being a day on which Caesar awoke before dawn is realized *when* the property λxJx of being a day on which Caesar awoke (at all) is realized".

Lumping the semantical steps 4 - 6 together we take (2) and (2*) to assert that the ratio of the number of "favorable" days on which Caesar awoke before dawn to that of the "admissible" days on which Caesar awoke (at all) is low, say, ≤ 0.3. Expressed more technically: (2*) is true in a model M iff

$$P(ext^{M}(\lambda xHx) \,/\, ext^{M}(\lambda xJx)) \leq 0.3,$$

with the usual proviso.

Our next example is adduced by Lewis (1975) to show "that the entities we are quantifying over, unlike times, may be distinct although simultaneous":

(3) Riders on the Thirteenth Avenue line seldom find seats.

The interest of this example, we claim, is due to the fact that the open sentences will have to contain *two* distinct free variables and, consequently, the corresponding λ-predicates will be *binary*. We propose the following analysis of (3):

Step 1.

Target condition:

(3a) x is a person & y is a time & x finds a seat on the Thirteenth Avenue line at y; in symbols: Sxy

Reference condition:

(3b) x is a person & y is a time & x is a rider on the Thirteenth Avenue line at y; in symbols: Rxy

Step 2.

Target λ-predicate:

(3c) λxySxy

Reference λ-predicate:

(3d) λxyRxy

Step 3.

We suggest formalizing (3) as

(3*) Seldom-When(λxySxy, λxyRxy)

the reading of which is as follows: "it is *seldom* the case that the binary 'seat-finding' relation λxySxy between persons and times is realized *when* the binary 'riding' relation λxyRxy among persons and times is realized".

Going through the semantical steps 4 - 6 we arrive at a model M such that (3*) is true in M iff

$P(ext^M(\lambda xySxy)/ ext^M(\lambda xyRxy)) \leq 0.3$ (say), provided that the cardinality of $ext^M(\lambda xyRxy) > 0$.

Note here that these extensions will be *binary relations* on the domain of the model, i.e. sets of *ordered pairs* of persons and times, or, in Lewis's terminology, sets of *cases*, whereby he means "tuples" or "sequences of coordinates" - in perfect agreement with the standard interpretation of predicates of degree > 1 in the semantics of predicate logic. We may add that Lewis's own discussion of (3) - his (9) - convincingly shows that any attempt to handle (3) on the basis of merely *one*-place predicates of *times* is doomed to failure.

We now deal with an example involving open sentences in *three* distinct free variables and, consequently, *three*-place λ-predicates:

(4) A man who owns a donkey always beats it now and then.

Step 1.

Target condition:

(4a) x beats y now and then during z; in symbols: Bxyz

Reference condition:

(4b) x is a man & y is a donkey & z is a period of time & x owns y throughout z; in symbols: Oxyz

Step 2.

Target λ-predicate:

(4c) λxyzBxyz

Reference λ-predicate:

(4d) λxyzOxyz

Step 4.

As a formal translation of (4) we then propose

(4*) Always-When(λxyzBxyz, λxyzOxyz)

the reading of which should by now be straightforward.

From a semantical point of view (4) and (4*) assert that *the ratio of* the number of "favorable admissible" cases, i.e. triples of men, donkeys and periods such that the man *both* owns the donkey throughout the period *and* beats it now and then during the period, *to* that of all the "admissible" cases, i.e. those triples where the man just owns the donkey throughout the period, *is equal to* 1. Technically speaking, (4) is true in a model M iff

$$P(ext^{M}(\lambda xyzBxyz)/ext^{M}(\lambda xyzOxyz)) = 1,$$

with the usual proviso.

5. *A comparison-of-frequency operator*

Consider

(5) John often goes to church for a Catholic.

We suggest that the analysis of (5) might involve *two pairs of* target and reference conditions in the following way:

1st target condition:

(5a) x is an occasion & John goes to church on x; formally, Jx

1st reference condition:

(5b) x is an occasion & on x John has an opportunity to go to church & John is a Catholic on x; formally, OJx

2nd target condition:

(5c) x is an occasion & Innocentius goes to church on x; formally, Ix

2nd reference condition:

(5d) x is an occasion & on x Innocentius has an opportunity to go to church & Innocentius is a "normal" or "typical" Catholic on x; formally, OIx

Then we form the corresponding λ-predicates λxJx, λxOJx, λxIx and λxOIx and propose the following formalization of (5):

(5*) MoreOftenThan(λxJx / λxOJx , λxIx / λxOIx)

the reading of which is roughly as follows:
"The property λxJx of being an occasion on which John (a Catholic) goes to church *is realized in relation to* the property λxOJx of being an occasion on which John has an opportunity to go to church *more often than* the property λxIx of being an occasion on which Innocentius (i.e. a "normal" or "typical" Catholic) goes to church *is realized in relation to* the property λxOIx of being an occasion on which Innocentius has an opportunity to go to church".

Semantically, we take (5) and (5*) to assert that, relatively to a given model, the conditional probability of the extension of the predicate λxJx *given* that of λxOJx *is greater than* the conditional probability of the extension of the predicate λxIx *given* that of λxOIx. Expressed in more technical jargon, (5*) is true in a model M iff

$$P(ext^{M}(\lambda xJx) \;/\; ext^{M}(\lambda xOJx)) >$$
$$> P(ext^{M}(\lambda xIx) \;/\; ext^{M}(\lambda xOIx));$$

provided that the cardinality of $ext^M(\lambda xOJx)$ as well as that of $ext^M(\lambda xOIx)$ are both distinct from 0.

Note also that, relatively to a suitable model M, we might well have that $ext^M(\lambda xOJx) = ext^M(\lambda xOIx)$, i.e. the set of occasions on which John has an opportunity to go to church might well be identical to the set of occasions on which the typical Catholic Innocentius has such an opportunity. This *need* not be so, though.*

MoreOftenThan is then a four-place or "tetradic" operator on predicates of the same degree. In our Appendix below we have chosen to define it in terms of another, primitive, tetradic operator of this kind, AtLeastAsOftenAs; see sec. 7 below.

---

* Hans Kamp has raised the following objection to our analysis of example (5). According to the analysis, (5) is true iff John (a Catholic) utilizes his opportunities to go to church *to a higher degree than* Innocentius (the typical Catholic) utilizes *his* ones. Now, this condition is fulfilled in a situation where John has, say, just one opportunity to go to church which he indeed utilizes, whereas in the same situation Innocentius has several such opportunities and does not utilize them all. According to the proposed analysis of (5), (5) will be true in the situation described (the first conditional probability being = 1, the second not); but, as Kamp argues, (5) is definitely *not true* in that situation, hence, our analysis is inadequate. -

A way of countering this objection is to point out that it is rather doubtful whether we are really able to judge of the question whether (5) is true, or is false, in the situation at hand. In order to be able to make a meaningful comparison between John and Innocentius here, we have to presuppose that the number of their respective opportunities to go to church is *sufficiently high* - this applies to both of them. But this requirement is obviously not met in the situation envisaged by Kamp; hence, we suggest, his objection ceases to apply to our analysis. This is one way of getting around the difficulty; another is, of course, not to use the notion of opportunity at all in the analysis of (5), but to base it on some different conception. At this juncture we have to leave the matter open for further debate.

6. Languages of FA

A *language* L of the *logic* FA of *frequency adverbs* or, in short, a FA-*language*, is a structure made up of the following disjoint *basic syntactic categories:*

I. A denumerable set Var_L of *individual variables*. Syntactic or meta-linguistic notation: $x, y, z, x_1,\ldots,x_n,\ldots$.

II. A denumerable set $Cons_L$ of *individual constants*. Syntactic notation: $a, b, c, a_1,\ldots,a_n,\ldots$.

III. For each nonnegative integer n, a denumerable set Pr^n of n-place *predicate letters*. Syntactic notation: P^n (n = 0,1,2...).

IV. *Logical constants*.

(a) For each nonnegative integer n, a fixed *tautologous* n-place *predicate* T^n(n = 0, 1, 2,...).

(b) *Sentential connectives:* –, &, v and → for, respectively, negation, conjunction, disjunction and material implication (in terms of which the symbol ↔ for material equivalence is defined in the usual way).

(c) *Quantifiers* (over individuals): ∀ ("for each") and ∃ ("for some").

(d) A symbol for *identity* (among individuals): =.

(e) A predicate-forming variable-binding operator λ for *relational abstraction*.

(f) Nine dyadic sentence-forming *adverbs of frequency*, viz.

Always-When	VeryOften-When	VerySeldom-When
Never-When	Often-When	Seldom-When
Sometimes-When	FairlyOften-When	FairlySeldom-When

(g) A four-place *comparison-of-frequency operator:* AtLeastAsOftenAs.

By simultaneous recursion (or induction) we now single out two categories of symbol strings, viz. the set $Sent_L$ of (well formed) *sentences of* L, and for each positive integer n = 1, 2,..., the set $Pred^n_L$ of n-place *predicates of* L:

(0) T^0 as well as all members of Pr^0_L are in $Sent_L$.

(1) For all a, b in $Cons_L$, (a = b) is in $Sent_L$.

(2) If Q^n is in $Pred^n_L$ and $a_1,\ldots,a_n$ are (not necessarily distinct) members of $Cons_L$, then $Q^n a_1,\ldots,a_n$ is in $Sent_L$ (n > 0).

(3) If A, B are in $Sent_L$, then so are –A, (A & B), (A v B) and (A → B).

(4) If x is in Var_L, a is in $Cons_L$, and if A is in $Sent_L$ and contains a but not x, then $\forall x A^x/a$ and $\exists x A^x/a$ are in $Sent_L$. (Here A^x/a is the result of replacing every occurrence of a in A by one of x).

(5) If Q^n and R^n are in $Pred^n_L$ (n > 0), then

Always-When(Q^n, R^n)
VeryOften-When(Q^n, R^n)
Often-When(Q^n, R^n)
FairlyOften-When(Q^n, R^n)
FairlySeldom-When(Q^n, R^n)
Seldom-When(Q^n, R^n)
VerySeldom-When(Q^n, R^n)
Never-When(Q^n, R^n)
Sometimes-When(Q^n, R^n)

are in $Sent_L$.

(6) If Q^n, R^n, S^n and U^n are in $Pred^n_L$ (n > 0), then AtLeastAsOftenAs(Q^n/R^n, S^n/U^n) is in $Sent_L$.

(7) T^n as well as all members of Pr^n_L are in $Pred^n_L$ (n > 0)

(8) If $x_1,\ldots,x_n$ are *distinct* members of Var_L, if $a_1,\ldots,a_n$ are *distinct* members of $Cons_L$, and if A is in $Sent_L$ and contains $a_1,\ldots,a_n$ but none of $x_1,\ldots,x_n$, then $\lambda x_1,\ldots,x_n A^{x_1}/a_1,\ldots,{}^{x_n}/a_n$ is in $Pred^n_L$ (n > 0). (Here $A^{x_1}/a_1,\ldots,{}^{x_n}/a_n$ is the result of replacing every occurrence of a_i in A by one of x_i, for each i such that $1 \leq i \leq n$.).

(9) Nothing is in $Sent_L$ or in $Pred^n_L$ (n > 0) except by virtue of the rules (0) - (8) above.

7. *Some definitions, readings and notational conventions*

It appears from clause (5) of the recursive definition just given that our adverbs of frequency Always-When, VeryOften-When,..., Sometimes-When are *binary* or *dyadic* operators which form sentences in L out of *pairs of* n-place predicates of L. These L-sentences can be understood in accordance with the following paradigm: read Always-When(Q^n, R^n) as "it is *always* the case that (the relation) Q^n is realized *when* (the relation) R^n is realized", and analogously for the remaining eight frequency adverbs. We now introduce a series of matching unary, i.e. one-place, operators as follows.

Let Q^n be in $Pred^n_L$ $(n > 0)$. We then lay down the definitions:

D1. ALWAYS Q^n =df Always-When(Q^n, T^n)

D2. VERY OFTEN Q^n =df VeryOften-When(Q^n, T^n)

.

.

.

D9. SOMETIMES Q^n =df Sometimes-When(Q^n, T^n).

These definitions then make use of our fixed tautologous n-place predicates T^n in an obvious way.

We can also use our nine dyadic frequency adverbs to obtain definitions of the following series of *frequentative quantifiers*, where $x_1, \ldots, x_n$ are assumed to be *distinct* members of Var_L not occurring in Q^n or in R^n $(n > 0)$:

D10. ForAll $x_1, \ldots, x_n$-When($Q^n x_1, \ldots, x_n$, $R^n x_1, \ldots, x_n$) =df Always-When(Q^n, R^n)

D11. ForVeryMany $x_1, \ldots, x_n$-When($Q^n x_1, \ldots, x_n$, $R^n x_1, \ldots, x_n$) =df VeryOften-When(Q^n, R^n)

.

.

.

D14. ForFairlyFew $x_1, \ldots, x_n$-When($Q^n x_1, \ldots, x_n$, $R^n x_1, \ldots, x_n$) =df FairlySeldom-When(Q^n, R^n)

.

.

.

D18. ForSome $x_1, \ldots, x_n$-When($Q^n x_1, \ldots, x_n$, $R^n x_1, \ldots, x_n$) =df Sometimes-When(Q^n, R^n)

The quantifiers just defined could also be made more "absolute" by having T^n do duty for R^n in the manner illustrated by our definitional series D1 - D9.

Again, L-sentences of the form AtLeastAsOftenAs(Q^n/R^n,S^n/U^n) may be read somewhat as follows: "Q^n is realized in relation to R^n *at least as often as* S^n is realized in relation to U^n". In terms of this primitive tetradic operator we can now define the following operators, where Q^n,R^n,S^n and U^n are in $Pred^n_L$ and $n > 0$:

D19. AtMostAsOftenAs(Q^n/R^n, S^n/U^n) =df AtLeastAsOftenAs(S^n/U^n, Q^n/R^n).

D20. EquallyOftenAs(Q^n/R^n, S^n/U^n) =df AtLeastAsOftenAs(Q^n/R^n, S^n/U^n) & AtMostAsOftenAs(Q^n/R^n, S^n/U^n).

D21. $\text{MoreOftenThan}(Q^n/R^n, S^n/U^n)$ =df
$-\text{AtLeastAsOftenAs}(S^n/U^n, Q^n/R^n)$.

D22. $\text{LessOftenThan}(Q^n/R^n, S^n/U^n)$ =df
$-\text{AtLeastAsOftenAs}(Q^n/R^n, S^n/U^n)$.

In view of the paradigm given above the readings of these new locutions are obvious.

Finally, in connection with L-sentences of the kind introduced by clause (2), we shall sometimes write $\{Q^n\}a_1,\ldots,a_n$, thus enclosing the predicate in braces. This convention facilitates reading when Q^n is formed by the use of λ (see clause (8) and cf. Prawitz (1965) p.63 f.).

8. *Finite Laplacean frames*

First, a preliminary explanation of some familiar terminology. If D is a non-empty set, then D^k (where k is a positive integer, i.e. k = 1,2,3,...) is to be the set of all ordered k-tuples of elements of D or, as we might as well put it, the k-th Cartesian cross-product of D with itself. Also, D^1=D so that ordered 1-tuples are identified with their sole terms.

By a *finite Laplacean frame* we shall now understand an ordered pair $\langle D, \{c_k\}_{k \text{ is a positive integer}}\rangle$ where

(i) D is a non-empty and finite set ("domain of individuals"), and

(ii) $\{c_k\}$ is a family of cardinality functions indexed by the set of positive integers; which means that, for each k = 1,2,3,..., $c_k : \wp D^k \to \text{Nat}$ is a function from the power set of D^k into the set Nat of all natural numbers (= {0,1,2,...}) satisfying the following conditions, for all $X,Y \subseteq D^k$:

(a) $c_k(X) = 0$ iff $X = \emptyset$

(b) if $X \subseteq Y$, then $c_k(X) \leq c_k(Y)$, and

(c) if $X \cap Y = \emptyset$, then $c_k(X \cup Y) = c_k(X) + c_k(Y)$.

Also, we require any c_k to be such that

(d) for all x in D^k: $c_k(\{x\}) = 1$ (or, in other words, that for all $d_1,\ldots,d_k$ in D: $c_k(\{\langle d_1,\ldots,d_k\rangle\}) = 1$).

The intuitive import of $c_k(X)$, for any $X \subseteq D^k$, is then obviously the *cardinality of* X (or the *number of elements in* X).

Definition. Let $\langle D, \{c_k\}_{k \text{ is a positive integer}}\rangle$ be any finite Laplacean frame, let X and Y be any subsets of D^k such that $Y \neq \emptyset$ so that, by (ii)(a), $c_k(Y) \neq 0$. We define the k-*th conditional probability of* X *given* Y, in symbols: $P_k(X/Y)$, by stipulating that

$$P_k(X/Y) = \frac{c_k(X \cap Y)}{c_k(Y)} .$$

Again, we define the k-*th "absolute" probability of* X, $P_k(X)$, by requiring that

$$P_k(X) = P_k(X/D^k) = \frac{c_k(X \cap D^k)}{c_k(D^k)} \left(= \frac{c_k(X)}{c_k(D^k)}, \text{ since } X \subseteq D^k\right).$$

Note that $P_k(X)$ is well defined because, by (i) above, $D \neq \emptyset$ so that for each $k = 1,2,3,\ldots$, $D^k \neq \emptyset$ and, by (ii)(a), $c_k(D^k) \neq 0$.

Exercise. Prove that, as defined, the probability measures $P_k(\ /\ )$ and P_k are real-valued functions on $\wp D^k \times (\wp D^k - \{\emptyset\})$ and on $\wp D^k$, respectively with the following familiar properties, where $X,Y,Z \subseteq D^k$ and $Y \neq \emptyset$:

P1. $0 \leq P_k(X/Y) \leq 1$.

P1.1 $0 \leq P_k(X) \leq 1$.

P2. $P_k(D^k) = P_k(D^k/D^k) = 1$.

P3. $P_k(-X/Y) = 1 - P_k(X/Y)$.

P3.1 $P_k(-X) = 1 - P_k(X)$.

P4. If $X \cap Z = \emptyset$, then $P_k(X \cup Z/Y) = P_k(X/Y) + P_k(Z/Y)$.

P4.1. If $X \cap Z = \emptyset$, then $P_k(X \cup Z) = P_k(X) + P_k(Z)$.

P5. $P_k(X/Y) = \dfrac{P_k(X \cap Y)}{P_k(Y)}$.

9. *FA-models and truth conditions*

Let L be any FA-language. By a FA-*model for* L we mean a triple $M = \langle D, \{c_k\}_{k = 1,2,3,\ldots}, V\rangle$ such that

(i) $\langle D, \{c_k\}_{k = 1,2,3,\ldots}\rangle$ is a finite Laplacean frame,

(ii) V is a *valuation of* L *on* the domain D in the sense of a one-place assignment-function satisfying the following conditions:

(a) $V(a) \varepsilon D$, for each a in $Cons_L$.

(b) $V(P^O) \varepsilon$ the set of truth-values $\{1,0\}$, for each P^O in Pr_L^O.

(c) $V(P^n) \varepsilon$ the cartesian product D^n, for each P^n in Pr_L^n where $n > 0$.

Let $M = \langle D, c_k\ k = 1,2,3,\ldots, V\rangle$ be any FA-model for L. By simultaneous induction we now define two notions, viz. $\models^M$, i.e. *truth in* the model M, and ext^M, i.e. *extension* (or denotation) *in* the model M, where the former concept applies to the members of $Sent_L$ and where the latter applies to the members of $Pred_L^n$ ($n > 0$). Thus, we read the locution $\models^M A$ as "the sentence A is true in the FA-model M" and we read the locution $ext^M(Q^n)$ as "the extension in M of the n-place predicate Q^n". The recursive definition runs as follows (cf. the various clauses in the definition of $Sent_L$ and of $Pred_L^n$ ($n > 0$) above):

(0) $\models^M T^O$

(0.1) $\models^M P^O$ iff $V(P^O) = 1$, where $P^O \varepsilon Pr_L^O$

(1) $\models^M (a = b)$ iff $V(a)$ is identical to $V(b)$

(2) $\models^M Q^n a_1,\ldots,a_n$ iff $<V(a_1),\ldots,V(a_n)> \varepsilon\ ext^M(Q^n)$, where $n > 0$

(3) $\models^M -A$ iff it is not the case that $\models^M A$

(3.1) $\models^M (A \& B)$ iff both $\models^M A$ and $\models^M B$

The truth conditions (3.2) and (3.3) for L-sentences having v and → as their main connective are then familiar.

(4) $\models^{M = <D, \{c_k\}, V>} \forall x A^x/a$ iff $\models^{<D, \{c_k\}, V'>} A$,

for each valuation V' on D such that $V' =_a V$ (see explanation below). Here A is assumed to be any L-sentence containing the individual constant a but not the individual variable x; in other words, A, a, x are to satisfy the hypothesis of the formation rule (4) in section 6 above.

(4.1) $\models^{M = <D, \{c_k\}, V>} \exists x A^x/a$ iff $\models^{<D, \{c_k\}, V'>} A$,

for some valuation V' on D such that $V' =_a V$ (see explanation below). Here A, a, x are to satisfy the hypothesis of the formation rule (4) in section 6 above.

Clauses (5) - (5.8) below are throughout to be understood with the following proviso: "provided that $ext^M(R^n) \neq \emptyset$ ($n > 0$)".

(5) $\models^{M}$ Always-When(Q^n,R^n) iff $P_n(ext^M(Q^n)/ext^M(R^n)) = 1$

(5.1) $\models^{M}$ VeryOften-When(Q^n,R^n) iff $P_n(ext^M(Q^n)/ext^M(R^n)) \in$
$\in [0.9,\ 1]$

(5.2) $\models^{M}$ Often-When(Q^n,R^n) iff $P_n(ext^M(Q^n)/ext^M(R^n)) \in$
$\in [0.7,\ 1]$

(5.3) $\models^{M}$ FairlyOften-When(Q^n,R^n) iff $P_n(ext^M(Q^n)/ext^M(R^n)) \in$
$\in\ ]0.5,\ 1]$

(5.4) $\models^{M}$ FairlySeldom-When(Q^n,R^n) iff $P_n(ext^M(Q^n)/ext^M(R^n)) \in$
$\in [0,\ 0.5[$

(5.5) $\models^{M}$ Seldom-When(Q^n,R^n) iff $P_n(ext^M(Q^n)/ext^M(R^n)) \in$
$\in [0,\ 0.3]$

(5.6) $\models^{M}$ VerySeldom-When(Q^n,R^n) iff $P_n(ext^M(Q^n)/ext^M(R^n)) \in$
$\in [0,\ 0.1]$

(5.7) $\models^{M}$ Never-When(Q^n,R^n) iff $P_n(ext^M(Q^n)/ext^M(R^n)) = 0$

(5.8) $\models^{M}$ Sometimes-When(Q^n,R^n) iff $P_n(ext^M(Q^n)/ext^M(R^n)) \neq 0$

Note that there is nothing sacrosanct about the particular choice of values for P_n in the series of clauses (5.1) - (5.6) but that, on the other hand, a certain kind of symmetry is to be preserved if one makes changes in these values in the series.

(6) $\models^{M}$ AtLeastAsOftenAs$(Q^n/R^n,\ S^n/U^n)$ iff
$P_n(ext^M(Q^n)/ext^M(R^n)) \geq P_n(ext^M(S^n)/ext^M(U^n))$; provided that $ext^M(R^n) \neq \emptyset$ and $ext^M(U^n) \neq \emptyset$ $(n > 0)$.

(7) $ext^M(T^n) = D^n$ $(n > 0)$

(7.1) $ext^M(P^n) = V(P^n)$, for all P^n in Pr^n_L $(n > 0)$

(8) $ext^M(\lambda x_1,\ldots,x_n A^{x_1/a_1,\ldots,x_n/a_n}) = |A|^M =$
$\{<d_1,\ldots,d_n> : d_1,\ldots,d_n$ are in D, $V^{\bullet}(a_1) = d_1$ and ...
...and $V^{\bullet}(a_n) = d_n$, and $\models^{<D,\{c_k\},V^{\bullet}>} A$, for some $V^{\bullet}$ on D such that $V^{\bullet} \overline{\overline{\ \ \ \ }}_{a_1\ldots a_n} V\}$;

see explanation below. Here A, $a_1,\ldots,a_n$, $x_1,\ldots x_n$ are to satisfy the hypothesis of the formation rule (8) for λ-expressions in section 6 above.

Explanation. Let V, $V^{\bullet}$ be valuations of L on a non-empty domain D, and let a be any member of $Cons_L$ /let $a_1,\ldots,a_n$ be distinct members of $Cons_L$/. We say that $V^{\bullet}$ *differs from* V *at most with respect to* a /*with respect to*

$a_1,\dots,a_n$/, in symbols: $V =_a V$ / $V \underset{a_1,\dots,a_n}{=\!=\!=} V$/,

iff

(i) $V'(P^n) = V(P^n)$ for each P^n in Pr^n_L where $n \geq 0$, and

(ii) $V'(b) = V(b)$ for each b in $Cons_L$ that is distinct from a /from each of $a_1,\dots a_n$/.

This completes our simultaneous inductive definition of $\models^M$ and ext^M.

Let A be any L-sentence, and let $M = \langle D, \{c_k\}_{k=1,2,3,\dots}, V\rangle$ be any FA-model for L. We say that M is *appropriate for* A just in case the concept of truth in M ($\models^M$) is defined for A. We ought to be able to verify that M is appropriate for A iff (i) there is no subformula of A having any of the forms Always-When(Q^n,R^n),... (down to)... Sometimes-When(Q^n,R^n) which is such that $ext^M(R^n) = \emptyset$, *and* (ii) there is no subformula of A of the form AtLeastAsOftenAs$(Q^n/R^n, S^n/U^n)$ such that $ext^M(R^n) = \emptyset$ or $ext^M(U^n) = \emptyset$.

Given this notion of appropriateness, we say that an L-sentence A is FA-*valid* iff $\models^M$ A for each FA-model M such that M is appropriate for A. Note that this conception of validity is slightly unorthodox in the following two respects: (a) FA-models are built on *finite* sets, viz. Laplacean frames, and (b) we have to restrict ourselves to such FA-models as are appropriate for L-sentences.

Bibliographic References

Lewis, D., 1975, Adverbs of Quantification, in E. Keenan (ed.)*Formal Semantics of Natural Language*. Cambridge University Press, pp.

Prawitz, D., 1965, *Natural Deduction*. Stockholm: Almqvist & Wiksell.

TENSES AND ASPECTS AS FUNCTIONS ON VERB-PHRASES*

Emmon Bach, University of Massachusetts/Amherst

0. *Introduction*

Practically all tense logicians that I've read have assumed that tenses and aspects are to be interpreted as sentence operators (one exception: Bäuerle, 1977). And there is a good deal of motivation for this idea. To take just one example, consider one test offered by Thomason and Stalnaker (1973), for helping to determine whether adverbials are sentential or not:

> Criterion 2: Only if an adverb is a sentence modifier can it give rise to quantifier scope ambiguities in simple existential or universal sentences.

Their example for this test is the following:

(1) Frequently, someone got drunk -- Someone got drunk frequently.

By this criterion, *frequently* must be a sentence adverb. By the same reasoning we could show that the present perfect is a sentence level operator if (2) is scopally ambiguous:

(2) Every president has resigned.

In Montague's PTQ (Paper 8 in Montague, 1974) sentences like (2) are claimed to be ambiguous. By direct generation (S17) we get a version like (3), by "quantifying in" (S14), we get one like (4):

(3) $H\ \hat{P} \bigwedge x\ [\mathit{president}'(x) \rightarrow P\{x\}]\ ({}^{\wedge}\mathit{resign}')$

(4) $\hat{P} \bigwedge x\ [\mathit{president}'(x) \rightarrow P\{x\}](\hat{x}_3\ H\ \mathit{resign}\ (x_3))$

It's easy to see that to retain this account in the theory of English presented in PTQ, the tenses/aspects *must* be sentential operators.

A consequence of this general way of looking at tenses/aspects can be seen in the very rules of PTQ: there are six totally independent ways of putting together subjects and verb phrases, one for each of the six combi-

* Dedicated to the memory of Michael Bennett.

nations of negation, present perfect, *will* and nothing (i.e. present tense) which are allowed by PTQ. We can make a typical linguist's argument against Montague's treatment. It is a complete accident, on that account, that English sentences in the six combinations end up showing the same basic structure: *NP(Aux)VP;* the distribution could just as well be this:

(5) John leaves.
    Doesn't leave John.
    Won't leave John.
    John has left.
    Hasn't left John.
    John will leave.

The grammar of English would be just as simple if (5) were the correct distribution. Note also that things just get worse as we add other tenses and modals.

What is it about PTQ that requires this sort of treatment? I will argue that the reason for this state of affairs is the view that subject NP's are functions taking IVP's as arguments. If we drop this idea and revert to an analysis given by Montague in "Universal Grammar" (Paper 7 in Montague, 1974) we can accomodate the facts about tenses and aspects in English in a much more natural way. The theory of UG, which has recently been argued for strongly by Keenan and Faltz (1978) within a different framework, takes intransitive verb phrases as denoting functions from (intensions of) NP-type things (sets of properties) to truth values. Once we take this step, we can reconsider the status of tenses and aspects.

My paper goes in three stages. In Section 1, I'll sketch some of the evidence for the UG, Keenan/Faltz analysis (henceforth 'UG/KF'). In Section 2, I'll show how this permits us to treat tenses/aspects as functions defined on IVP's and give rules that result in a system exactly equivalent to that of PTQ. In Section 3, I'll speculate in a more informal way about a more thoroughgoing revision that doesn't end us up exactly where PTQ does, and in Section 4 I'll consider extensions of PTQ under the new framework.

1. Why IVP's are of category t/T.

In this section, I'll consider evidence for the notion that intransitive verb-phrases should be assigned a higher-type category than *t/e*. It should be pointed out at the outset that PTQ has no category of tensed IVP's, since the tenses and aspects are introduced syncategorematically. So the view that tensed IVP's at least are of category t/T is quite consistent with PTQ and amounts only to adding a higher category (t/T) and spelling out the surface forms of English via an intermediate stage. For a number of arguments for the more general idea that IV's themselves are of this category I refer to the monograph by Keenan and Faltz (1978). I'll confine myself here to a few ways in which this view solves a number of problems in an elegant way. To them should be added the considerations about tense and aspect that make up the bulk of this paper.

1.1 English constituent structure.

On the assumption that conjunction should be defined for all or almost all and only genuine syntactic categories of English (or "major" categories), the UG/KF view predicts that tensed IVP's should be conjoinable. They are:

(6) Harry left at three and is here now.

(7) John lives in New York and has always lived there.

Note that these assumptions and facts are incompatible with a popular view of English sentences as having this structure:

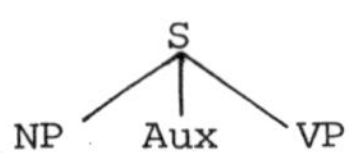

1.2 Raising verbs

Verbs like *seem* or phrases like *seem to me* seem semantically to work like propositional operators. That is, a sentence like (8) feels like it ought to have the indicated structure in its interpretations (cf. Wotschke, 1972):

(8) A unicorn seems to be approaching.

seem' (ϕ)

where ϕ corresponds to the proposition expressed by (9):

(9) A unicorn is approaching.

It's hard to get this in the PTQ system. But if we take the view that IVP's are functions, it's easy. We can let *seem* take IV's with *to* and assign it this interpretation:

(10) $\lambda P \lambda \mathcal{P}\ seem' (\mathcal{P}\{P\})$

(This assumes that we retain PTQ's analysis of IVP's *without tenses* as of category *t/e*.)

1.3 Syntactic category theory.

One of the problems of Montague's syntax in PTQ is that there seems to be no natural connection between various subcategories of verbs: IV's are of category *t/e*, TV's of category *(t/e)/T*. If we regard IV's as of category *t/T* there is a unified way of classifying at least these two types together, and explaining why for example adverbial modifiers can often go with either type (c.f. Keenan and Faltz, 1978).

Moreover, it begins to look as if we will be able to establish connections between verbs and other categories. For example, verbs and prepositions share some properties. In the UG/KF system they share the property of being functions of category *X/T*.

Finally, we may begin to get a handle on why verbs are different from other kinds of things classified as *t/ne* in PTQ. (Keenan and Faltz provide a number of arguments of this general sort).

1.4 Montague phonology.

Finally, let me mention an argument that comes from a rather surprising source: the study of phonology.

Recent work in phonology (Selkirk, ms., Liberman & Prince, 1973) has shown that it makes a lot of sense to think of the phonology of a language as including a system that specifies various domains along a hierarchy: the syllable, the foot, the word, the phonological phrase, etc. (This work is in part a revival and vindication of a lot of older ideas, e.g. those of Kenneth Pike). It turns out that Montague grammar -- or more precisely categorial grammar -- offers an interestingly different way to think about such problems.

Among the rules of English phonology are principles that tell us how to organize sequences of words into phonological phrases. In a categorial framework it appears that we can state a very general principle of phrasal organization (for English and some other languages):

> Phonological phrases are the maximal chunks that you get by putting functions together with arguments on their "right".

This principle tells us for example that prepositions will go with their NP's, that a determiner will go together with its common noun phrase, that a transitive verb will go together with its object NP and so on. The principle seems to work just right for such disparate processes as the assignment of stress to English phrases, the organization of phrases for the purposes of assigning intonations (at the next level up), for defining the domain of French liaison, and Italian syntactic gemination.

Given this principle, the two views of subject NP's set forth in PTQ and UG/KF make opposite predictions. In fact, the UG/KF view seems correct: liaison never happens from a subject noun phrase to the tensed VP, nor does gemination in Italian take place here, and English phrasing respects the independence of the subject NP. (Work in progress by myself and Deirdre Wheeler is devoted to these questions, much oversimplified here.)

2. *Redoing PTQ*

In this section, I will sketch the simplest possible modification of PTQ. I retain the category assignments of PTQ, but add a new category of tensed verb phrases. In order to make everything work out right it's necessary to make a distinction between verbal and non-verbal predicate types. The system is very "surfacy" in that it generates auxiliary elements directly in place without recourse to obligatory rules like Affix Hopping (cf. Lapointe, ms.).

Let us first add a new category for tensed IVP's: $P_{t/T}$. There are no basic members of this category. Expressions in the category are formed in the first place by a rule that takes PTQ's IV's ($P_{t/e}$) and changes the tenseless main verbs of the phrase to their present tense singular forms. The new phrases are to be interpreted as functions from intensions of NP's (sets of properties) to truth values. We have two choices: one is to give the truth conditions for these functions directly (as do Keenan/Faltz, Gazdar); equivalently we can interpret each such tensed IVP as $\lambda\mathcal{P}\mathcal{P}\{{}^\wedge\gamma\}$ where γ translates the old IV.

This rule will give us tensed IVP's like *walks, is a fish, sees a unicorn.* We can now treat the various tensed auxiliaries of English as functions from the old IV's to members of the new category $P_{t/T}$. To do this right, however, it's necessary to remove *be* from the category of transitive verbs. It has long been known (Chomsky, 1957; more recently Akmajian and Wasow, 1975) that the tensed forms of *be* act like auxiliaries. That is, we have paradigms like these:

(11) John doesn't walk.
John isn't a fish.
*John doesn't be a fish.
John walks, doesn't he?
Mary is Bill, isn't she?
*Mary is Bill, doesn't she?

So let's introduce a syntactic distinction between verbal IV's and copular IV's (letting the latter be $P_{t////e}$, P_{COP} for short).

We need to distinguish four classes for Auxiliaries. First, we have *will* and *won't* which can go directly with verbal and copular IV's; call them Aux-1. Second, we have *doesn't*, which can go only with verbal IV's; call them Aux-2. Third, we have the tensed forms of *be*, *is* and *isn't*, which take term phrases (or other things in a fuller fragment); call them Aux-3. Finally, we have the tensed forms of *have*, *has* and *hasn't*, which take both verbal and copular IV's in their participial forms.

The rules for forming tensed IVP's with auxiliaries are straightforward. For the first three we just concatenate the Aux element with the appropriate kinds of predicates. For the present perfect we have to form the participial version of the predicate. Translations for the Aux elements except for *is* and *isn't* all have the form $\lambda P \lambda \mathcal{P} X \mathcal{P}\{P\}$, where X = W, ¬W, H, ¬H, ¬, For *be* we simply reconstruct Montague's interpretation. We let *be* itself form copular predicates by combining it with a term phrase. (This is all spelled out formally in Appendix A.).

The new system allows us to correct a minor mistake implicit (I think) in PTQ. PTQ allows us to conjoin arbitrary IVP's (by rule S12): *walk* or *be a fish*, for example. But what is the present tense third person negative of this phrase? The answer is that there is none: **John doesn't walk and be a fish*. The best we can do is *John doesn't walk and isn't a fish*. As suggested above, we do want to conjoin tensed IVP's, so we can add a rule to do this. And we can formulate rules that will allow *John will walk and be a fish* but exclude the bad sentence above. (This problem was pointed out to me by Barbara H. Partee.). I'll defer further discussion of the conjoining rule to the next section.

This might all appear unduly complex, but please note that the rules of Appendix A spell out much of what is buried in Montague's phrases "third singular present" (S4) and the corresponding phrases of S17. Moreover, an explicit definition of *main verb* allows us to correct a mistake in PTQ (which gives **John walks and talk*; cf. Bennett, 1974, and also the more careful formulation of UG with explicit definitions of the relation *main verb of*).

Finally, it's pretty straightforward to extend the fragment by adding new members to the Aux categories, for example, the other modals of English.

3. *Revising PTQ.*

In this section I want to explore a more radical revision of PTQ's type assignments along the lines suggested by Keenan and Faltz.[1] That is, we let the old IV's themselves be mapped into expressions (or interpretations) of type <<s,f(T)>, t>. I take *walk* to denote a function from intensions of sets of properties to truth values. For an extensional verb like *walk* we can say that the value of *walk'(^NP')* at a world and time is 1 just in case *walk'*$_*$ is in *NP'* at that world and time. In the formalism of PTQ we replace the category IV by a new category $P_{t//T}$ -- distinct from tensed IV's which stay in $P_{t/T}$.

On this view, then, the translation of *John walks* will be just *walk'(^j*)*. The interpretations of the Auxiliaries will be more simply represented as just $\lambda \mathbb{P}\ X\ \mathbb{P}$, where $\mathbb{P}$ ranges over things of type <s, f(t/T)>, X as above = W, ¬W, H, ¬H, ¬ but these operators are redefined. One immediate advantage is this: we can add *have* to the fragment (in it's infinitive form) to get things like *will have walked*. On this view the Auxiliaries are more truly VP-operators. What are we to say about their interpretations?

In the logic we eliminate all clauses introducing W, H, ¬ as sentence operators and replace them by these definitions:

> If $\gamma \varepsilon$ $ME_{<<s,f(T)>,\ t>}$ then so are $H\gamma$, $W\gamma$, $\neg\gamma$.

And similarly we want to replace the clauses giving the denotations of expressions involving the old sentence operators by ones like these:

> $H\gamma$ is that function *h* defined for NP intensions such that $h(\alpha)$ at i, j is 1 iff there is j', j' < j, and $\gamma(\alpha)$ is 1 at i, j', and similarly for W and ¬.

Let's now think about the rule for conjoining tensed IVP's as in these sentences.

(12) John has walked and is a fish.

(13) A woman has walked and will run.

---

1 And apparently independently by Gazdar, 1979. Again, let's note that this was the system of *UG*. Hoepelman, 1978, follows the same system.

The syntactic rule is simple:

If $\gamma, \delta \varepsilon P_{t/T}$, then so are γ *and* γ *or* δ

A first stab at a translation rule fails:

If γ, δ translate as γ', δ' respectively, then γ *and* δ and γ *or* δ translates as $\lambda\mathcal{P}[\gamma'(\mathcal{P}) \wedge \delta'(\mathcal{P})]$ and $\lambda\mathcal{P}[\gamma'(\mathcal{P}) \vee \delta'(\mathcal{P})]$ respectively.

This rule would work fine for (12) but would make (13) come out as equivalent to (14).

(14) A woman has walked and a woman will run.

(This was one of the stumbling blocks of the old meaning-preserving conjunction reduction transformation.). My judgment is that (13) has no reading like (14). So we need something fancier, say, this:

If γ, δ translate as γ', δ' respectively, then γ *and* δ and γ *or* δ translates as
$\lambda\mathcal{P}\mathcal{P}\{\hat{x}\ \gamma'(\hat{P}\ P\{x\}) \wedge \delta'(\hat{P}\ P\{x\})\}$ and $\lambda\mathcal{P}\mathcal{P}\{(\hat{x}\ \gamma'(\hat{P}\ P\{x\}) \vee \delta'(\hat{P}\ P\{x\})\}$
respectively.

This rule has the effect of reconstructing the "quantifying in" interpretation. This rule will have the same effect as the more direct interpretations made available in systems like those of Keenan/Faltz (1978) and Gazdar (1979) where Boolean operations on a whole bunch of categories besides sentences are allowed. The interpretation of (13) on this account will now come out equivalent to this formula:

(15) $\bigvee x$ [woman' (x) $\wedge$ H walk' (x) $\wedge$ W run' (x)]

I take this to be correct.[2]

In favor of this view of tenses and negation as basically VP operators in English is the following fact. There are no simple ways in English of giving negations or tense operations on conjoined sentences, in striking contrast to the simple syntax of negation and tenses in logics. Thus, in treatises on tense logic, it is necessary when discussing examples to resort to higher verbs: *it is not the case that p and q, it will be the case that p or q,* etc. (We'll see that this fact doesn't hold for time adverbials: *Yesterday John left and Mary arrived.*).

---

2 In the discussion at the conference, Lauri Karttunen and Ed Keenan disagreed with my judgments about such sentences, offering examples that seemed to show that you do need to provide for interpretations like (14). Note that if this is correct we can provide a very strong argument for the analysis adopted here, in the case of intensional expressions such as the following sentence: A secretary is being sought and will be required to know three languages. One reading for this sentence (like the second interpretation above) is available via the usual route of quantifying in. But the other reading cannot be captured by an interpretation of the first sort given. Apparently, all we can say is that the interpretation is some function from NP intensions to truth values (Keenan pointed out to me that this is generally true for conjunctions of intensional expressions). So the inability to state a generalization about such conjunctions turns out to be an argument for the general point!

4. *Extending PTQ.*

What we've done so far isn't really a very good test of our theory. To really check it out we have to consider the harder problems, to wit, the progressive, the past tense, and time adverbials. This section will be devoted to an attempt. Clearly, in a paper of this length, I can only scratch the surface.

4.1 *The progressive.*

The best treatment of the English progressive that I know of is due to Frank Vlach (ms. a.). Vlach argues persuasively (as have others) that an adequate analysis of the progressive must take into account something like the classification of certain English expressions developed by Vendler, Kenny, and others. According to this view English expressions must be divided into those having to do with *states*, *processes*, *accomplishments*, and *achievements*. I say "expressions" advisedly, since one problem we must address is this: what kinds of things are to be divided into these classes?

Recall that Vendler and Kenny talked in the first place in terms of *verbs*, interpreted liberally to be sure, since they included things like *push a cart*, already some sort of verb phrase. Dowty (1972)[3] pointed out that the nature of the object and other things in the verb phrase made a crucial difference and that the subject had to be taken into account as well. For example, (16) is an achievement predicate, (17) a process predicate, (18) an achievement sentence, (19) a process sentence:

(16) ...find a unicorn

(17) ...find unicorns

(18) I discovered the village (? all summer).

(19) Tourists discovered the village (all summer).

For such reasons, Dowty, Vlach and others have concluded that the domain of the progressive must be the sentence.

Let me note once more that this flies in the face of English syntax. Writers like Akmajian and Wasow (1975) have given evidence that the progressive ought to be the element of the English tense/aspect system that is farthest "down" in the verb phrase system. The progressive is then a major test of our theory, since if the progressive has to be a sentence operator, then everything in the system, since "higher", must also be.

3 Following Verkuyl (1971).

I don't want to try to provide a full theory of the progressive here. I agree with Vlach that an adequate theory must take at least this fourfold classification into account. I am solely concerned here with showing that we can work out a way of looking at the classification, within the framework proposed here, which does not require that the progressive be a sentence operator. For simplicity, I'll confine myself to three cases: (i) sentences in which the nature of the object affects the nature of the predicate; (ii) cases where one kind of adverbial (hopefully this case will generalize to many) affects the nature of the predication, (iii) cases where the nature of the subject affects the nature of the predication. The first and the last cases were illustrated already. The second case can be illustrated by these sentences:

(20) Max pushed a cart (process)

(21) Max pushed a cart for an hour (accomplishment on one reading)

(Hoepelman, 1976, has discussed some of these examples).
The facts to be explained can be summarized thus:

(i) If an object (subject) in the singular makes a nonprocess predicate (sentence) then replacing the object (subject) by a mass noun or bare plural allows the interpretation of the predicate (sentence) as a process expression.

(ii) If certain adverbials (e.g. *for an hour*) are added to process predicates (sentences) then the result is interpreted as an accomplishment or achievement.

Intuitively, the difference between processes and events (as I prefer to call both of the other types of accomplishments and achievements) is this: an event is a one-time kind of thing, it has a beginning and middle and an end (if it is an accomplishment, if it is an achievement its beginning is its end and it has no middle, cf. Mourelatos, 1978, and Gabbay and Moravcsik, this conference/volume), processes just go on and may never have begun or never end. Events are countable, processes aren't. Processes are like mass nouns, events like count nouns. Suppose we had an ontology in which there were entities like events and processes. Then we could say various things this: if x and y are processes of the same kind and x abuts y then there is a process z of the same kind of which x and y are parts (call this *additivity*). Clearly, events aren't like this. Similarly, if x is a process, we can often subdivide x into parts that are also processes of the same sort. But with events this is never true: if x is an event of a certain sort (*building a cabin*, *finding a unicorn*), then no proper part of x is also an event of that sort. Call this *indivisibility*. Considerations like these, by the way, convince me that it is futile to attempt to reconstruct such notions in terms of properties of moments or intervals of time (as Montague tried to do for events in his paper "On the nature of certain philosophical entities", Paper 6 in Montague, 1974). We certainly can't say that if *John build a cabin* is a property of an interval of time, then there is no proper subinterval of that interval of which *John build a cabin* holds; similarly, there can be several strikings of Shem by Shaun that are simultaneous but different (I owe this last example to Terry Parsons). In fact, I believe that our notions of time itself are built up from more primitive notions about events and the like and their relationships.

Now, still intuitively, it seems obvious that if you put together a verb like *find* with a bare plural or mass noun the result is going to be a process predicate: it is additive and not indivisible. Similarly if you put together a plural subject with a verb like arrive the result will be a process. And if you put together an expression like *for an hour* (or e.g. a goal expression like *to the store*) with a process predicate like *run* or *push a cart* you will now have an event predicate or sentence: pushing a cart for an hour, running to the store, do have beginnings, middles, and ends, are indivisible and not additive.

How are we to understand this classification? There are two aspects to the problem. One is what we might call the *ethnometaphysical* problem. I believe that this problem is best attacked by providing an ontology in which there are things like events (instantaneous and protracted), processes, and states. We may not be able to say what they are exactly, but we can say something about their properties, as I have tried to do above. (Work in progress is devoted to this problem, where I try to reconstruct our notions about time on the basis of more primitive notions about events and the like and their relationships, following a line initiated by Whitehead (1920), Russell (1956) and currently being pursued by Hans Kamp. The second problem we might call the *descriptive* problem: how are we to represent all of this in an explicit account of English syntax and semantics. For example, one might try to build into a fragment of English purely syntactic distinctions corresponding to the metaphysical ones. I believe that such an attempt would be misguided, since it would be treating the classification as purely artibrary, like German gender. In effect, we would be saying that there could be a language just like English, say Shmenglish, in which *find a unicorn* and *push a cart* would mean exactly what they mean in English but had just the opposite properties with respect to the progressive, time adverbials, and the like. So the facts are at least semantic. How should they be reflected in a semantics for English? One way would be to state meaning postulates individually for each verb, since the hub of the interpretation does seem to be the verb (or one might do some kind of decomposition of the meanings of the verbs, à la Jackendoff, 1976; or Dowty, 1976; Hoepelman, 1978). I think that even this is too much, so I'll take a Cresswellian stance and say that it's just part of our understanding of the function denoted by *find* that tells us that *find a unicorn* corresponds to a set of events (cf. Cresswell's remarks on *walk*, 1978).[4]

With this as background, let's introduce the progressive as an operator on verbal phrases. I'm only going to try to give an account of the semantics, and that only in part, and not try to spell out the syntax. Let's use a shorthand for the differing truth conditions that must be given for the different types of expressions: condition P for the case of processes, conditions E-1 and E-2 respectively for the cases of achievement (instantaneous events) and accomplishments (protracted events) (protracted events) (say, along the lines of Vlach, ms. a.).

> We add Prog(γ) to our list of well-formed expressions in the logic, where γ is a meaningful expression of type $<s,f(t/T)>$ *and* Prog(γ) is also. Prog(γ) is that function *h* such that h(^NP') = 1 under conditions P, E-1, E-2, according as γ(^NP') is a process, an instantaneous event, or a protracted event.

4.2 The Past tense.

I endorse the view that there is a semantic difference between the past tense in English and the present perfect. (As a matter of ordinary usage, it must be admitted that the two are often interchangeable.). The difference is that in the present perfect the truth conditions explicitly quantify over a time in the past, while a past tense sentence without a time adverbial must be understood under some contextual assignment of an interval or point in the past. Without such an assignment a past tense sentence has something of the status of a sentence with an unbound pronoun in it (cf. Partee, 1973). Suppose you ask

---

4 Barbara Partee has pointed out to me that the view followed here, taken together with Vlach's treatment of the progressive is quite problematical for a compositional semantics, since it makes it impossible to give an explicit statement of truth conditions for sentences in the progressive. This observation forces us to one of three conclusions: (1) we must include meaning postulates or some other device of the sort rejected here; (2) the semantics of the progressive can't be given compositionally, i.e. it requires reference to matters of knowledge and belief; or (3) there is a uniform truth condition for the progressive, but we just haven't found it yet.

me *Did Mary leave?* Unless it's clear from the context what period you are talking about I can't answer the sentence Yes or No. Some contexts might yield something equivalent to a present perfect sentence. Thus we do get a kind of pragmatic entailment:

If Mary left, then Mary has left.

By now it should be fairly obvious what the *form* of my treatment of the past tense is going to be. It is going to be some kind of modifier or operator on IV phrases (or our new type). One way to capture the context dependence of the past tense just mentioned is to think of the past tense as adding a variable over intervals or points to the interpretation of a sentence. In the next section I'll treat time adverbials as elements that can bind such a variable. So the first question we might ask is this: should this variable be a variable over points or intervals? It seems as if both kinds of adverbials can specify the context of a past tense sentence:

What happened at 3 o'clock? Mary left.
What happened yesterday? Mary left.

At 3 o'clock is a typical point-time adverbail, *yesterday* a typical interval adverbial.

Vlach (ms. b.) has argued that the semantics of tense and time in English (and undoubtedly other languages) needs three somewhat independent notions: truth *in* an interval, *at* an interval, and *for* (throughout) an interval. I believe this is correct. For example, sentences like *Mary builds a cabin* or *Mary finds a unicorn* are true at intervals (the latter an instant), sentences like them can be true in intervals (iff they are true at some subinterval of the interval): *Mary built a cabin in 1972*. Neither kind of sentence can be true *for* an interval. Sentences about processes and states can be true *at* as well as *for* intervals. Note, however, that we can reduce *in* to *at* in the special case of an instant. I believe that the proper view of the past tense variable is as one ranging over intervals and that truth for past tense sentences should be thought of in terms of the *in* relation. In this way we can explain why the past tense variable can be bound either by a point-time adverbial like *at 3* or a proper interval variable like *yesterday*. Thus, I think that the correct semantics of the past tense is rendered almost exactly by the English phrase *in the past*.

The way I propose to do this is by introducing a distinguished variable over intervals in the translation of past tense sentences, in some manner yet to be spelled out. A pragmatic theory in the sense of Stalnaker (1978) will assign values to this variables from context if the variable remains unbound. In the next section I'll suggest that certain time adverbials can act to bind this distinguished variable.

I was hung up for a long time by the fallacious assumption that the use of a single distinguished variable would require giving it the same value throughout a sentence. On that view a conjunction of sentences in the past tense would require the same contextually specified value for the variable in each conjunct. To avoid this consequence, I worked out a very elaborate system using an infinite set of indexed past tense variables in the syntax, somewhat like Montague's indexed variables for pronouns. But clearly, parts of a sentence can themselves help to establish the context, and hence the context can change throughout a sentence, as von Stechow (1977) has argued. Consider the following sentences:

(22) Mary got married and got pregnant.

If we interpret this sentence as containing two occurrences of a free variable over intervals and if we allow the first part of the sentence to change the context then the value of the free variable can also change. On the other hand, addition of a point time adverbial can bind (indeed must bind on the normal interpretation) both occurrences of the variable:

(23) At 3 o'clock (on such and such a day) Mary got married and got pregnant.

This requires a somewhat unusual wedding ceremony to be sure.).

Dowty (1977) suggested that the past tense be considered to get into sentences when time adverbials are added to them. I am going to consider the past tense as *being* in some sense an adverbial, but in line with the general approach of this paper as a modifier at the IVP level, and as just suggested, as an interval adverbial. So the semantics of a past tense sentence is going to be just like that of a sentence with an interval adverbial in it. Suppose I is an interval, then we can say this:

In $I(\gamma)$ is that function h defined for NP intensions such that $h(\alpha) = 1$ at i, j iff $\gamma(\alpha) = 1$ in I at i, j.

Now let's translate verb phrases in the past tense as containing a distinguished variable I_P. A pragmatic rule would take a sentence with this variable in it and interpret it from context. What we need from the context for such an interpretation is the set of ordered pairs $\langle I, j \rangle$ such that I is an interval and j a moment and $I < j$.

4.3 *Time adverbials.*

Are time adverbials like *yesterday, at 3, when I arrived,* sentence operators or verb-phrase operators? My answer is yes. In a sentence like (24) *yesterday* is acting like a sentence modifier, in (25) as an IVP modifier (or possibly also as a sentence modifier):

(24) Yesterday, Mary left and John arrived.

(25) Mary left yesterday.

Let's let such expressions be interpreted primarily as IVP modifiers. We can handle interval adverbials like *yesterday* and point adverbials like *at 3* with the machinery set up so far. They can receive these interpretations:

yesterday: $\lambda \mathbb{P}\, \lambda I_p\, \mathbb{P}$ (*yesterday*')

at 3 : $\lambda \mathbb{P}\, \lambda I_p\, \mathbb{P}$ (*at-3*')

And we can provide a rule that lets each such adverbial act also as a sentence operator, binding the distinguished past tense variable:

yesterday: $\lambda p\, \lambda I_p\, p$ (*yesterday*')

at 3 : $\lambda p\, \lambda I_p\, p$ (*at 3*)

All we need do is say that if I is a moment, than *in I* reduces to *at I*.

Examples (24) and (25) work like this:

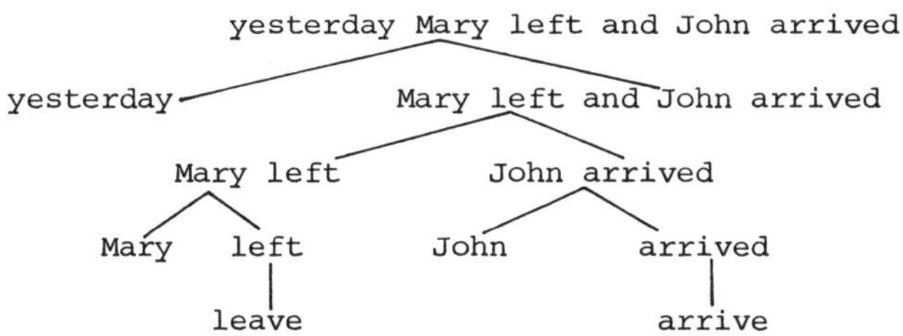

Translation: $\lambda p \lambda I_p p(\textit{yesterday}')\lambda_p\ \lambda I_p P(\textit{yesterday}')(\hat{}[\mathrm{In}\ I_p(\textit{leave}')(\hat{}m^{*})] \wedge [\mathrm{In}\ I_p\ (\textit{arrive}')(\hat{}j^{*})])$.

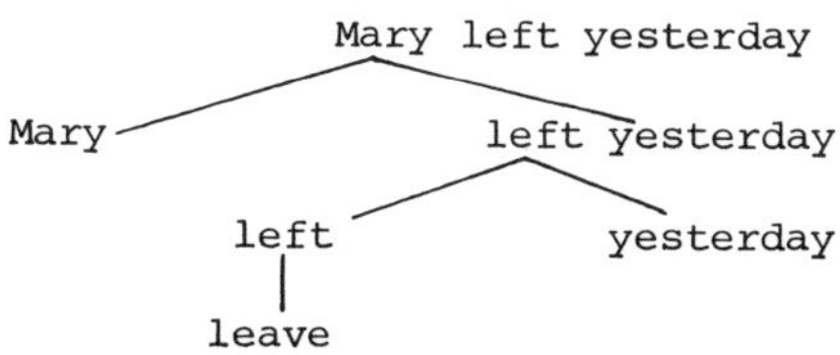

Translation: $[\lambda \mathbb{P} \lambda I_p\ (\textit{yesterday}')\mathbb{P}\ (\mathrm{In}\ I_p\ (\mathrm{leave}'))](\hat{}m^{*})$

It remains for us to say something about durational adverbials like *for three hours*, *between 3 and 4*. (I'm excluding the modal temporals with explicit *for* as in the natural interpretation of *Mary got up for three hours*, which are a whole other story.). Given that we have interpreted the past tense as an *in*-type adverbial, how can we connect it up to durational adverbials?

I believe that the answer to this question is that durational adverbials do not function to bind the past tense variable at all and that we must think of a sentence like (3) not as somehow resulting from the application of a durational adverbial to a past verb phrase (or sentence) but rather as the past of a sentence with a durational in it already:

(26) Mary ran for an hour.

In Priorian language (3) asserts that it was the case that Mary 'runs' for an hour rather than it is the case for an hour that Mary ran. But that story and the right interpretation of frequency adverbials takes us into mysteries of the present tense that I am not prepared to discuss in this paper.[5]

---

5 I would like to thank Irene Heim and Barbara H. Partee for discussion and critical comments on this paper. Many of the details of the paper were worked out or initiated in discussions with Rick Saenz. Naturally, none of these friends should be held responsible for any blunders on my part. Much of my preliminary research on tenses and aspects was carried out at the Center for Advanced Studies in the Behavioral Sciences, in part under a grant from the National Endowment for the Humanities.

Appendix A: A minimal revision of PTQ

1. Remove *be* from B_{TV}.
2. Delete S4 and S17.
3. Add these categories:

   t/T: the category of tensed verb phrases

   t////e: the category of non-verbal (copular IV's (in *be*) (abbreviation: COP) and the categories whose basic members are as follows:

   $B_{COP/T}$: {*be*}

   B_{Aux-1} : ((t/T)/COP): {*will, won't, ...*} (true modals)

   B_{Aux-2} : ((t/T)/(t/e)): {doesn't, ...} (tensed forms of *do*)

   B_{Aux-3} : ((t/T)/T): {*is, isn't, ...*} (tensed forms of *be*)

   B_{Aux-4} : ((t/T)/(t/e)): {*has, hasn't, ...*} (tensed forms of *have*)

4. Assume basic forms of pronouns are him_i.
5. Assume the definitions of these relations:

   main verb(s) of, main term phrase(s) of

6. Define the following subfunctions:

   NOM(α) where α is a term-phrase = the result of replacing all main term phrases of α that have the form him_i, by he_i, otherwise leave them the same.

   PRES(γ) where γ is in P_{IV} = the result of replacing the main verb(s) of γ by their present tense singular forms.

   EN(γ) where γ is in P_{IV} or P_{COP} = the result of replacing the main verb(s) of γ by their past participle forms.

7. Add these rules:

   S99 : (lexical): If γ is in P_{Aux-1}, then γ is in P_{Aux-2}.
   Tr : γ'.

   S100 : If γ is in P_{IV}, then PRES(γ) is in $P_{t/T}$.
   Tr : $\lambda\mathcal{P}\mathcal{P}\{\hat{}\gamma'\}$

   S101 : If γ is in $P_{COP/T}$ and α is in P_T, then $F_6(\gamma,\alpha)$ is in P_{COP}.
   Tr : $\gamma'(\hat{}\alpha')$

S102 : If γ is in P_{Aux-1}, δ is in P_{COP}, then $F_6(\gamma,\delta)$ is in $P_{t/T}$.

Tr : $\gamma'(\hat{}\delta')$

S103 : If γ is in P_{Aux-2}, δ is in P_{IV}, then $F_6(\gamma,\delta)$ $P_{t/T}$.

Tr : $\gamma'(\hat{}\delta')$

S104 : If γ is in P_{Aux-3}, α is in P_T, then $F_6(\gamma,\alpha)$ is in $P_{t/T}$.

Tr : $\gamma'(\hat{}\alpha')$

S105 : If γ is in P_{Aux-4}, δ is in $P_{IV} \cup P_{COP}$, then $F_{100}(\gamma,\delta)$ is in $P_{t/T}$

where $F_{100}(\gamma,\delta) = \gamma EN(\delta)$

Tr : $\gamma'(\hat{}\delta')$

S106 : If γ is in $P_{t/T}$ and α is in P_T, then $F_{101}(\gamma,\alpha)$ is in P_t

where $F_{100}(\gamma,\alpha) = NOM(\alpha)$

Tr : $\gamma'(\hat{}\alpha')$

New translations:

will	:	$\lambda P \lambda \mathcal{P}\ W \mathcal{P}\{P\}$
won't	:	$\lambda P \lambda \mathcal{P} \neg W \mathcal{P}\{P\}$
doesn't	:	$\lambda P \lambda \mathcal{P} \neg \mathcal{P}\{P\}$
has	:	$\lambda P \lambda \mathcal{P}\ H \mathcal{P}\{P\}$
hasn't	:	$\lambda P \lambda \mathcal{P} \neg H \mathcal{P}\{P\}$
is	:	$\lambda \mathcal{P}_1 \lambda \mathcal{P}_2\ \mathcal{P}_2 \{\hat{x} \mathcal{P}_1 \{\hat{y}[{}^{\vee}x = {}^{\vee}y]\}\}$
isn't	:	$\lambda \mathcal{P}_1 \lambda \mathcal{P}_2 \neg \mathcal{P}_2 \{\hat{x} \mathcal{P}_1 \{\hat{y}[{}^{\vee}x = {}^{\vee}y]\}\}$

References

Akmajian, A. and Wasow, T., 1975, The constituent structure of VP and AUX and the position of the verb BE. *Linguistic Analysis* 1: 205-45.

Bäuerle, R., 1977, Tempus, Temporaladverb und die temporale Frage. Universität Konstanz: *Sonderforschungsbereich 99 "Linguistik"* 13, 16.

Bennett,M., 1974, Some extensions of a Montague fragment of English. Unpublished doctoral dissertation, UCLA.

Cresswell, M.J., 1978, Prepositions and points of view. *Linguistics and Philosophy* 2: 1-41.

Dowty, D., 1972, Studies in the logic of verb aspect and time reference in English. Unpublished doctoral dissertation, The University of Texas, Austin.

Dowty, D., 1976, Montague grammar and the lexical decomposition of verbs. In Barbara H. Partee, ed., *Montague Grammar* (New York: Academic Press).

Dowty, D., 1977, Toward a semantic analysis of verb aspect and the English 'Imperfective' progressive. *Linguistics and Philosophy* 1: 45-79.

Gazdar, G., 1979, Constituent structures. (Unpublished paper).

Hoepelman, J., 1976, Mass-nouns and aspects, or: Why we can't eat gingercake in an hour. *Amsterdam Papers in Formal Grammar* 1: 132-153.

Hoepelman, J., 1978, The treatment of activity verbs in a Montague-type grammar, a first approximation. In Franz Guenthner and Christian Rohrer, eds., *Studies in Formal Semantics* (Amsterdam: North-Holland).

Keenan, E.L. and Faltz, L.M., 1978, Logical types for natural language. *UCLA Occasional Papers in Linguistics* 3.

Jackendoff, R., 1976, Toward an explanatory semantic representation. *Linguistic Inquiry* 7: 89-150.

Lapointe, St., 1978, Subcategorization and co-occurrence restrictions. (Unpublished manuscript).

Liberman, M. and Prince A.S., 1977, On stress and linguistic rhythm. *Linguistic Inquiry* 8: 249-336.

Montague, R., 1974, *Formal Philosophy*. Ed. by Richmond Thomason. New Haven.

Mourelatos, A.P.D., 1978, Events, processes, and states. *Linguistics and Philosophy* 2: 415-434.

Partee, B.H., 1973, Some structural analogies between tenses and pronouns in English. *Journal of Philosophy* 70: 601-609.

Russell, B., 1956, *Logic and knowledge*, Allen & Unwin, London.

Selkirk, E.O., (forthcoming) The role of prosodic categories in English word stress. (To appear in Linguistic Inquiry.).

Stalnaker, R., 1978, Assertion. In Peter Cole, ed., *Syntax and Semantics* 9: *Pragmatics* (New York: Academic Press).

Stechow, A., von, 1977, Occurrence-interpretation and context-theory. Universität Konstanz: *Sonderforschungsbereich 99 "Linguistik"* 14.

Thomason, R.H. and Stalnaker R.C., 1973, A semantic theory of adverbs. *Linguistic Inquiry* 4: 195-220.

Vlach, F., (forthcoming, ms. a.) The semantics of the progressive. (Paper presented in the 1978 Brown University conference on tense and aspect, to appear in the proceedings.).

Vlach, F., (ms. b.) The semantics of tense and aspect. (Draft, January 1979).

Whitehead, A.N., 1920, Concept of Nature. Cambridge: University Press.

POINTS AND PERIODS[1]

Johan van Benthem, Filosofisch Instituut / Rijksuniversiteit Groningen

1. *Introduction*

In traditional tense logic statements are evaluated at *moments*; i.e., at points in time (cf. [12]). More recently, however, *intervals* (periods of time) have been proposed as the appropriate indices of evaluation (cf. [5]). In fact, intervals had been used already - even within traditional tense logic - in the study of *events*; being their temporal "substrata", so to speak (cf. [1]).[2] In this paper the connection between the point-approach and the period-approach will be discussed, both from a mathematical and a locical point of view.

This possible choice between moments and intervals of time is interesting for various linguistical and philosophical reasons. Although "point" would seem to be the more elementary concept theoretically, it comes later in the factual experience of time; "periods" being primary. There is a clear analogy with geometry in this respect: extended bodies come before points in our spatial experience, though not in the resulting theory of space.[3] It would be tedious to argue in general terms for or against one approach.[4]

---

1 At the Stuttgart conference Hans Kamp has presented a contribution rather similar to the present one. Moreover, the subject of this paper turned out to have a long intellectual pre-history, going back to A.N.Whitehead and B.Russell & N.Wiener (cf. [16], [14]). Further comments provided by several participants have been collected in some of the following notes.

2 More involved temporal substrata for events have been proposed by Gabbay & Moravscik. In this paper only the simplest case is studied.

3 There is a contemporary logico-philosophical program in which the theory of space is developed in the order of experience; viz. P.Lorenzen's "proto-geometry" (cf.[9]). This could be viewed as an attempt to bridge the gap between the common sense world (the world of natural language?) and the world of science.

4 Note that, at least, the issue cannot be settled simply by invoking natural language as a kind of impartial arbiter. (Cf. the argument about "the wedding moment" given in [2].).

Instead, both will be considered here.[5]

The usual procedure in tense logic is to go "from points to periods". One constructs periods as point-sets; e.g., using interval topologies on some time axis. But mathematics also provides for the converse direction, "from periods to points", by means of the Stone representation which introduces points as maximal filters of intervals (cf. [4]). This method forms the main theme of this paper.[6] It is related to the Henkin method for completeness proofs in intensional logic, as given in [8].[7] In addition to such *mathematical* correspondences, there is also the question of *logical* connections between statements (in some tense-logical language) at one level and "similar" statements at another. Here the situation is much less clear, and the present paper has nothing more to offer than some tentative remarks.

Finally, a remark about the exposition. The main line of reasoning is relatively immune to additions or changes, it is hoped. This means, e.g., that it could be modified to accommodate richer conceptions of events than the one presented here. In order to make that main line as accessible as possible it has been given in a heuristic fashion, following the suggestions of Lakatos in [7]. People preferring a more conventional deductive style will have to do some logical reconstruction.

---

5 Just for the record, I sympathize with the interval approach in semantics. That is not to say, however, that all arguments advanced against the traditional point-approach are valid, or even fair.

6 In the area of modal logic, this subject was investigated quite thoroughly in [6], which develops the duality theory of modal algebras and Kripke models. This theory could be transplanted to tense logic;but there would be essential differences with the treatment given here. (One would need "tense *operations*" in event structures, instead of having a precedence relation.)

7 For the benefit of readers interested in broad historical outlines, one could sum up years of logical research in one equation, viz. HENKIN = STONE + LINDENBAUM. (Cf. [13].).

2. *From moments to events*

2.1 *Time structures induce interval structures*

A *time structure* $\mathcal{T} = \langle T, < \rangle$ consists of a non-empty set T of points of time and a binary relation $<$ of temporal precedence which is *transitive* (i.e., $\forall t_1, t_2, t_3 \in T (t_1 < t_2 < t_3 \Rightarrow t_1 < t_3)$) and *irreflexive* (i.e., $\forall t \in T\ t \not< t$). *Linearity* (i.e., $\forall t_1, t_2 \in T (t_1 = t_2 \vee t_1 < t_2 \vee t_2 < t_1)$) is not required. From such structures various kinds of interval may be constructed; "open-ended", "closed" and the like. To fix our thoughts, let us consider the general class of *convex* subsets of T; i.e., those sets $X \subseteq T$ for which

$$\forall t_1, t_2 \in X\ \forall t_3 \in T (t_1 < t_3 < t_2 \Rightarrow t_3 \in X).$$

These may be thought of as "uninterrupted" stretches of time. Clearly, T itself is convex, and so is the empty set. All singleton subsets {t} ($t \in T$) are convex too. (If $t < t' < t$, then $t < t$; which contradicts irreflexivity.[8])

For standard reference in the sequel, the following three time structures are useful. All have "smaller than" as their precedence relation.

ZZ = $\langle Z, < \rangle$: the integers:

$\mathbb{Q}$ = $\langle Q, < \rangle$: the rationals:

IR = $\langle R, < \rangle$: the reals:

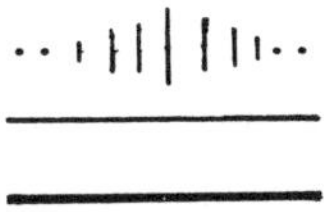

If the reader wants a non-linear example, to avoid being prejudiced in the checking of proofs, it is useful to keep the binary tree in mind with the usual precedence ordering:

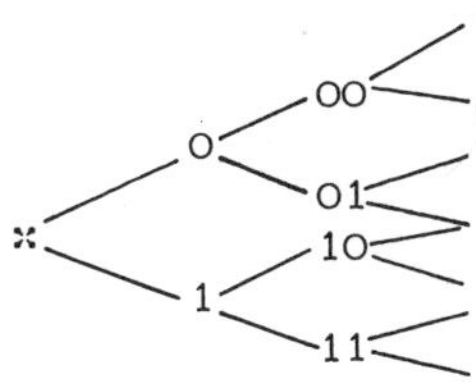

E.g., the non-trivial convex subsets of Q are all subsets of the following forms.

8 If $<$ is not transitive, then singleton subsets need not be convex. Cf. the structure $\langle \{0,1\}, \{\langle 0,1\rangle, \langle 1,0\rangle\}\rangle$

Open intervals:

			$(q_1, q_2 \varepsilon R!)$
(q_1,q_2)	$=_{def}$	$\{q\varepsilon Q \mid q_1 < q < q_2\}$	$(q_1 < q_2)$
$(q_1,+\infty)$	$=_{def}$	$\{q\varepsilon Q \mid q_1 < q\}$	
$(-\infty,q_2)$	$=_{def}$	$\{q\varepsilon Q \mid q < q_2\}$	

Closed intervals:

			$(q_1, q_2 \varepsilon Q)$
$[q_1,q_2]$	$=_{def}$	$\{q\varepsilon Q \mid q_1 \leqslant q \leqslant q_2\}$	$(q_1 \leqslant q_2)$
$[q_1,+\infty)$	$=_{def}$	$\{q\varepsilon Q \mid q_1 \leqslant q\}$	
$(-\infty,q_2]$	$=_{def}$	$\{q\varepsilon Q \mid q \leqslant q_2\}$	

Half-open intervals ("pins"):

$(q_1,q_2]$	$=_{def}$	$\{q\varepsilon Q \mid q_1 < q \leqslant q_2\}$	$(q_1 \varepsilon R, q_2 \varepsilon Q;\ q_1 < q_2)$
$[q_1,q_2)$	$=_{def}$	$\{q\varepsilon Q \mid q_1 \leqslant q < q_2\}$	$(q_1 \varepsilon Q, q_2 \varepsilon R;\ q_1 < q_2)$.

Note that convex subsets of Q have either one element (the closed intervals [q,q] are the singleton subsets) or (countably) infinitely many.

Any intersection of convex sets is itself convex. Unions of convex sets need not be convex, however; witness the rational interval $(0,1) \cup (1,2)$. Such unions correspond to "repetitive" events. Note that, by the distributive laws, any intersection of such unions is itself a union of (intersections of) convex sets. For any time structure $\mathcal{C} = \langle T, < \rangle$, the resulting set of subsets of T forms a *distributive lattice* (cf.[4]) with respect to the operations $\cap$ and $\cup$. It also has a smallest element $\emptyset$ and a greatest element T.

Example: The finite unions of convex subsets of ZZ are precisely all finite and cofinite subsets of Z. ($X \subseteq Z$ is *cofinite* if Z-X is finite.).

Complements played no role in the preceding. This is because not every event need have a "complementary" event covering exactly the remaining periods of time. In particular cases, the above lattice may be closed under complements, however. This happens, e.g., in the previous example - as is explained by the following result.

Proposition:
If $<$ is linear in $\langle T, < \rangle$, then the complement of a finite union of convex subsets of T is itself a finite union of convex subsets of T.

In fact, there is even a related result without linearity.

Proposition:

The complement of any convex set is itself a finite union of convex sets.

Proof:

Suppose that $X \subseteq T$ is convex. It suffices to show that both

$$X_1 =_{def} \{t \varepsilon T \mid \neg \exists t' \varepsilon X\ t < t' \ \&\ t \notin X\}$$

and

$$X_2 =_{def} \{t \varepsilon T \mid \exists t' \varepsilon X\ t < t' \ \&\ t \notin X$$

are convex, since

$$X_1 \cup X_2 = T - X.$$

First, if $t_1, t_2 \varepsilon X_1$ and $t_1 < t_3 < t_2$, then $t_3 \notin X$. (Otherwise, t_1 would have a $<$-successor in X, contradicting the fact that it is in X_1.). Moreover, if $\exists t' \varepsilon X\ t_3 < t'$, then $\exists t' \varepsilon X\ t_1 < t'$ (by transitivity): again a contradiction. It follows that $t_3 \varepsilon X_1$.

Next, if $t_1, t_2 \varepsilon X_2$ and $t_1 < t_3 < t_2$, then $t_3 \notin X$. (Otherwise, we would have $t_3 < t_2 < t_4$ for some $<$-successor t_4 of t_2 which belongs to X, and hence $t_2 \varepsilon X$, since X is convex. This contradicts the fact that $t_2 \varepsilon X_2$.). Moreover, by transitivity, t_3 has a $<$-successor in X because t_2 has.

It follows that $t_3 \varepsilon X_2$. QED.

A simple application of De Morgan laws and distributivity now shows that the complement of any finite union of convex sets is itself a finite union of convex sets; whether $<$ is linear or not. (This observation is due to S.K. Thomason.).

When our lattice is closed under complements it may be turned into an ordinary *Boolean algebra* (cf. [4]). In the sequel, both complements and unions will be disregarded:[9] only intersections are needed.

The relation $<$ on $\mathcal{T}$ induces a precedence relation on the convex sets in an obvious way, yielding the following kind of structure.

Definition:

Let $\mathcal{T} = \langle T, < \rangle$ be a time structure. The *convex event structure* $C(\mathcal{T})$ is the quadruple $\langle C(T), \emptyset, \cap, < \rangle$, where $C(T)$ consists of all convex subsets of T and $<$ is the relation on $C(T)$ defined by

---

9 In an earlier draft, unions (and hence distributuve lattices) were used throughout. The present parsimony makes the paper more perspicuous, however, without changing anything essential.

$i < j$ if and only if $\forall t_1 \varepsilon i \forall t_2 \varepsilon j\ t_1 < t_2$.[10]

Generally speaking, certain properties of $<$ will be lost in the transition from $\mathcal{T}$ to $C(\mathcal{T})$. E.g., even when $<$ is linear on $\mathcal{T}$, it need not be linear on $C(\mathcal{T})$. A counterexample is provided by the incomparable intervals (0,2) and (1,3) in $C(\Phi)$. But, at least, $\langle C(T),<\rangle$ is a time structure: the induced relation $<$ remains transitive and irreflexive (provided one disregards minor anomalies involving the empty set).

Defining *inclusion* in the usual way:

$i \subseteq j$ if and only if $i \cap j = i$,[11]

one obtains the following properties of the structures $C(\mathcal{T})$.

(1) $\forall i,j \varepsilon C(T)(i \neq j \Leftrightarrow \exists k \neq \emptyset((k \subseteq i\ \&\ k \cap j = \emptyset) \vee (k \subseteq j\ \&\ k \cap i = \emptyset)))$.

[Sketch of proof: If $i \neq j$, say $t \varepsilon i-j$, then the convex set $\{t\} \neq \emptyset$, $\{t\} \subset i$ and $\{t\} \cap j = \emptyset$. If $k \neq \emptyset$, $k \subseteq i$, $k \cap j = \emptyset$, then $i \neq j$.]

(2) $\forall i,j \varepsilon C(T)(i \not< j \Leftrightarrow \exists k \neq \emptyset, l \neq \emptyset (k \subseteq i\ \&\ l \subseteq j\ \&\ \forall m \subseteq k, n \subseteq l\ (m \not< n \vee m = \emptyset \vee n = \emptyset)))$.

[Sketch of proof: If $i \not< j$, say $t_1 \varepsilon i, t_2 \varepsilon j$, $t_1 \not< t_2$, then take $\{t_1\}$, $\{t_2\}$ for k,l respectively. For the converse, note that if $k \subseteq i, l \subseteq j$, $k \not< l$, then $i \not< j$.]

(3) $\forall i,j,k \varepsilon C(T)((i \subseteq j\ \&\ j < k) \Rightarrow i < k)$

(4) $\forall i,j,k \varepsilon C(T)((i < j\ \&\ k \subseteq j) \Rightarrow i < k)$.

[These are obvious monotonicity requirements.]

(5) $\forall t_1,t_2 \varepsilon T(t_1 \neq t_2 \Leftrightarrow \exists i,j \varepsilon C(T)(t_1 \varepsilon i\ \&\ t_2 \varepsilon j\ \&\ i \cap j = \emptyset))$.

[This is a separation postulate comparable to Leibniz's Law of Identity.]

(6) $\forall t_1,t_2 \varepsilon T(t_1 < t_2 \Leftrightarrow \exists i,j \varepsilon C(T)(t_1 \varepsilon i\ \&\ t_2 \varepsilon j\ \&\ i < j))$.

(7) $\forall t \varepsilon T, i \varepsilon C(T)(t \not\varepsilon i \Leftrightarrow \exists j \varepsilon C(T)(t \varepsilon j\ \&\ j \cap i = \emptyset))$.

---

10 A different notion of induced precedence was proposed by Ruth Manor. One could set $i < j$ iff $i \cap j$ is convex and $i-j, j-i$ are non-empty sets such that $i-j$ precedes $j-i$ in our stricter sense. The effect of this is to make $<$ better-behaved as far as comparability of intervals is concerned.
Manor also takes interval boundaries to be "fuzzy"; a possibility which has not been considered here.

11 Another interesting approach would take $<$ and $\subseteq$ as basic relations in event structures, dropping the operation $\bullet$. Kamp takes still different primitives, viz. "precedence" and "overlapping". In a recent manuscript [11] , Needham uses the mereological "part of" relation as well as a "betweenness"-predicate (in terms of which precedence may be defined).

The reasons for mentioning these particular properties will be given in section 3.1. Then it will also become clear that one needs a more general concept of event structure than the one described here. In fact, all collections of sets based upon some time structure which satisfy "enough" of the above conditions are to be considered. For this purpose, it is useful to generalize the concept of a "time structure" itself.

Definition:

A *general time structure* $\mathcal{T}$ is a triple $\langle T,<,\mathcal{S}\rangle$ such that

(i) $\langle T,<\rangle$ is a time structure,

(ii) $\mathcal{S}$ is a set of subsets of T which contains $\emptyset$ and is closed under intersections,

(iii) $\mathcal{S}$ satisfies the above condition (1) (with "C(T)" replaced by "$\mathcal{S}$") and

(iv) for the relation < lifted to $\mathcal{S}$ like in the preceding definition, the above condition (2) holds.[12]

Each convex event structure $C(\mathcal{T})$ corresponds to the general time structure $\langle T,<,C(T)\rangle$. Note that the above properties (3), (4) hold for all general time structures.

In order to get the flavour of conditions (1), (2), we consider a few examples. Notice that in the above proofs for the case of convex sets, essential use was made of the presence of *singletons*. This is the case, e.g., in the general time structure consisting of IR together with all closed intervals. But they also hold for IR together with all open intervals; i.e., in the absence of singletons. [If i $\not<$ j, i.e., $t_1 \varepsilon i, t_2 \varepsilon j$ and $t_1 \not< t_2$, for certain t_1, t_2, then $t_2 \leqslant t_1$. Since j is open (in the topological sense), t_2 is not a boundary point of j, hence a subinterval l of j exists which is completely "to the left of" t_2.
Similarly, i contains a subinterval k completely "to the right of" t_1.
These two are the required intervals; for even l < k.] IR together with all finite unions of open intervals does not satisfy (1), however. Consider, e.g., the distinct members $(0,1) \cup (1,2)$ and $(0,2)$: no k as required exists.[13] remains true in this case.

Finally, the following concept will be needed.

---

12 This move from "standard" time structures to "general" time structures is only to be expected from the analogy with the duality theory of Kripke frames and modal algebras (cf. [6]), which has led to the introduction of so-called "general frames" into modal logic. (Cf. also [15].).

13 This does not mean that the Stone construction of section 3.1 cannot be applied to this structure. It only follows that $(0,1) \cup (1,2)$ and $(0,2)$ will become identified in the resulting general time structure.

(2) remains true in this case.

Finally, the following concept will be needed.

Definition:
Let $\mathcal{T} = \langle T, <, \mathcal{S}\rangle$ be a general time structure.
The *derived event structure* $i(\mathcal{T})$ *of* $\mathcal{T}$ is the quadruple $\langle \mathcal{S}, \emptyset, \cap, < \rangle$, with the relation $<$ on $\mathcal{S}$ defined like above.

2.2. *Matters of language*

The preceding section was devoted to the mathematical connection between point-structures and interval-structures. There is also a variety of logical questions concerning connections between (appropriate) languages interpreted at these two levels. Only a few obvious ones will be stated here.

To begin with, there are traditional questions concerning the language of propositional tense logic. The preceding section suggests several operations on time structures, and one might ask which tense-logical principles are preserved by them. E.g., consider $\mathbb{Q}$. Its tense-logical theory is axiomatized by the minimal tense logic K_t together with the axioms FT ("succession to the right"; i.e., $\forall t_1 \exists t_2\ t_1 < t_2$), PT ("succession to the left"; i.e., $\forall t_1 \exists t_2\ t_2 < t_1$), $Gp \rightarrow GGp$ (transitivity), $GGp \rightarrow Gp$ (denseness; i.e., $\forall t_1, t_2 (t_1 < t_2 \Rightarrow \exists t_3\ t_1 < t_3 < t_2)$) and $G((Gp \wedge p) \rightarrow Gq) \vee G(Gq \rightarrow Gp)$ ("linearity to the right"; i.e., $\forall t_1, t_2, t_3 ((t_1 < t_2\ \&\ t_1 < t_3) \Rightarrow (t_2 = t_3 \vee t_2 < t_3 \vee t_3 < t_2)))$, $H((Hp \wedge p) \rightarrow Hq) \vee H(Hq \rightarrow Hp)$ ("linearity to the left"; i.e., $\forall t_1, t_2, t_3 ((t_2 < t_1\ \&\ t_3 < t_1) \Rightarrow (t_2 = t_3 \vee t_2 < t_3 \vee t_3 < t_2)))$ (cf. [3]). Now consider the frame $BC(\mathbb{Q})$ consisting of all bounded convex intervals of Q as ordered by the above relation $<$. What is its tense-logical theory?[14]

---

14 Our conjecture is: K_t, FT, PT, $Gp \rightarrow GGp$ and $FGp \rightarrow GFp$ ("directedness"; i.e., $\forall t_1, t_2, t_3 ((t_1 < t_2\ \&\ t_1 < t_3) \Rightarrow \exists t_4 (t_2 < t_4\ \&\ t_3 < t_4))$, $PHp \rightarrow HPp$. Directedness is all that remains of linearity. Denseness disappears, witness situations of "immediate precedence" like that of (0,1) and (1,2). This, by the way, is the sense in which expressions like "the next moment" may be used consistently, while preserving the intuition that (the underlying) time is dense.

A similar question concerns the time structure $\langle C(Q),<\rangle$.

One would prefer a closer connection between the two levels, however. Let us follow the most simple-minded approach.[15] Consider a tense-logical model $\langle \mathfrak{T},V\rangle$ where $\mathfrak{T}=\langle T,<\rangle$ is a time structure and V is a *valuation* assigning subsets of T to the proposition letters of the language. A valuation C(V) on $C(\mathfrak{T})$ is to be found such that, for all tense-logical formulas ϕ , truth of ϕ at an interval i and truth of ϕ at the points $t\varepsilon i$ are related in some natural fashion. The following correspondence seems reasonable.

$$(\ast)\quad \langle C(\mathfrak{T}),C(V)\rangle \models \phi[i] \text{ if and only if } \langle \mathfrak{T},V\rangle \models \phi\ [t] \text{ for all } t\varepsilon i.$$

E.g., one might start by setting $C(V)(p)=_{def}\ \{i\varepsilon C(T)\,|\,i\sqsubseteq V(p)\}$; and possibly V could be required to satisfy some useful restrictions. Obviously, this procedure requires us to redefine the logical constants at the interval level, since there is no reason to suppose that non-truth of ϕ at i (i.e., ϕ is false at *some* $t\varepsilon i$) will coincide with truth of $\neg\phi$ at i (i.e., ϕ is false at *all* $t\varepsilon i$); except maybe in special cases. In other words, negation should be taken in a more intuitionistic sense, say like

$$\langle C(\mathfrak{T}),C(V)\rangle \models \neg\,\phi[i] \text{ iff } \langle C(\mathfrak{T}),C(V)\rangle \not\models \phi[j] \text{ for all } j\subseteq i.$$

But this implies that disjunction cannot be taken in its usual sense either. For, $\phi\vee\neg\phi$ is true throughout $\langle \mathfrak{T},V\rangle$ and hence it should be true in each interval, because of ($\ast$). One solution might be as follows:

$$\langle C(\mathfrak{T}),C(V)\rangle \models \phi\vee\psi[i] \text{ iff, for all } j\subseteq i, \text{ there exists } k\subseteq j \text{ such that either } \langle C(\mathfrak{T}),C(V)\rangle \models \phi[k] \text{ or } \langle C(\mathfrak{T}),C(V)\rangle \models \psi[k].$$

We leave further exploration of such possibilities to active and ambitious readers.

Several more sophisticated languages are available for interpretation at our two levels. Here we just mention the much more elaborate tense logic of [1], the ordinary tensed predicate logic of [10] and, of course, second order tensed predicate logic, which spans both levels at once.

---

15 One sense in which this approach is simple-minded is that one might just as well look for a *translation* τ from the language of propositional tense logic to some language at the interval level and prove ($\ast$) with "$\tau(\phi)$" on the left-hand side. This objection was made by Dov Gabbay. Still, the original approach may have heuristic value for a truth definition of the traditional language at the interval level.

3. *From events to moments*

3.1 *Representation by maximal filters*

Abstracting from the set-theoretic origin of the event structures in section 2.1, one arrives at the following concept of a more algebraic nature. It should be stated beforehand that the definition given here is "proof-generated" in Lakatos' sense (cf. [7]):the postulates were extracted from the proof of the main theorem below.[16]
Thus, the proof itself serves *as a heuristic device* to find out which conditions are needed for representation.[17]

Definition:

An *event structure* is a quadruple $\mathcal{E} = \langle I, O, \cdot, < \rangle$, where I is a non-empty set (of "intervals"), $O \in I$, $\cdot$ is a binary operation on I and $<$ is a binary relation, satisfying the following postulates (involving the inclusion relation $\subseteq$ defined by: $i \subseteq j$ iff $i \cdot j = i$):

(1) $\forall i \in I \; O \cdot i = O$

(2) $\forall i,j \in I \; i \cdot j = j \cdot i$ (commutativity)

(3) $\forall i,j,k \in I \; (i \cdot j) \cdot k = i \cdot (j \cdot k)$ (associativity)

(4) $\forall i \in I \in i \cdot i = i$ (contraction)

(5) $\forall i,j,k \in I \; ((i < j \;\&\; k \subseteq j) \Rightarrow i < k)$

(6) $\forall i,j,k \in I \; ((i \subseteq j \;\&\; j < k) \Rightarrow i < k)$ (monotony)

(7) $\forall i,j,k \in I \; ((i < j \;\&\; j < k) \Rightarrow i < k)$ (transitivity)

(8) $\forall i,j \in I \; (i < j \Rightarrow i \cdot j = O)$

(9) $\forall i,j \in I \exists k \in I \; (i \neq j \Rightarrow (k \neq O \;\&\; ((k \subseteq i \;\&\; k \cdot j = O) \vee (k \subseteq j \;\&\; k \cdot i = O))))$ (separation)

(10) $\forall i,j \in I \exists k,l \in I \; (i \nless j \Rightarrow (k \neq O \;\&\; l \neq O \;\&\; k \subseteq i \;\&\; l \subseteq j \;\&\; \forall m,n \in I \; ((m \subseteq k \;\&\; n \subseteq l) \Rightarrow (m \nless n \vee m = O \vee n = O))))$.

Now, we want to prove a representation theorem to the effect that any such event structure may be viewed as being generated by some underlying time structure. The obvious method in this field is that of the Stone Representation Theorem (cf.[4]), which introduces points as maximal filters on the

---

16 Hence the reader should not be discouraged by not grasping the point immediately: the proof will supply the motivation. But one could no doubt think up some kind of philosophical motivation at this stage already.

17 In an earlier draft, the full definition of "event structure" was only given after completion of the proof. This turned out to be confusing, even for quite intelligent readers; whence this concession to human frailty.

algebra. The idea is that each point separates the set of events into those which "cover" it and those which do not.[18] For any point, the set of its covering events is closed under intersections and extensions, while not containing O (i.e., it is a (proper) *filter*); and indeed it is maximal with respect to these closure properties. Let us, then, identify points with such maximal filters.[19]

The Stone construction starts from a given event structure $\mathcal{E} = \langle I, O, \cdot, < \rangle$. Define $\underline{U_I}$ to be the set of *maximal filters* on $\mathcal{E}$, i.e., those subsets u of I which satisfy

(i) $\forall i, j \in u \; i \cdot j \in u$
(ii) $\forall i \in u, j \in I (i \subseteq j \Rightarrow j \in u)$
(iii) $O \notin u$: i.e., u is a proper filter on $\mathcal{E}$, and
(iv) if $u \subseteq v \subseteq I$ and v satisfies (i),(ii),(iii), then $v = u$: i.e., u is maximal.

Next, set $\underline{u < v}$ if there exist $i \in u, j \in v$ with $i < j$.
This yields a time structure $\langle U_I, < \rangle$.
Moreover, the obvious map σ defined by setting

$$\sigma(i) =_{def} \{u \in U_I \mid i \in u\}$$

takes elements of I to subsets of U_I. This (hopefully) establishes an isomorphism between $\mathcal{E}$ and intervals of $\langle U_I, < \rangle$; though not necessarily with $C(\langle U_I, < \rangle)$.
The point is that we want to admit only "intervals" of the form $\sigma(i)$, which gives rise to a *general* time structure defined as follows.

Definition:

Let $\mathcal{E} = \langle I, O, \cdot, < \rangle$ be an event structure. The *Stone representation* $\mathfrak{t}(\mathcal{E})$ *of* $\mathcal{E}$ is the triple $\langle U_I, <, \{\sigma(i) \mid i \in I\} \rangle$, with $U_I, <$ and σ as described above.

Examples will be found in section 3.2 on duality.

---

18 Note that this is heuristics only. The whole enterprise should be taken seriously, in the sense that we do not "have" points any more, although we can philosophize about what they would be like. Otherwise, the procedure loses its philosophical significance.

19 It is an idle speculation if one maximal filter could "cover" more than one point; especially when one recalls Leibniz's principle of the identity of indiscernibles. In fact, Stone's method could be viewed as one mathematical implementation of Leibniz's idea.

First, an obvious requirement may be checked.

Lemma:

For any event structure $\mathcal{E}$, $\mathfrak{t}(\mathcal{E})$ is a general time structure.

Proof:

The assertions to be verified are the following.
(i) $\langle U_I, < \rangle$ is a time structure.
(ia) $<$ is transitive: Suppose that $u < v < w$; i.e., for certain $i \varepsilon u, j, j' \varepsilon v, k \varepsilon w$: $i < j$ and $j' < k$. Since v is a filter, $j \cdot j' \varepsilon v$. Clearly, $j \cdot j' \subseteq j$. (For, $(j \cdot j') \cdot j = j \cdot (j \cdot j')$ (by clause (2) in the definition of "event structure") $= (j \cdot j) \cdot j'$ (by clause (3) in that same definition) $= j \cdot j'$ (by clause (4) in that definition).). Because of clause (5), it follows that $i < j \cdot j'$. Clause (6) implies that $j \cdot j' < k$. Then, because of transitivity (clause (7)), $i < k$. Hence, by the above definition, $u < w$.
(ib) $<$ is irreflexive: Suppose that, on the contrary, $u < u$; i.e., for certain $i, j \varepsilon u$, $i < j$. By clause (8) in the definition of "event structure", this implies that $i \cdot j = 0$. But this contradicts the fact that u is a filter.
(ii) Conditions on $\{\sigma(i) \mid i \varepsilon I\}$.
(iia) $\{\sigma(i) \mid i \varepsilon I\}$ contains $\emptyset$: $\sigma(0) = \emptyset$, because no filter contains 0.
(iib) $\{\sigma(i) \mid i \varepsilon I\}$ is closed under intersections: σ even preserves intersections; i.e., for all $i, j \varepsilon I$, $\sigma(i \cdot j) = \sigma(i) \cap \sigma(j)$. To see this, note that for any filter u, $i \cdot j \varepsilon u$ if and only if both $i \varepsilon u$ and $j \varepsilon u$.
iic),(iid): As for the remaining two conditions (nrs (iii) and (iv) in the definition of "general time structure"), these could be established right here, using clauses (9), (10) in the definition of "event structure".[20] We prefer to take a short-cut, however. The next proof, which is independent from the present one, except for an inessential reference to the present cases (iia) and (iib), establishes an isomorphism that enables one to see directly that the required two conditions obtain in $\mathfrak{t}(\mathcal{E})$, given the analogous clauses (9), (10) for $\mathcal{E}$.

This completes the proof of the lemma. We are now ready for the main result. Recall the notion of "derived event structure" $\mathfrak{i}(\mathcal{T})$ introduced at the end of section 2.1.

---

20 E.g., to prove condition (iii), one would use the fact that, if $i \notin u$, then $i \cdot j = 0$ for some $j \varepsilon u$; u being maximal.

Theorem:

For any event structure $\mathcal{E}$, the function σ defined above is an isomorphism between $\mathcal{E}$ and $\mathfrak{it}(\mathcal{E})$.

Proof:

The requirements for σ to be an isomorphism are the following.

(i) σ is injective:

Suppose $i \neq j$. A maximal filter u is needed which is in $\sigma(i)-\sigma(j)$, or vice versa. Here is where clause (9) of the definition of "event structure" comes in. For, suppose that $k \neq 0$, $k \subseteq i$ and $k.j=0$. It suffices to find a maximal filter u such that $k \in u$. (Obviously, then, $i \in u$, but $j \notin u$.).

To prove that such a maximal filter exists, one applies Zorn's Lemma to the set of all filters containing k as ordered by set-theoretic inclusion.[21]

(ii) σ is surjective:

This holds by the definition of $\mathfrak{t}(\mathcal{E})$.

(iii) $\sigma(0)=\emptyset$:

This was shown above.

(iv) σ preserves intersections:

This was also shown above.

(v) σ preserves $<$:

One direction is easy. If $i < j$ and $u \in \sigma(i), v \in \sigma(j)$, then, by definition, $u < v$ (since $i \in u, j \in v$).

If, on the other hand, $i \nless j$, then $u \in \sigma(i), v \in \sigma(j)$ are to be found such that $u \nless v$. One way of getting this is provided by clause (10) in the definition of "event structure". For, take any two maximal filters u,v containing k,l respectively, with k,l as described in that clause. Suppose that $u < v$; i.e., that $r < s$ for some $r \in u$, $s \in v$. Consider r.k and s.l. Clearly, by clause (6), $r.k < s$ (because $r.k \subseteq r$),

and hence $r.k < s.l$ (because $s.l \subseteq s$; using clause (5)).

But, this contradicts the particular choice of k and l, given that $r.k \subseteq k$, $s.l \subseteq l$ and $r.k \neq 0$ (being in u), $s.l \neq 0$ (being in v). This concludes the proof of the theorem.QED.

---

21 The only non-routine feature here is to see that this set is non-empty. For this purpose, consider $\{i \mid k \subseteq i\}$.
If one were to spell out the argument to rock-bottom level, it would have to be verified - amongst others - that, if $k \subseteq m$ and $k \subseteq n$, then $k \subseteq m.n$. The required calculation boils down to: $k.(m.n)=(k.m).n=k.n=k$.

We will now discuss several heuristic features of the above proof. To begin with, the previous arguments supplied the ten conditions which a "good" event structure has to satisfy, in order to be representable. Now, the general notion of such a structure was inspired by the "derived event structures" $\mathfrak{i}(\mathcal{C})$ of section 2.1. Clearly, then, we want such derived event structures to satisfy these ten conditions unearthed here. It would have been most elegant if they satisfied all ten solely by virtue of their set-theoretic nature and the definition of $<$. This is indeed the case as far as the first eight conditions are concerned; but (9), (10) required a special stipulation. This explains the occurrence of conditions (iii),(iv) in the definition of "general time structure". These conditions, in their turn, force one to consider the cases (iic),(iid) in the proof of the above lemma.[22]

It may be seen from the above that the given conditions on an event structure are *necessary* as well as sufficient for representability. Still, not all conditions are equally essential. E.g., a structure satisfying (1), ...,(8) and (10) can, at least, be mapped homomorphically onto a derived event structure.

The above method can also be used to find out which additional conditions are (necessary and) sufficient for certain desired properties of $\mathfrak{t}(\mathfrak{X})$. E.g., in order for its relation to be *linear* one needs precisely "$\forall i,j \in I(i \cdot j=0 \Rightarrow i < j \vee j < i)$". And, if one wants the sets $\sigma(i)$ to be *convex* in the sense of section 2.1, then the suitable condition is "$\forall i,j,k,l \in I((i < j < k \;\&\; i \subseteq l \;\&\; k \subseteq l) \Rightarrow j \subseteq l)$". Thus the method may be adapted to a whole range of concepts of time.

---

22 The whole procedure is like straightening out a loop. In fact, this kind of construction work may make the *results* obtained seem less and less impressive. But, after all, this is the price one pays for playing the game according to Lakatos' rules: with the cards on the table.

The mentioned "irregularity" leads one to suspect that a neater result lies hidden somewhere; using fewer, or simpler, special stipulations.

3.2 Duality

The Stone construction established the following connection between the functors $\mathbf{i}$ and $\mathbf{t}$:

$\mathbf{it}(\mathcal{E})$ is isomorphic with $\mathcal{E}$ for each event structure $\mathcal{E}$. But $\mathbf{i}$ and $\mathbf{t}$ are not each other's inverses; not even up to isomorphism. For, $\mathbf{t}\,\mathbf{i}(\mathcal{T})$ need not be isomorphic to $\mathcal{T}$ itself, for arbitrary general time structures $\mathcal{T}$.[23]

Example:

Consider the general time structure $\langle Z, <, \mathcal{S}\rangle$, where $\mathcal{S}$ consists of all convex sets of integers. Since all singleton subsets $\{z\}$ belong to $\mathcal{S}$, one obtains maximal filters of the form $\{X \varepsilon \mathcal{S} \mid z \varepsilon X\}$ corresponding, in an obvious way, to each $z \varepsilon Z$. But the nest of all sets of the form $\{z \varepsilon Z \mid z > k\}$ (together with Z itself) is a maximal filter too, and hence $\mathbf{t}\,\mathbf{i}(\langle Z, <, \mathcal{S}\rangle)$ contains a "point at infinity"; which makes an isomorphism impossible. In fact, the latter structure is ZZ with an added smallest and greatest point.

Another example is provided by $\langle Q, <, \mathcal{S}\rangle$, where $\mathcal{S}$ consists of all bounded open intervals. It turns out that $\mathbf{t}\,\mathbf{i}(\langle Q, <, \mathcal{S}\rangle)$ is rather like the *real* number axis (with the bounded open *rational* intervals, of course).[24,25]

Finally, it turns out that $\mathbf{t}\,\mathbf{i}(\langle Z, <, \mathcal{S}\rangle)$, where $\mathcal{S}$ now consists of all *finite* subsets of Z, is in fact isomorphic with $\langle Z, <, \mathcal{S}\rangle$. This can be proved quite generally.

Proposition:

If $\langle T, <\rangle$ is a time structure and $\mathcal{S}$ is the set of all finite subsets of T, then $\mathbf{t}\,\mathbf{i}(\langle T, <, \mathcal{S}\rangle)$ is isomorphic with $\langle T, <, \mathcal{S}\rangle$.

---

23 Thus, when a person has derived some event structure from his (secret) conception of time, the Stone construction will not always tell others exactly which conception that is.

24 In fact $\mathbf{t}(\langle Q, <, \mathcal{S}\rangle)$ is a real number axis in which all rational points have been "duplicated", to form discrete jumps. The proof is involved, though not deep.

25 At one time, it seemed that this construction would also yield *non-standard* real numbers, e.g., infinitesimals, through the use of nests like $\{(0,1/n) \mid n=1,2,\ldots\}$ lying "arbitrarily close" to the nest $\{(-1/n,+1/n) \mid n=1,2,\ldots\}$, which goes to zero. But the latter nest is not maximal: in fact one could add the former to it!

Proof:

There are two assertions in one really. First, $\langle T,<,\mathcal{S}\rangle$ is a general time structure, thanks to the presence of all singletons. Then, each maximal filter on $\boldsymbol{i}(\langle T,<,\mathcal{S}\rangle)$ has a singleton set as its intersection; from which it is easy to see how to define the required isomorphism. QED.

Now, let us find more generally all fixed points of the operation $\boldsymbol{t}\,\boldsymbol{i}$; i.e., those general time structures $\mathcal{C}$ for which $\boldsymbol{t}\,\boldsymbol{i}(\mathcal{C})$ is isomorphic to $\mathcal{C}$ itself. There is an obvious mapping which should yield an isomorphism, viz. the one going from maximal filters to some (hopefully unique) point in their intersection:

$$\tau(u) =_{def} \bigcup\bigcap u.$$

Again one could write out an attempted proof and thereby discover which conditions $\mathcal{C}$ has to satisfy. This time we will just mention the results of such a procedure, for a general time structure $\langle T,<,\mathcal{S}\rangle$.

(R1) $\forall u\ \bigcap u \neq \emptyset$ ("saturation")

(R2) $\forall t_1,t_2 \in T(t_1 \neq t_2 \Leftrightarrow \exists i,j \in \mathcal{S}(t_1 \in i \ \&\ t_2 \in j \ \&\ i \cap j = \emptyset))$

[Cf. (5) of section 2.1.]

(R3) $\forall t_1,t_2 \in T(t_1 < t_2 \Leftrightarrow \exists i,j \in \mathcal{S}(t_1 \in i \ \&\ t_2 \in j \ \&\ i < j))$

[Cf. (6) of section 2.1.]

(R4) $\forall t \in T\, \forall i \in \mathcal{S}(t \notin i \Leftrightarrow \exists j \in \mathcal{S}(t \in j \ \&\ j \cap i = \emptyset))$

[Cf. (7) of section 2.1.]

R1 and R2 ensure that τ is well-defined and injective. R3 takes care of preservation of $<$. R4 is needed in order to show that τ is surjective.[26]

Note that R2,R3 and R4 hold in all general time structures derived from convex event structures. In fact, the presence of all singletons in $\mathcal{S}$ is a sufficient - though by no means a necessary - condition for their validity. (E.g., R2 and R3 are also satisfied in $\mathbb{Q}$ with the open intervals; though R4 is not: consider 0 and (0,1).). If all sets in $\mathcal{S}$ are finite, then R1 will hold as well. (This explains the above proposition.) But, when $\mathcal{S}$ contains infinite sets, R1 will normally fail; witness the above examples. Still, there are non-trivial examples of general time structures satisfying all four requirements; e.g., the real numbers with all bounded closed intervals.[27] Let us call structures satisfying R1,R2,R3 and R4 *expressive*. Then we have arrived at the following result.

---

26 Using it one shows that, for each $t \in T$, $\{i \mid t \in i\}$ is a maximal filter.

27 In this case, saturation follows from the Bolzano-Weierstrass theorem.

Theorem:

For any general time structure $\mathcal{T}$, $\mathfrak{t}\,\mathfrak{i}\,(\mathcal{T})$ is isomorphic with $\mathcal{T}$ if and only if $\mathcal{T}$ is expressive.

A further proof of this statement is omitted here.

One would like to extract more information from the duality between event structures and general time structures as sketched here. A reasonable wish concerns a correlation of semantical notions involving the one with corresponding notions involving the other. E.g., if $\mathcal{T}_1$ is a substructure of $\mathcal{T}_2$ (in the usual sense), then the mapping $i \mapsto i \cap T_1$ defines a homomorphism from $\mathfrak{i}(\mathcal{T}_2)$ onto $\mathfrak{i}(\mathcal{T}_1)$. Unfortunately, various unexpected difficulties arise here. E.g., if $\mathcal{E}_1$ is a substructure of $\mathcal{E}_2$ (again in the usual sense), then the mapping $u \mapsto u \cap I_1$ should yield a function from $\mathfrak{t}(\mathcal{E}_2)$ into $\mathfrak{t}(\mathcal{E}_1)$ possessing some pleasant properties (cf. [6]). Nothing of the sort happens, however. $u \cap I_1$ need not even belong to $\mathfrak{t}(\mathcal{E}_1)$! (Simple examples show that $u \cap I_1$ may fail to be maximal in $\mathcal{E}_1$, even though u is a maximal filter in $\mathcal{E}_2$.) This subject clearly requires a more careful investigation.

3.3 Languages again

The above results establish mathematical equivalences between event structures and time structures. This does not answer all *logical* questions, however. E.g., is it possible to relate truth of tense-logical formulas, when evaluated at intervals, to truth of such formulas when evaluated at the moments in these intervals? Such questions need much further thought, especially in view of the fact that explicit truth definitions with respect to intervals (instead of moments) are scarce. Here we will consider just one particular proposal, and point out its difficulties.

Suppose that tense-logical formulas are to describe (possibly complex) events, in a language with logical operators $\wedge, \vee, \neg$, F and P. Let $\mathcal{E}$ be an event structure. The usual algebraic evaluation consists in assigning elements V(p) of I to proposition letters p and then lifting this assignment V to a function $\bar{V}$ from the set of all formulas into I; with clauses like $\bar{V}(A \wedge B) = \bar{V}(A) \cdot \bar{V}(B)$. This procedure requires the introduction of special unary operations to provide an interpretation for F and P, however. Instead, we experiment with an ordinary truth definition, keeping in mind that each formula A is to describe a set of intervals which forms, in some suitable sense, a "complex event". This may be taken to mean that such a set is closed under the inclusion relation defined in terms of $\bullet$ in section 2.1.[28,29] Let us call such sets *hereditary*. For a start, we take a function V assigning hereditary subsets of I to the proposition letters. This yields the obvious starting clause:

(i) $\langle \mathfrak{E}, V\rangle \models p\ [i] \Leftrightarrow i \varepsilon V(p)$

The clauses for $\wedge$ and $\vee$ are as always; e.g.,

(ii) $\langle \mathfrak{E}, V\rangle \models A \wedge B\ [i] \Leftrightarrow \langle \mathfrak{E}, V\rangle \models A[i]\ \&\ \langle \mathfrak{E}, V\rangle \models B\ [i]$

As for negation, we want $\neg A$ to be true at those intervals which are "completely free from A":

(iv) $\langle \mathfrak{E}, V\rangle \models \neg A\ [i] \Leftrightarrow \forall j \varepsilon I (\langle \mathfrak{E}, V\rangle \models A\ [j] \Rightarrow i \bullet j = 0)$

Finally, the treatment of F and P is traditional again; e.g.,

(v) $\langle \mathfrak{E}, V\rangle \models FA\ [i] \Leftrightarrow \exists j \varepsilon I (i < j\ \&\ \langle \mathfrak{E}, V\rangle \models A\ [j])$.

A straightforward induction shows that each set $V(A) =_{def} \{i \varepsilon I \mid \langle \mathfrak{E}, V\rangle \models A\ [i]\}$ is hereditary. The proof is rather like that for similar statements in intuitionistic logic.[30]

It is a plausible guess that this concept of truth may be related to that in $\mathfrak{t}(\mathfrak{E})$, provided that a suitable assignment is chosen in the latter structure. One such assignment is as follows: $\underline{t(V)}(p) =_{def} \{u \mid u \cap V(p) \neq \emptyset\}$. Thus, $\langle \mathfrak{t}(\mathfrak{E}), t(V)\rangle$ becomes an ordinary tense-logical model; and one would like to prove, at least, that

(a) if $\langle \mathfrak{E}, V\rangle \models A\ [i]$ and $i \varepsilon u$ (intuitively: "$u \varepsilon i$"), then $\langle \mathfrak{t}(\mathfrak{E}), t(V)\rangle \models A\ [u]$.

---

28 This is just one notion of "complex event", of course. There are also types of event which are typically not preserved in passing from an interval to its subintervals.

29 In the original draft, in which event structures contained an additional operation + (union), there was also a requirement that the sets V(A) be closed under +. In mathematical terms, then, these sets would be (possibly improper) *ideals* on $\mathfrak{E}$.

30 In the presence of +, the arguments can easily be extended. E.g., if $\langle \mathfrak{E}, V\rangle \models \neg A\ [i]$ and $\langle \mathfrak{E}, V\rangle \models \neg A\ [j]$ and $\langle \mathfrak{E}, V\rangle \models A\ [k]$, then $i \bullet k = 0 = j \bullet k$, whence $(i \bullet k) + (j \bullet k) = 0 = (i+j) \bullet k$ (by distributivity); and hence $\langle \mathfrak{E}, V\rangle \models \neg A\ [i+j]$.
An additional condition is needed to take care of the tense operators, however.

Some reflection shows that the following equivalence is more promising:

(b) $\langle \mathfrak{t}(\mathfrak{X}), t(V)\rangle \models A\,[u] \Leftrightarrow V(A) \cap u \neq \emptyset$.

Clearly, (b) implies (a). But, (b) would require, amongst others, that, for each formula A and each maximal filter u, $V(A) \cap u \neq \emptyset$ or $V(\neg A) \cap u \neq \emptyset$. And it is not too hard to show that this may not always be the case. Thus, the above guess may very well turn out to be wrong. In brief, our logical discussion of possible reducibility of truth has remained inconclusive.

In view of the simplicity of our sample truth definition, the general prospects for reduction look gloomy.

References

Åqvist, L. & Guenthner, F., 1978, "Fundamentals of a Theory of Verb Aspect and Events within the setting of an improved Tense Logic", in: F.Guenthner & C.Rohrer (eds.): *Studies in Formal Semantics,* North-Holland, Amsterdam.

van Benthem, J.F.A.K., 1977, Tense Logic and Standard Logic, *Logique et Analyse* 80, pp.395-437.

van Benthem, J.F.A.K., 1977, Tense Logic, Second-Order Logic and Natural Language, to appear in the *Proceedings of the Tübingen Workshop on Formal Semantics.*

Birkhoff, G., 1967, "Lattice Theory", *American Mathematical Society Colloquium Publications,* vol.25, Providence.

Cresswell, M.J., 1977, "Interval Semantics and Logical Words", in: C.Rohrer (ed): *On the logical Analysis of Tense and Aspect,* TBL Verlag Gunter Narr, Tübingen, pp.7-29.

Goldblatt, R.I., 1976, "Metamathematics of Modal Logic", *Reports on Mathematical Logic* 6, pp.41-77,and 7, pp.21-52.

Lakatos, I.M., 1976, "Proofs and Refutations", Cambridge University Press, Cambridge.

Lemmon, E.J. & Scott, D., 1977, "An Introduction to Modal Logic", edited by K.Segerberg, Blackwell, Oxford.

Lorenzen, P., & Schwemmer, O., 1975, "Konstruktive Logik, Ethik und Wissenschaftstheorie", B.I.Wissenschaftsverlag, Mannheim.

Needham, P., 1975, "Temporal Perspective", *Filosofiska Studier* 25, Uppsala.

Needham, P., "Time, Modality and Aspect: Some Problems of Predicating Intervals", manuscript.

Prior, A.N., 1967, "Past, Present and Future", Clarendon, Oxford.

Rasiowa, H. & Sikorski, R., 1970, "The Mathematics of Metamathematics", Polish Scientific Publishers, Warszawa.

Russell,B., 1956, "Logic and Knowledge", edited by R.C.Marsh, Allen & Unwin, London.

Thomason, S.K., 1972, "Semantic Analysis of Tense Logics", *The Journal of Symbolic Logic* 37, pp.150-158.

Whitehead, A.N., 1929, "Process and Reality" Cambridge University Press, Cambridge.

VERBS, EVENTS, AND THE FLOW OF TIME

Dov Gabbay/Julius Moravcsik

Verbal forms are used typically in English to denote what takes place in time. The temporal can be located in the past, in the future, or at the point of utterance itself. What takes place in time may be something instantaneous, or an ongoing process, or the completion of activity. In a language like English we find a complex system of tense, aspect, and temporal modifiers that makes the variety of temporal references possible. Our aim is to uncover this complex network, and give it a formal representation.

We shall show that the full force of temporal reference of verb phrases is the product of the interplay of four systems: a) Semantic verb phrase categorization b) Tense c) Aspectual semantics and d) The semantic categorization of temporal modifiers. Having presented each of these systems, we shall show how the results can be combined in a formal syntactic and semantic representation of the relevant fragment of English. Though our results cover only English, we believe that the kind of formal framework that we employ can be used to account for tense and aspect in other natural languages as well, thus providing a framework for discovering interesting linguistic formal universals.

Consider the following sentences:

(1) They are reading their lesson.

(2)* They are knowing their lesson.

(3) She arrives on Tuesday.

(4) She arrives on Tuesday every week.

A formal representation such as ours will have to account for facts of the following sort: why is (1) well formed but (2) unacceptable? (I.e. why does 'read' take the progressive while 'know' does not?) why is it that (3) has as one of its reading the simple future, while (4) does not admit such a reading at all? (I.e. why does the addition of the temporal modifier 'every week' change the semantical force of the sentence to which it is added?).

In the course of answering such questions one can see by comparing (3) and (4) that we cannot build up the temporal semantics of sentences molecularly; i.e. by assigning a semantic object to each of the relevant linguistic units regardless of linguistic context. The addition of a linguistic element can change the meaning assignment of other elements in a sentence.

Philosophers and logicians working on similar problems previously have differed on what kinds of objects to assign as denotations to verb phrases. Basically, there are three options: One can take verbs to be denoting events while the relevant noun phrases denote objects - one is then left with the problem of linking the two domains. Again, one can take events as the basic domain, and attempt to interpret objects "adverbially" - i.e. as modifications of events. Finally, one can take objects as the basic domain, and interpret, e.g., a verb like 'walk' as denoting (at a certain time, in a certain way) all those who walk. Our paper is neutral with regard to this issue; our four systems and the resulting formal representations can be grafted unto any of these three types of semantics. Our aim includes delineating sharply what is to be regarded as falling into each of the four systems. Only in this way can we hope to build a formal structure within which significant universals can be expressed.

Thus, e.g. we do not cover facts due to lexical peculiarity. That "sleeping slowly" does not entail interrupted sleep while "sleeping badly", according to some, does, is irrelevant to our analysis.

Again, aspect has been treated as the conceptual waste-basket by those analyzing tense. But we mean to include only matters relevant to the interpretation of sub-phrases, and bearing on matters of temporal reference while having explicit syntactic representation. Thus, e.g. the fact that

0.0. people walked across the lawn

admits of two readings (habitual and narrative) while

0.1 3 people walked across the lawn

does not admit of the habitual reading is irrelevant to our analysis. (This is one of the many ways in which our analysis differs from Verkuyl (1976)).

We are not trying to account for everything that has been labelled "aktionsarten".

I. *Some previous efforts*

Three previous efforts in philosophy relate to the issues with which our paper deals. But the main aim of authors like Vendler and Kenny[1] has been different from ours. Their classification of verbs into categories like "performance" or "achievement" is meant to serve as the foundation for a theory of action rather than as a cornerstone of a formal representation of tense and aspect. Semantic differences that are crucial for a theory of action (e.g. factors affecting intentionality, and causal effect) are not always the same as the ones needed for the interpretation of tense and aspect.
For example, for Vendler's purposes it is important to separate a verb like 'start' that indicates the inception of an action from a verb like 'find' which indicates the completion of a process, and both of these from a verb like 'see' which though instantaneous, does not indicate inception or completion. From the point of view of a system of tense and aspect, however, notions like inception or end are irrelevant. What matters is that some verbs denote instantaneous events (and thus 'see' and at least one sense of 'start' fall into the same category) while others, such as 'walk' denote events with duration. Again, within the context of a theory of action Vendler characterizes 'walk' as denoting continuous activity. But from the point of view of a tense and aspect representation it is very important that what we characterize as walking can admit of small gaps; a few momentary stops - to catch one's breath - will not deprive the activity of being called in the ordinary sense continuous walking. As we shall see, the possibility of these small gaps does affect tense and aspectual interpretation.

Along the same lines, in Kenny's work verb phrases like 'find' and 'build a house' are treated as relevantly similar, since both indicate result and imply activities leading up to the result as well as strategy, plan, and intention. But from the point of view of our system the two phrases are very different. For we can say:

(5) She was building her house for some time.

but not

(6)* She was finding her house for some time.

---

1 Z. Vendler "Verbs and Times" *Philosophical Review* 56 (1957) pp.143-160. A. Kenny *Action, Emotion, and Will* New York, 1963, chpt.8.

it will be shown that the two expressions fall into different semantic categories; and this classification will explain why the temporal modifier 'for some time' is compatible with one verb phrase but not the other.

Thus Vendler and Kenny differ from us both in aim, and also in scope insofar as they restrict their analysis to action-verbs while we consider all verb phrases taking tense and aspect, regardless of whether they denote actions or events such as rusting, melting, etc.

A recent paper by A.Mourelatos[2] is much closer to our enterprise both in scope and in spirit. Mourelatos' interest is aspect and tense regardless of its impact for action theory; furthermore, his threefold distinction of event, process, and state is fairly close to the ontology that we shall assume. Still, his interests are not in the formal representation of a fragment of English, and though he says much that is useful on aspect, he does not have the conception of the four distinct semantic systems - sketched above - interacting so as to give the right readings to the relevant set of English sentences.

II. The Ontology

We posit as the underlying reality for our analysis of tense and aspect a world of entities and temporal as well as atemporal properties. E.g. an atemporal property is that of being an odd number. There are also properties that are atemporal relative to certain things; e.g. the property of being a human is such that if something has it, it has it necessarily and throughout its temporal career. Finally, there are temporal properties such as being sick, walking, seeing, being a lawyer, etc. We construe this world as teeming with potentialities, possibilities, as well as constraints of necessity; in short, we take a realist attitude towards the modalities. We posit also temporal intervals and points.

One might ask: what are entities? Our answer would parallel Dana Scott's reply in *Advice on Modal Logic* to the analogous question about individuals. Having raised the question, he drops it with the comment that it is "too good" a question to be answered. Modal logic has made great advances without being

---

2 Alexander P. Mourelatos "Events, Processes, and States" *Linguistics and Philosophy* 2 (1978) pp.415-434. The first draft of our paper was prepared in early 1978; Mourelatos' work came to our attention as we were making revisions in our first draft.

able to say what an individual is. We think that similar advances should be possible in tense and aspect representation without being able to say what a particular is.

There is no expression in English that would cover all of the different types of instantiations of temporal properties. Thus we will label this genus simply as *instantiations of temporal properties*. Some of these instantiations, e.g. being sick, knowing, being solid, etc. are *states*. A state is an instantiation of a temporal property P of a thing x such that i) Px holds for some duration; thus it cannot be instantaneous. ii) Px holding over a certain duration of time does not imply certain specific changes in x. E.g. being healthy does not imply specific changes in the subject while an instantiation that is not a state, for example the event of walking does imply such changes (movements of the legs, etc.). iii) When Px holds over a period of time, it does so without any gaps, or interruptions. E.g. if someone knows something, then forgets, and then knows it again, we say that he rediscovered the relevant item; we count two states of knowing. Likewise, if someone is sick on a day, recovers, and becomes sick again, we say that the person was sick twice during the day. We can represent a state E=P(x) semantically as follows. Let (a,b) be the interval of time in which the state E holds

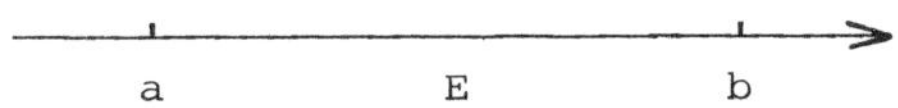

then (a,b) itself is considered as the semantic object associated with E. Since (a,b) has a definite length (i.e. it is not a single point) the property (i) of a state - namely duration is adequately represented. Property (iii) is also reflected, since there are no gaps. Property (ii) can also be represented but this will be done when we deal with the semantics of processes. The semantic representation of processes will account for change via a device that will not be available for the semantics for states, as they have no change.

Non-states, therefore, are either without duration, or if they have duration they imply specific changes, or at least allow gaps in what is nevertheless counted as one instantiation. We shall call these instantiations *events*. Among events with duration we distinguish *mere events* from *processes* on the ground that processes are made up of a series of changes that culminate in a state (e.g. outcome, or the production of something) not reached at any time while the process takes place. During the course of the mere event of walking we reach the state of having walked over and over again, but when the process of, e.g., writing a book reaches the state of having a book completed, this is not a state reached at any time while the process lasts.

It is clear that processes allow "gaps", i.e. the same process need not go on continuously. E.g., the process of cleaning up the yard can be interrupted by a friend or a neighbour, and even the most dedicated artist might take some time out for sleep when composing a poem. Less obviously, the same holds for mere events. E.g. 'she was yelling at him for some time' can be true over a certain period of time, even if the agent stops for a few seconds; at least to catch her breath, if for nothing else.

To give a semantic representation of events, we use again, intervals of time. In this case we must take account of the possibility of gaps and also change. We imagine an interval (a,b) in which the entire event takes place from beginning to end.

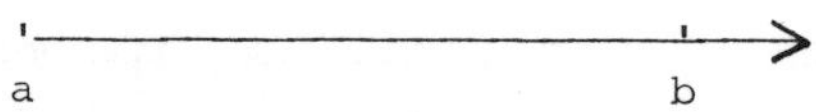

Since we allow gaps, there will be, inside (a,b) subintervals of non-activity. Let us take for example (x,y) as the only gap

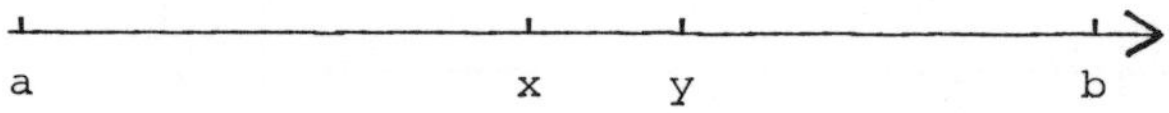

So we must list, besides (a,b) also the intervals (a,x),(y,b) in which activity occurred. So the semantic representation of an event (which has duration) will be composed (1) of one big interval in which the entire event occurs and (2) a list of subintervals in which the actual activity occurs. Thus

$$\begin{matrix} (1) & , & (2) \end{matrix}$$
$$\left[\begin{matrix} (ab) & ; & \begin{matrix}(a,x)\\ (y,b)\end{matrix}\end{matrix}\right]$$

The above analysis is adequate for events without change. In case we have change we must find a way to reflect this too in our semantic representation.

The idea we use is already needed to solve the following difficulty. Suppose we have a state E occurring in (a,b). So its semantic representation is (a,b) itself. Suppose also we have an event F (e.g. yelling) occuring at (a,b) *without gaps*. Its semantic representation will be [(a,b), (a,b)]. How can we tell the difference? Technically there is a difference. One is represented as (a,b) and one as [(a,b), (a,b)] , but this is a trick of form.

Is there a true semantical difference? The answer is yes. The interval (a,b) is only one instantiation of E and of F. Since E is a state, every instantiation of E will be an interval (c,d) while, across the flow of time, it is conceivable that some instantiations of F, being an event, will have gaps. Thus when we look at all instantiations of E and F they will *not* be the same.

This same idea will help us to represent change. Given F_1 and F_2 both events, both occurring between the temporal points a and b. Thus

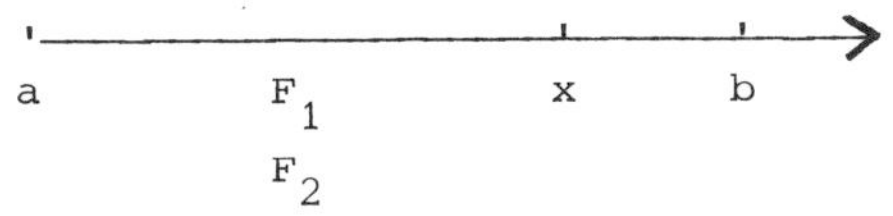

F_1 involves change, F_2 does not. E.g. F_1 = *building a house*; F_2 = *yelling*. We tell the difference semantically because when the change is completed at b, we begin with a new state E and at no point x between a and b do we have the state E.

This is not true for F_2. It may be that for this particular instantiation of F_1 the above does not happen but it will happen for other ones.

Among instantaneous events we distinguish non-repetitious events, represented by points, from repetitious ones, such as hitting, knocking, etc. represented by intervals containing sets of points.

Within this scheme there are no instantaneous states; but this is more a matter of conceptual decision rather than that of facts. The decision reflects the fact that from the point of view of a tense and aspect semantic there is no significant difference between a momentary instance of rest, and a momentary instance of seeing; or for that matter, momentary events marking the beginning or ending of some durational event or state. This is not to deny that there may be significant semantic differences among the verbs in question within theories of biology, action, etc.

Our ontology is, then, made up simply of time points, time intervals, and changes. The categorization can be summed up in the following scheme:

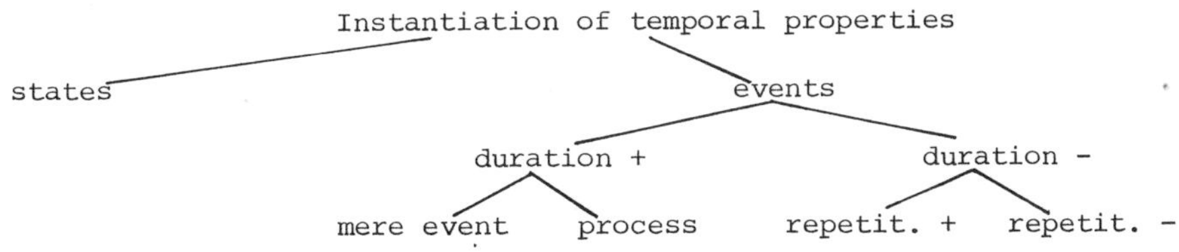

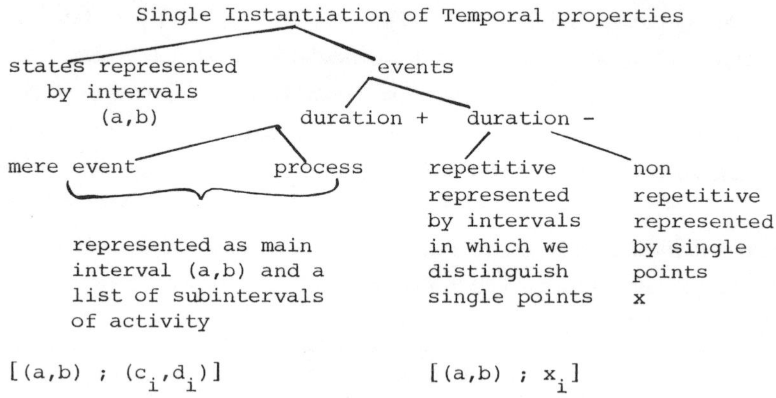

The difference between mere events and processes is brought out by the way different instantiations behave across the time line.

III. Ontology and the Semantics of V and -VP

The articulation of reality outlined in the previous section is mirrored by semantic types into which certain English verbs (V) and verb phrases (VP) fall. Many elements within a verb phrase will play no significant role in the semantic analysis from the point of view of tense and aspect. E.g. , manner adverbials like 'slowly' do not play any role except for lexical peculiarities; 'walks' and 'walks slowly' have the same tense and aspect semantics. Again, certain NP's and PP's are irrelevant. E.g., there is no relevant difference between 'gives' and 'gives to Tom'. Other NP's and PP's, however, do make a difference; e.g. the difference between 'write' and 'write a book', or 'walk' and 'walk to Rome'. We assume that a systematic demarcation of the relevant classes of NP's and PP's can be effected, but we shall not attempt this within the confines of this paper.

Furthermore, what we classify are, strictly speaking, not VP's but various senses, or uses, of VP's. E.g., though 'see' denotes instantaneous events in its characteristic use, it has a use in which it does not, as in

(7) She is seeing him often.

The fact, however, that 'see' can denote instantaneous events forming a whole series, over a time interval, is not a reason for ascribing yet another sense to it. Thus in

(8) He will see her.

and

(9) They saw the opera last night.

'see' has the same meaning. To be sure, (9) implies that they saw the whole opera; and the opera takes time. It does not follow from this that (9) implies the existence of non-instantaneous events of seeing. (What is instantaneous here is simply thus represented in the model. This causes no metaphysical claims about instantaneous seeing).

To take another example, 'start' and 'begin' have a sense in which they denote instantaneous events. But when combined with verbs or verb phrases that denote events or processes the initiation of which is conventionally associated with certain routines, they have a sense in which they denote events with duration, thus taking also the progressive. E.g. "he is starting to run" does not make any sense, except when used to say the equivalent of "he is learning how to run"; but "he is starting to warm up the car" (or "cleaning the yard", etc.) does denote a routine assumed by convention or well ingrained habit; i.e. pumping the gas pedal, reaching for the ignition, etc.

For purposes of brevity, however, we shall continue to talk of the classification of VP's. As one would expect from the ontological classification, *state VP's* are expressions like 'know', 'is healthy', etc. They denote the continuous instantiation of temporal properties over a time interval, without allowing gaps, and without the implication of specific changes required by the state. This characterization of this VP class, together with the characterization of the progressive aspect to be given below, explains why state VP's do not take the progressive; for the progressive operates on periods of actual goings on, excluding gaps, or pauses. But since state VP's do not allow for gaps, the semantic structure needed for the application of the progressive is simply not there.

Another class of VP's are the non-repetitious instantaneous event VP's such as 'start', 'see', etc. Since the progressive requires denotations over duration (or interval), it is clear why these VP's cannot take the progressive. Repetitious instantaneous event VP's do take the progressive; that aspect marking an interval containing the repetitions. E.g. "he was knocking on the door for some time". This is shown in the diagram of section II. Whatever is represented by [(a,b),(a,b)] can take PROGRESSIVE.

Verbs like 'walk', 'push', 'write', etc. belong to the class of mere event VP's; i.e. the ones with duration that do not qualify as process VP's. Negatively, this implies that they do not lead to a terminal state that is different from the states characterizing partial stages of events. Positively, it implies that the events denoted by these VP's bring with them specific changes. A walking person moves his legs, a woman pushing a cart is moving her arms, etc. Some cessations of these events count as terminations of events, while others do not. This was already illustrated in the previous section in the case of yelling. Other obvious examples include a walk with momentary halts, standing on the corner with an occasional rest on the bench, etc. Even if in fact such an event takes place without "gaps", it is always possible that it could have included gaps; and since we define the semantic objects corresponding to these VP's as functions across possible states (or "world" if one insists), the crucial feature of the possibility of gaps is included in the formal representation. What kind of gap counts as cessation and what kind is simply part of the event, depends on the lexical item in question, and is determined by human interest, scientific theory, etc.

One might speculate why states should be continuous and events allow gaps. The answer seems to be twofold. First, since states do not entail specific changes, they are "static" instantiations; hence one pause counts for as much as any other. Since events imply change, what counts primarily is the completion of certain changes; if these completions are interrupted momentarily, that does not seem to matter. Secondly, unlike states, events have ordinarily as parts of their explanations teleological factors; after all, with the addition of appropriate NP's event VP's can be turned into process VP's ('write' + 'write a book'; 'walk' + 'walk to Rome'; 'change' + 'change into so-and-so', etc.). Hence matters other than sheer duration will become relevant and criterial for human categorizations.

There are two special classes of VP's that do not entail specific changes, but still behave syntactically like event VP's, and qualify semantically by allowing gaps. One of these is the class of verbs describing position; e.g. 'sit', 'stand', 'lie', etc. The peculiarity of these verbs was noted already by Aristotle who placed them in a special class within his theory of Categories. The other class can be called: verbs of attention, e.g. 'watch', 'look', 'wait', and 'hope'. That the latter admit of gaps is obvious. 'she hopes to get her diploma next June' can be true over a certain period of time even if the person in question cannot be said to be hoping every second for the diploma. Likewise 'John watched for fires from the tower' does not exclude John stepping away from the lookout post for a few seconds. Attention is analogous to movement or change; it is something active and dynamic. Hence the similarity to events rather than to states. When used with the progressive, these verbs denote actual happenings, as in:

(10) He was waiting for the train when a former friend bumped into him.

Positions seem like states, but they are the result of movement or activity; hence, presumably, the event VP-like behaviour of the relevant class of verbs. The possibility of gaps is indicated by the fact that

(11) The statue will stand on the square.

is compatible with the statue being removed during the future interval indicated for a day for, e.g. cleaning.

Finally, we have the class of process VP's; they differ from event VP's only in their perfect form describing a state that did not hold at any non-final stage of the process, i.e. the relevant series of changes.

We can extend the count + - distinction, used to classify NP's, to VP's as well. (For discussion see Gabbay + Moravcsik "Sameness and Individuation, *J. of Philos.*, 1973).

State and non-process durational VP's are count- terms, thus resembling those NP's that we call mass terms. For given a state *S* many subparts of it are also S; furthermore, many states that are S when taken together add up to an S. Not all parts of a state are necessarily also the instantiation of the same property as the state itself; with states, and especially events, there may be minimal parts counting as the instantiation of the property the state exemplifies. But, clearly, if someone is sick over a period of time, then he is in that state also over most if not all of the parts of that state. Again, if someone is walking over a period of time, then there will be sub-events of that event which are also instances of walking.

It is not true that any two states of sickness add up to a state of sickness. E.g., it makes sense to say: "he was sick twice last year", as we saw already. Likewise, the parallel is not complete between mass nouns (or noun phrases) and event VP's either; for not any two events of walking count as a walking; e.g. we can say: "she took two walks while on vacation."

But the contrast is worth making, since process VP's exhibit very strong count + behaviour; for the state of, e.g. "having written a book" does not hold for *any* of the parts of the process that culminates in this result. We have a clear and precise arithmetic for process VP's, while we have only partial arithmetics for state and mere event VP's.

Our analysis can resolve the problem mentioned by Dowty 1977 (Towards a Semantic Analysis of Verb Aspect and the English Imperfective Progressive *Linguistics and Philosophy* 1, pp.45-77).

Since *drawing a circle* is a process VP from *John was drawing a circle* it does not follow that *John drew a circle*. *Pushing a cart*, however, is not a process VP, hence from *John was pushing a cart* it does follow that *John pushed a cart*.

Schematically, we arrived at the following semantic classification of VP's needed for tense and aspect representation.

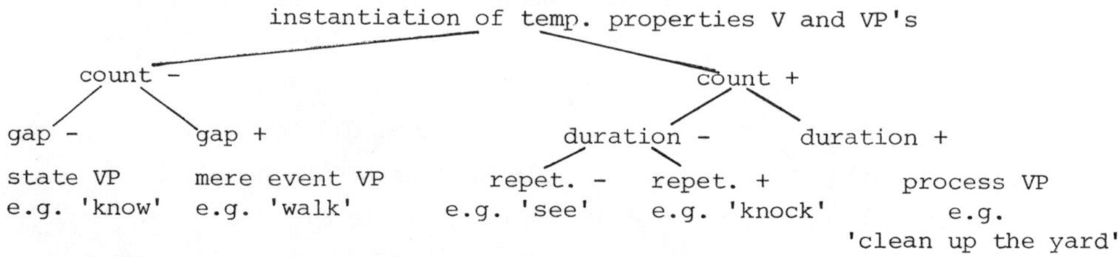

A comparison of the ontology and the VP classification shows that the capturing of the verb-semantics requires more than a mere ontology of things, properties, temporal intervals and temporal instances. Our characterization of process, event vs. state, etc. brought into play such notion as agency, causation, completion of processes or development. This is hardly surprizing; the semantics of natural languages will reflect partly the influence of concepts and factors that relate not simply to metaphysical structures but to general conditions of human activity, and human knowledge - factors that help to define what Wittgenstein called forms of life.

The classification presented will enable us to explain - together with the tense and aspect semantics to follow - why state VP's do not take progressive, the interpretation of the perfect, and the force of the combination of perfect and progressive as well as the selectional restrictions on that combination.

IV Algebra of Events

We saw above that single instantiations of states or events can be semantically represented by intervals or combinations of intervals. Thus for example an instantiation of *walk* can be represented, for a particular case happening e.g. today by

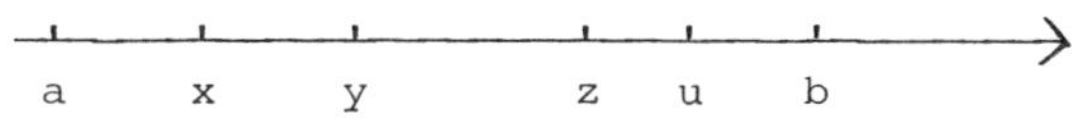

$$\begin{bmatrix} (a,b) \;;\; & \begin{matrix} (a,x) \\ (y,z) \\ (u,b) \end{matrix} \end{bmatrix} \qquad \text{walking occurs in } (a,b)$$

and actual walking in (a,x), (y,z), (u,b), (x,y) and (z,u) are gaps.

The above is a single instantiation of *walk* let us say the one that occurs between a = 12.00 hours Jan.1, 1979 and b = 13.00 Jan.1,1979. There may be other instantiations of *walk* on other days. The semantic object associated with *walk* denoted by ||walk|| is the *set of all instantiations of walk*. We can similarly talk about ||know||, ||knock|| etc. As we remarked before, the difference between a process and non-process durational event is brought out not necessarily by the properties of a single instantiation of the event but by the way the entire set of instantiations is related to other events. To give an example take *yell* = E and *write a book* = F. We know that with every instantiation of E in (a,b) say, there is E^{*} = *having yelled* such that we have

i.e. +E reads 'E is valid', $+E^{*}$ says 'E^{*} is valid'.
With F we have F^{*} = *having written a book* such that with every instantiation we have

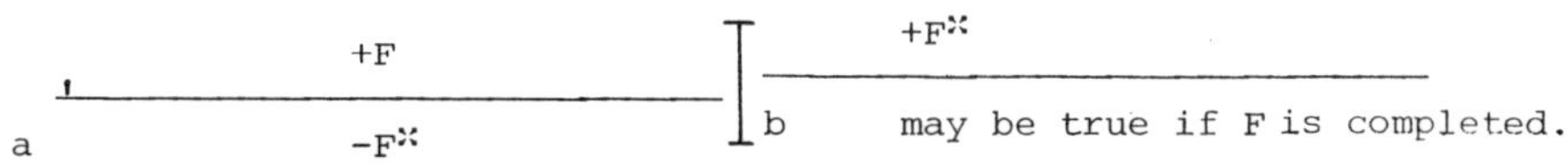

$-F^{*}$ says 'F^{*} is not valid'.

Thus properties of events are brought out by using equations with $\pm F$, $\pm F'$ etc. in what we call the algebra of events.

Let F_1, F_2 be two events. If F is an event such that every instantion of F_1 followed immediately by an instantiation of F_2 gives an instantiation of F we write

$$F_1 + F_2 \Rightarrow F$$

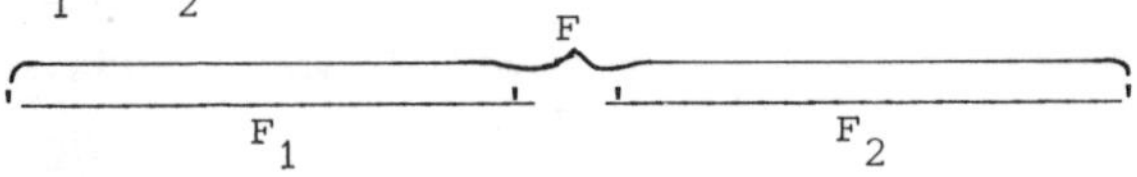

e.g. *Turning ignition key + stepping on the gas pedal = starting car.*
If every instantiation of F is brought about by first instantiating F_1 and then F_2 we write

$$F \Rightarrow F_1 + F_2$$

Different kinds of events satisfy different equations. E.g.

$$||walk|| \Rightarrow ||walk|| + ||walk||$$

Since every walk can be decomposed into two walks. On the other hand it is not true that $||sick|| + ||sick|| \Rightarrow ||sick||$. Given an event F, we can look at $||F||$, this is a family of instantiations of F. Take on such instantiation, say

$$[(a,b),\ (c_i,d_i)]$$

(Assume F has instantiations of this form). With F we can associate the set F^{*} of points of completion of F. We can also associate the sets $F^{\underline{a}}$ = all the intervals (a,b) in all instantiations of F, and $F^{\underline{b}}$ = all intervals (c_i,d_i) of actual activity of all instantiations of F. The triple $(F^{\underline{a}},F^{\underline{b}},F^{*})$ is all the information we need on F for the evaluation of prog., perf. and other tense operators. Verbs like *know* or *win* have no $F^{\underline{b}}$. For verbs like *knock*, $F^{\underline{a}}$ contains the intervals in which there were groups of knockings and $F^{\underline{b}}$ contains those points around which there are cumulative knockings.

By laying down different definitions on this structure, we can formally characterize the V or VP classes introduced informally in the previous section.

For instantaneous VP $F^{\underline{a}} = F^{*}$; and $F^{\underline{b}} = F^{\underline{a}}$ for repetitous ones:
$F^{a} = (F^{*},\ F^{*},\ F^{*})$.
For state verbs or verb phrases: $F^{\underline{a}} = (F^{*}\ F^{*}\ F^{*}\ F^{*})$ (continuous)
For mere event VP's $F^{\underline{b}} \subset F^{\underline{a}}$, possibly $F^{\underline{a}} \neq F^{\underline{b}}$.
For process VP's: no two $F^{\underline{a}}$'s are identical with a 3rd $F^{\underline{a}}$.

V. *Tenses and the Asymmetry of Time*

We postulate three tenses, past, present, and future, without committing ourselves to the thesis that in all natural languages there must be distinct syntactic elements marking solely these temporal reference. E.g., if it is true that in English there is no "pure future", that is compatible with our claim that reference to future time is one of the semantic aspects of VP's with future auxiliary el.

We postulate two aspects; the continuous, or progressive, and the complete, or perfect. Between the system of temporal references and that of aspect we have a cross-classification. This can be represented, with illustrations, in the following table:

	present	past	future
simple	sings	sang	will sing
PROG	is singing	was singing	will be singing
PERF	has sung	had sung	will have sung
PERF+PROG	has been singing	had been singing	will have been singing

The evidence to be presented suggests that from the point of view of the English tense system time is asymmetrical insofar as we have to represent the future as branching time. Consider:

(12) The parcel will arrive tomorrow.
(13) The parcel is going to arrive tomorrow.
(14) The parcel is arriving tomorrow.
(15) The parcel arrives tomorrow.
(16) The parcel will be arriving tomorrow.

The difference between (12) and (16) seems to be that (12) is prediction while (16) is a mere statement of what will happen in the normal course of events. If (12) has this predictive force, then the formal representation will have to include branching times for the future.

(15) seems to imply deliberate planning while (16) lacks this force. Thus we are more likely to say: "According to our plans, the parcel will arrive tomorrow" rather than: "according to our plans the parcel will be arriving tomorrow".

(14) expresses what is planned for the future; we would not use this form if we were talking about something that is certain to happen. Hence we do not say: "the sun is rising at 5 o'clock tomorrow.". We are likely to use a form like (13) when we express the belief that something is going to happen tomorrow because we see it starting now. Thus, e.g., we might say:

(17) This planet is going to burn itself out in 20 years.

These readings can vary with different lexical items but what is clear is that branching future is required.

VI. Aspect

As stated above, we posit two aspects; the progressive indicates continuous actual activity or happening, while the perfect singles out completion and relates that to the point of evaluation.

Within our system the progressive (PROG) is an operator that works on //VP//, i.e. the semantic object assigned to a VP. As we saw, this has the structure of ($F^{\underline{a}}$, $F^{\underline{b}}$, F^{*}). Since the progressive signifies actual goings on, it picks out the F^{b} element, and relates it to the point of evaluation or some other, contextually given, point of reference. Thus

$$PROG//VP// = //PROG//VP//// = //F^{b}, F^{a}, F^{*}//$$

E.g., 'he is walking' is true iff at the point of evaluation the agent in question is actually engaged in the act of walking. The similar point is made by the use of a process VP and with past reference in

(18) She was reading a book when the phone rang.

This interpretation can account for the difference between

(19) She walks in these hills.
(20) She is walking in these hills.

The simple present expresses either the historical present, or habitual activity. But (20) is a natural answer to a question like: "What is she doing right now?"

Certain activities at times assume ceremonial or institutional significance. In these cases what would be usually gaps disappear. E.g. "she is baby-sitting tonight" or "John is sitting at this plan" can be time even if what is counted usually as the activity is interrupted.
Furthermore, there may be several layers of activity e.g. F = *building a castle* may have $F^{\underline{a}}$ = 15th century and $F^{\underline{b}}$ = (1440 - 1460) but actual work was done only in the summers and further, only in the mornings. So really we can add $F^{\underline{c}}$, ", $F^{\underline{d}}$ etc. Our claim is that one always looks at two nested activities and no more, while maintaining a given point of view.

While the simple past is neutral between picking out the actual occurrence or an event with possibly gaps in it - i.e. our $F^{\underline{a}}$ or $F^{\underline{b}}$, the PROG must pick out the $F^{\underline{b}}$ element. This does not come out in pairs like:

(21) He worked on his paper yesterday.
(22) He was working on his paper yesterday.

since the simple past does not restrict the reference to the F^{b} element, though - since the reference is to F^{a} - F^{b} is included. But the difference is brought out sharply by pairs like:

(23) She was working on her paper when the phone rang.
(24)* She worked on her paper when the phone rang.

In (23) the ringing of the phone is related to a period of actual work, i.e. F^{b} in our representation; hence the need for the PROG.

The same situation obtains when one compares the simple future with the future progressive, as shown by:

(25) When the chandelier falls, she will be singing her great aria.
(26)* When the chandelier falls, she will sing her great aria.

Our formal representation of the different VP classes explains why state VP's cannot take the progressive. For PROG operates on the F^{b} element, and there is no such element in the semantic structure of a state VP. It becomes also clear how PROG operates on dur.-VP's; it cannot go with the instantaneous ones, for there is no F^{b} element, to be contrasted with F^{a}; and F^{b} element is created in the case of the dur.-repetitious VP's by taking a series of points. This corresponds to the intuitive fact that PROG indicates extended goings-on, hence it cannot operate on elements that do not allow for extended goings-on.

We turn now to the perfect.

Perf. F true at x iff x is in the future of a point of completion from F^{x}. We are not taking perf. F to be F* because e.g. *He has finished his book* does *not* allow that he finished just now! E.g. in

(27) He has cleaned up the yard

we state that the cleaning up of the yard was completed at some point in the past of the point of evaluation. PERF thus defined works in the intuitively obvious way on instantaneous VP's.

The difference between the perfect and the simple past can be seen in pairs like:

(28) He worked hard all through January.
(29)* He has worked hard all through January.

Since in the semantic structure of state VP's all we have is a series of * elements, there will be no basic difference between simple past and perfect, except that the latter relates the past state to the present, i.e. the point of evaluation. Our formal representation does not say what the nature of the relation to the present is; this is where the human element enters the semantics. The nature of the relation can depend on extra-linguistic context, human interests, etc.

Finally, we turn to the case in which the perfect operates on a semantic object which is the product of PROG having already operated on //VP//. Examples of this are:

(30) She has been walking all this time.
(31) He has been writing a book since last week.

It is obvious that there will be no perfect-progressive for state and duration-VP's since these lack the progressive in the first place.

Turning to perfect progressive let perf(prog E) be the set of all points x such that for some $t<x$, $(t,x) \varepsilon F^{\overset{b}{=}}$. Thus *He has been reading his book* is true at x iff he was reading until now, but is not necessarily reading now. Here we see a difference between the above and *He is reading his book*. The latter implies reading now.

As can be seen, perf(prog F) regards each point of $F^{\overset{b}{=}}$ as if it were a point of completion for the activity of prog.

From this analysis it does not follow that the activity referred to will not continue beyond the point of evaluation. E.g., (30) is compatible with the agent continuing the walking beyond the present. But the sentence leaves the option of continuation open.

Alternate suggestion to be explored: PERF-PROG takes as the F^b what is normally the F^a element, and gives the point of evaluation as *. This seems to account for possibility of "gaps" in what is described by sentences like (30) and (31); but the F^b element has to enter into the analysis somehow; else how can we block sentences like:

(32) She has been being sick.
(33)* He has been seeing the opera.

Thus the analysis of VP-classification, together with this analysis of tense and aspect accounts for the combinations not permissible in English and gives the intuitively natural reading for the relevant class of sentences.

Examples involving quantifiers in the object NP, and negation in the VP could also be given, but we do not see that these cause any special problems for our analysis. E.g.,

(34) He was dancing with all of the girls at the ball.

is analyzed like any process VP; 'dancing' is a mere event VP, but 'dancing with all of the girls' is a process VP.

VII. Temporal Modifiers

Different temporal modifiers apply to different semantic objects among the ones generated so far. This can be seen by examples such as

(35) They worked on this paper for some time.
(36)* They have worked on this paper for some time.
(37) They cooked the dinner in an hour.
(38)* They were cooking the dinner in an hour.

The differences are intuitively obvious. A modifier like 'in an hour' indicates the temporal duration it took to complete something; hence it cannot apply to a VP in progressive, since - as we have shown - the semantic object specified will be a F^b element, and not an F^* element. Thus what we need is a semantic classification of temporal modifiers, and a formal analysis of the different classes. If the analysis is adequate, it - together with the other components described in this paper - should account for the types of sentences blocked in English, and give us the right readings for sentences containing temporal modifiers.

Syntactically, temporal modifiers fall into two classes; some - like 'before', 'after', - are sentential modifiers, while others are generated within a VP or PRED constituent.

It turns out that semantically, not all temporal modifier classes do the same job. Some are sentential operators, others give us a point or interval of reference and only a restricted group of temporal modifiers have a strictly adverbial role, - analogous to that of manner adverbials - operating on either the F^a or the F^b element. The examples considered so far, 'for some time' and 'in an hour' are true adverbials, while 'on January 25, 1956' is not. Furthermore, some expressions, like 'yesterday', have a sense in which they function as adverbials, and one in which they give a period as reference .

The following gives an overview of our time modifier classification system:

A) *True Modifiers*	1) Durationals. E.g. 'for some time', 'for a while'
	2) Containers. E.g. 'within an hour', 'in a day'
	3) Instantaneous. E.g. 'right now'
B) *Referentials*	e.g. 'on June 25, 1928', 'in January', 'since', 'until'
C) *Sentential Modifiers*	1) Repetitous. E.g. 'every week'
	2) Relationals. E.g. 'after'.

We shall survey these now, one by one, giving their semantic roles, and pointing out how their semantic restrictions fall out of our analysis.

Durationals include 'for some time', 'for a while', 'for 3 hours', and also a sense of 'yesterday', 'today', etc. These are correlated semantically to functions that take an F^a or F^b and take it to an interval that is a part of the $F^{\underline{a}}$ or $F^{\underline{b}}$ referred to. E.g.:

(39) They were reading for some time.

asserts that there was an interval (contextually defined) which was part of the period in which the reading took place.

Since durationals need to operate on an $F^{\underline{a}}$ that is not simply an * element (or on an F^b) therefore they cannot modify repetition - duration-VP's. Furthermore, they can modify a VP that has progressive applied to it, since the resulting semantic object is an $F^{\underline{b}}$. But they cannot operate on something in perfect form, since - as we have shown - PERF deletes everything except the last * element. Hence the ungrammaticality of

(40)* They have written a book for 3 hours.

The only exception to the last provision is the perfect form of a state VP. Hence the grammaticality of

(41) She has known that for some time.

But, as our analysis showed, in the case of state VP's the semantic object is a series of * elements; hence the possibility of locating an interval within such a series. Durationals are blocked only in cases in which there is a distinguished * element.

Containers include phrases like 'within an hour', 'in a day'; the paraphrase of these modifiers is given by expressions like: "it took a day (hour, year, etc.) to do so-and-so. Thus, e.g.

(42) They cleaned up the yard in an hour.

These modifiers take a $F^{\underline{\underline{a}}}$ or $F^{\underline{\underline{b}}}$ element that preceded a distinguished ⁙ element i.e. the ⁙ signifies completion and give a durational specification of such element.

Intuitively, this means that the modifier must apply to an F^{a} that has a ⁙ as its completion. Hence they cannot modify a VP that is either in the progressive or the perfect, since the correlated semantic operations yield a ⁙ element or an F^{b}; and their semantic objects are not appropriate for container-modification. Thus these modifiers modify VP's that are without the two aspects.

Furthermore, since they must modify durations with ⁙ as their completions, they must modify process VP's; they cannot modify state or duration-VP's; nor can they modify mere event VP's, since, as we saw, these do not have a completion state that would not have held also for various stretches of $F^{\underline{\underline{a}}}$. Hence the ungrammaticality of

(43)⁙ They have seen within an hour.
(44)⁙ She walked in a day.

Instantaneous Modifiers. This class includes 'right now', 'instantly', etc. These modifiers take an $F^{\underline{\underline{a}}}$ or $F^{\underline{\underline{b}}}$ or F⁙ element, and place within it, or identify with it, the ⁙ element designating the present or some other contextually identified moment. Since an ⁙ element is the minimal unit in this semantic analysis, our system would predict that except for lexical idiosyncrasies, instantaneous modifiers should be compatible with any VP-class and with both aspects. And, in fact, some instantaneous modifier modifies some member of each VP-class of our classification, in each permissible aspect-form. Hence we get sentences like:

(45) They knew instantly who the rider was.
(46) She has completed the lesson just now.
(47) She is kissing him right now.

The various restrictions on which instantaneous modifier can go with which member of a given VP-vlass is a matter of specific lexical meanings, and is handled in our system by meaning postulates.

This completes the survey of the true modifiers. We now turn to temporal modifiers that simply indicate the time at which a certain instantiation of a temporal predicate takes place.

Referentials. These include 'on the 26th of May', 'in 1948', 'in January' and a sense of 'yesterday', 'today', as well as 'since' and 'until'. Since these are not true adverbial modifiers, we expect them to be neutral with regard to VP-classification and aspect, except with respect to their indicating a point of time or a period.

The only qualification needed is with regards to 'since' and 'until'. These mark beginning and termination respectively, and thus can modify only sentences in which the VP is duration + and not in PERF; i.e. does not receive an ⁎ element as the semantic object.

Finally, we have *sentential modifiers*. These include repetition marking modifiers such as 'every week', and well as sentential connectives such as 'before', 'after', and 'while'. Since they modify a whole sentence, we expect them to be neutral with regards to the classifications of VP and aspect sketched so far. This is indeed the case with repetition markers; thus we get sentences like

(48) My mother came to visit me every day.

Needless to say, the period within which the repetitions take place is determined by context, linguistic or otherwise.

There are various restrictions on the combinations of types of sentences that can be linked by 'before', 'after', and 'while'; but the detailed description of these restrictions, though expressible within our semantic framework, is beyond the scope of this paper.

Can we combine in a sentence any two temporal modifiers? That there must be some restrictions can be seen by the ungrammaticality of

(49)⁎ He was working in his garden for some time in a day.

One possible hypothesis is that temporal modifiers can be combined only if one is contained in the other. (We are indebted to Prof. Rohrer for this suggestion.). Thus, the following is grammatical:

(50) The session began on January 24, 1956, at 3 o'clock.

This, however, is not the only permissible type of combination. Other combinations are possible as long as the modifiers selected belong to two distinct classes from among the three basic semantic types (true adverbial, referential, sentential) that formed the backbone of our analysis of temporal modifiers. Thus, e.g. the following are grammatical:

(51) She worked in her garden for some time until the sun went down.
(52) They quarreled repeatedly a long time ago.

Finally, our analysis can explain ambiguities like the one in

(53) He was working on his paper yesterday.

If we take 'yesterday' to have the sense of a durational modifier, then the sentence says that the period of yesterday was a part of the ongoing activity of the work on the paper; if 'yesterday' is taken in the referential sense, then the sentence says that a period of paper writing can be located within the period of yesterday. In the first case, 'yesterday' has only the VP in its scope; in the second case it has the whole sentence within its scope.

VIII. Syntax and Semantics

Having outlined the semantic interpretation of verb classes, tense, aspect, and temporal modifiers, we shall conclude by showing how this semantics can be fitted to a syntactic analysis within the framework of transformational grammar. We are not committed to the thesis that semantic rules must mirror syntactic rules; the way they do in the grammars of formal languages. Still, it is interesting to see that in this case there is a fairly close correlation between the semantic system outlined and the syntactic analysis presented in "The category of AUX in Universal Grammar" by Akmajian, Steele, and Wasow (forthcoming in *Linguistic Inquiry*). The following is a slightly revised version of their constituent analysis:

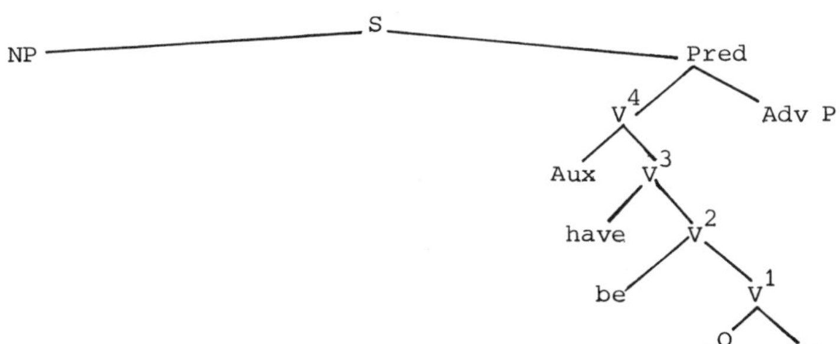

Our first syntactic category to which we assign a semantic object is either their V^0, or V^1. We have divided the appropriate VP genus into five classes. Thus according to our analysis, we start with

V^1a - //X'//
V^1b - //X''//   etc. for all five VP-classes.

The V^2 node of the tree presented corresponds to our applying the PROG, if any, to the semantic object assigned to the VP-class under analysis. Thus

V^2 - //PROG V^1//

The V^3 node corresponds to V^2 and PERF. Hence

V^3 - //PERF PROG V^1// or //PERF V^1 // if there is no Prog.

At this stage we evaluate the semantic object obtained at one of the tenses - keeping in mind that we might need more than one future operator. The tense element would be syntactically under the AUX category. Hence $//V^3//$ evaluated at past present or future corresponds to V^3+Aux = V^4.

Under the AdvP heading we distinguished eight distinct temporal modifiers. Thus we compute the semantic object corresponding to PRED by applying to whatever $//V^4//$ we obtained the temporal operators found in the sentence. Thus we have

AdvP', AdvP'', etc. up to 8. And //PRED// = Adv1...8 applied to $//V^4//$

Note that since the same semantic object can emerge at different stages of the analysis, we need to carry the whole derivation in order to obtain the correct semantic interpretation of any given predicate phrase with tense, aspect, and temporal modifier in it.

Even if this analysis turns out to be in need of correction with respect to some of the details, it should provide the basic framework for tense and aspect analysis and the semantics of temporal reference in English. Our work leads to two important questions. First, what modifications are required for the handling of temporal reference in other natural languages; i.e. how much of this framework represents universal grammar and semantics? Secondly, how many of the notions referred to in this analysis resist formal treatment? (E.g. the human element determining contextually the meaning of 'for a long time', etc.).

If our work sheds light on these questions, it will have served its purpose.

Bibliography

The following works have been consulted for examples but analyse the data differently.

Comrie, B., 1976, *Aspect,* C.U.P.
Huddlestone, K., 1969, Some Observations on Tense and Deixes in English, *Language,* vol.45.
Reichenbach, H., 1947, *Elements of Symbolic Logic.*
Verkuyl, H.J., 1972, *On the compositional nature of Aspect,* D.Reidel.
Verkuyl, H.J., 1976, 'Interpretive Rules and the Description of the Aspects'. In: *Foundations of Language,* 14, pp.471-503.

Note

The interested reader may wish to consult the following additional references. (H.J. Verkuyl's paper below is a reaction to the present work).

Aqvist, L., 1977, 'On the analysis of some accomplishment and activity verbs in: C.Rohrer (ed.), *On the analysis of Tense and Aspect,* Tübingen, Teubner Verlag, pp.31-65.
Hoepelman, J., *Verb classification and the Russian Aspect, a formal treatment.* D.Reidel, to appear.
Kamp, H., Events, Instances and temporal references, in A. von Stechow (ed.). *Semantics from different points of view,* J.Springer-Verlag, 1979.
Verkuyl, H.J., On the proper classification of events and verb phrases, to appear.

We wish to thank Emmon Bach, A. Galton, Alice ter Meulen, R. Oehrle, Christer Platzack, Helmut Schnelle, E. Traugott, Tom Wasow, and our audience at Tel Aviv University for useful suggestions; needless to say, they are not responsible for the results.

ON THE MASS-COUNT DISTINCTION AND THE FRENCH IMPARFAIT AND PASSÉ SIMPLE

J.Hoepelman, C.Rohrer

I. Introduction

In Vendler's well known four-fold classification of verbs one can put the accomplishments and achievements in one main category, and the activities and states in another, on the basis of certain syntactic and semantic similarities. This division seems to have a parallel in the Mass - Count distinction. There exists moreover a peculiar interaction between verbs of Vendler's four types and Mass or Count expressions. The connection between the classification of the verbs in two main groups and the imparfait - passé simple distinction has not gone unnoticed in traditional French grammars. P. Imbs writes: "Aussi bien l'emploi du passé simple est-il étroitement lié à la notion d'événement, dont l'exacte saisie est essentielle pour le juste emploi de ce temps". The same view is expressed by R. Martin: "L'aspect ponctuel-perfectif lie indissolublement le passé simple à la notion d'evenement". It is interesting to compare these remarks with Mourelatos' classification of accomplishments and achievements as *events* (Mourelatos, 1978).

In this paper we will discuss some of the phenomena of the mass - count distinction and present a semantics for mass and count expressions which is a restricted and modified version of (Hoepelman, 1976).
We will also list some of the phenomena, which accompany the use of the imparfait and the passé simple. Then we will present a new semantics for the imparfait and the passé simple, which provides a formal basis for the intuition that the imparfait has affinity to mass expressions and the passé simple to count expressions. Our proposal also gives a treatment of the infinitive and opens new perspectives for the handling of the progressive and the generic present.

II. *Mass Terms, Count Terms and Indefinite Plurals*

Of course, the first thing to be remarked in connection with mass terms (vs. count terms), is that one cannot use them in combination with cardinal count nouns.

(1) **three waters*

(2) **three wines*

(1) and (2) are unacceptable - unless one reinterprets "water" and "wine" as "kind of water" and "kind of wine". The French translations of (1) and (2) are unacceptable as well:

(3) **trois eaux*

(4) **trois vins*

That, on the other hand, count terms can occur with cardinal count nouns is too obvious to be corroborated by examples. In the second place one remarks that mass terms in object position with adverbs like "in an hour", whereas one can use them with expressions like "for an hour". It is, however, not acceptable to use count terms with "for an hour", in a non-repetive reading, and quite normal to use them with "in an hour":

(5) *John drinks water for an hour.*

(6) **John drinks water in an hour.*

(7) **John eats two apples for an hour.*

(8) *John eats two apples in an hour.*

French:

(9) *Jean boit de l'eau pendant une heure.*

(10) **Jean boit de l'eau en une heure.*

(11) **Jean mange deux pommes pendant une heure.*

(12) *Jean mange deux pommes en une heure.*

Thirdly, it must be realized that there are several ways to use mass terms, which are not equivalent. It is interesting to compare the use of the articles and of the partitive "de" in French with the corresponding uses in English in this connection. That there are at least three different uses of mass terms can be demonstrated by means of the following examples.

(13) *John drinks water.*

(14) *Water is indispensable.*

(15) *Water boils at 100° C.*

French: (16) *Jean boit de l'eau.*

(17) *L'eau est indispensable.*

(18) *L'eau bout à 100° C.*

We can paraphrase (13) as (19):

(19) *John drinks some water.*[1]

However, (14) and (15) cannot be paraphrased like that.
(14) does not mean the same as "some water is rare", and (15) not the same as "some water boils at 100° C". (15) can be paraphrased as (20).

(20) *All water boils at 100° C.*

We cannot do this with (13) and (14): (13) is not "John drinks all water", and (14) is not "All water is indispensable". The same phenomenon can be observed in connection with indefinite plurals:

(21) *The anteater eats ants.*

(22) *Ants are indispensable.*

(23) *Ants are fond of sugar.*

French:

(24) *Le fourmilier mange des fourmis.*

(25) *Les fourmis sont indispensables.*

(26) *Les fourmis aiment beaucoup le sucre.*

(21) can be rendered as (27):

(27) *The anteater eats some ants.*

It is not correct to render (22) as "Some ants are indispensable". Neither can one paraphrase (23) as "Some ants are fond of sugar". (23) can be considered to be "All ants are fond of sugar", but (22) is not equivalent to "All ants are indispensable, nor is (21) equivalent to "The anteater eats all ants".

So we see that there are at least three different uses of mass terms and indefinite plurals in English, which morphologically are all realized in the same way, however. In Dowty (1972) it has been noticed that Vendler's

---

1 Sentences (13), (14), (15) have a strong generic flavour. Therefore, rendering (13) as (19) might seem doubtful. It seems more acceptable to render "John is drinking water" with the present continuous as "John is drinking some water". However, if we construct the generic in (13) as "Normally when John drinks, then he drinks water", then it seems to be acceptable to paraphrase this as "normally when John drinks, he drinks some water". We will neglect the peculiarities of the generics here.

categorization of verbs is not as stable as it seems at first sight. Mourelatos (1978) too stresses this point.[2] If one has an activity verb like "to push", one just has to add a directional adverb like "to London" to obtain an accomplishment:

(28) **John pushed the cart in an hour.*

(29) *John pushed the cart for an hour.*

(30) *John pushed the cart to London in an hour.*

(31) **John pushed the cart to London for an hour.*

Accomplishments like "to close" seem to be able to change into activities:

(32) *John closed the door in ten minutes.*

(33) **John closed the door for ten minutes.*

(34) **John closed doors in ten minutes.*

(35) *John closed doors for ten minutes.*

The division mass - count is not so stable either. Out of count expressions one can form noun expressions and reversely. Mourelatos (op.c.) seems to be completely right in supposing that we have here a very fundamental fact of the ontology of natural languages. In English we have pairs like:

	Count	*Mass*
(36)	a chicken	chicken
	a fish	fish

French:

un poulet	du poulet
un poisson	du poisson

We also have more bizarre cases like English:

	Count	*Mass*
(37)	a missionary	missionary
	an encyclopaedia	encyclopaedia

French:

un missionaire	du missionaire
une encyclopédie	de l'encyclopédie

These examples are less bizarre in examples like the following:

(38) *I liked stuffed missionary, said the cannibal.*

(39) *I have eaten encyclopaedia all my life, said the wise bookworm.*

It seems safe to assume that in principle any count noun can be converted into a mass expression. Of course in a formal semantics we'll need a mechanism which allows us to do this.

---

2 Compare Hoepelman (1979).

Conversely, one can turn a mass expression into a count expression by means of a measure phrase, like "a liter of". We have seen that we don't have " *John drank water in an hour", but we do have

(40) *John drank a liter of water in an hour.*

As was to be expected the following is not acceptable:

(41) **John drank a liter of water for an hour.*

We observe the same in French. We have

(42) *Jean but un litre d'eau en une heur*

but not

(43) **Jean but un litre d'eau pendant une heure.*

These examples, by the way, make clear that one cannot consider "water" to mean "quantity of water", because one can have "John drank water for an hour", but not

(44) **John drank a quantity of water for an hour.*

The fact that one can form a mass term out of any count term makes the principle of physical decomposition less plausible than it seems to be if one takes into consideration only homogeneous substances like "water" or "gold". This principle has been formulated by Bennett as follows (1974,p.257):

> "This is to say that given any subset of the set of all quantities of a substance, there exists a quantity of that substance in the set which is the sum of the members of the subset".

If we now take the substance "gingercake", we might take those quantities of this substance which are just ginger. The sum of these quantities will be a quantity of ginger, not a quantity of gingercake.

The main idea of our analysis is, that "water", as it occurs in (14) (but not in (13) or (15)) is the name of an individual concept of a special kind, just like "John" is the name of an individual concept. The extension of the individual concept "water" and any point (i,j), where i is a world and j a moment of time, is the union of all things which are water, at (i,j). In other words: at each (i,j) there is a set of individuals to which the predicate "water" can rightly be applied. To be able to form the union of these individuals at (i,j), we must consider them as *sets*. This means, that our ontology has to be built up out of just sets. We will have no atoms. As the set theoretical basis for this ontology we will use Zermelo-Fraenkel's set theory. We can now form superindividuals (like the thing which is the union of all things which are water at a certain point (i,j)), and parts of individuals (as we might need to explain a sentence like "first I'll take a little bit of John, and then I'll have a sip of Mary" - spoken by Count Dracula).

In particular, after having made the superindividual "water", we can again speak about its subparts - leaving it open whether all these subparts are "water" or not - and about subparts of subparts. Of course we will need some additional machinery to make names out of predicates, and predicates out of names. In the next paragraph we will give a very short Montague fragment for English, in which we will introduce a few additional articles which enable us to make these transitions. These articles are not realized in the sentences which are constructed after the last rule (an "editing rule") has been applied. In the light of the use of articles in connection with mass terms in French, it will seem more plausible to introduce such articles.

The sense of "water" in (14) ("water is indispensable") will be connected with the article C_{*} (where the fact that the article does not appear on the surface is indicated by the subscript "$_{*}$").

As said before, we interpret (14) as to mean that the *individual concept* "water" has the property of being indispensable. The notion "water" in (13) ("John drinks water") will be connected with the article S_{*}, which doesn't appear on the surface either. (13) does not mean, that John drinks the individual concept "water". It rather means that John drinks a part of the extension of this concept. Moreover, for all the parts of this part, John drinks them as well. This last condition is the so called *condition of distributivity*. It should not be interpreted as saying that all the subparts of some subpart of water (at a certain point) are water - this would lead to the well known problems with the subparts of dirty water - but rather as saying that one and the same thing happened to some subpart of water and to all its subparts as well (where it does not play a role whether these subparts are water or not). It is the condition of distributivity which makes a sentence like (6) *"John drank water in an hour") unacceptable: If John drinks a subpart of water in an hour, it cannot be the case that he drinks all the subparts of this subpart in an hour.

Finally, the notion of "water" occurring in (15) ("water boils at 100°C") will be connected with the unrealized article $\forall_{*}$. (15) can be interpreted as saying that all subparts of water have the property of boiling at 100°C.

In the intensional logic we will introduce an operator "o" which makes an individual expression "oP" out of any predicate expression "P". At a point (i,j) the extension of oP is the union of all the things which are P at (i,j). Reversely, the operator "sm" makes predicate expressions "sma" out of individual expressions "a". At each (i,j) sma denotes the characteristic function of the set of all subsets of a. For example: if P translates "water" (as a predicate) then oP denotes the individual concept consisting of the function from points (i,j) to the union of all things (sets), which are water at (i,j). $sm^{\vee}o\ P(^{\vee}x)$ means, that $^{\vee}x$ is a subset of this individual. Notice that x is not necessarily identical with any of the sets which originally made up the individual "water" (at (i,j)).

As the formula translating "John drinks water" we propose (45), where Q translates the phrase "John drinks", and "$\vee$" is Montague's extension operator (Montague, 1974).

(45) $(\exists x)(\forall y)(sm^{\vee}o\ water'(^{\vee}x)\ \&\ sm^{\vee}x(^{\vee}y) \rightarrow Q(^{\vee}y))$

For "Water boils at 100°C" we have a formula of the form (46),

(46) $(\forall x)(sm^{\vee}o\ water'(^{\vee}x) \rightarrow R(^{\vee}x))$

where "R" represents "boils at 100°C". For "water is indispensable" we propose a formula like (47):

(47) $S(o\ water')$

where "S" represents "is indispensable".

Regarding the indefinite plural, we proceed in almost the same way. The term "ants", for instance in sentence (22) ("ants are rare") is considered to be a *name*. Its denotation is the individual concept which as its extension at each point (i,j) has the set of all individuals to which at (i,j) the predicate "ant" can truly be applied. Notice the difference with the mass term "ant", which can equally well be formed and has as its denotation the individual concept which has as its extension at each (i,j) the union of all things (which are sets) to which the predicate "ant" can rightly be applied at (i,j).

The following Montague fragment is not meant to provide a description which is exhaustive in any way. It is rather set up to give us the opportunity to demonstrate the transitions from count to mass expressions and reversely and to develop a semantics with an adequate ontology.

III. A small Montague Fragment for English count terms, mass terms and indefinite plurals.

The following fragment of a Montague grammar presupposes some knowledge of Montague (1974). We have somewhat extended the categorial grammar and had to complicate the rules of the categorial grammar accordingly.

1. Syntax of a fragment of English.

Let e and t be two objects. $\mathbb{N}$ the set of natural numbers.

Def. 1. The set CAT of categories or indices of expressions is the set such that

i) for any n ε $\mathbb{N}$: (n,e) and (n,t) ε CAT.

ii) for any n,m ε $\mathbb{N}$: if A and B ε CAT, then (n,m,A,B) ε CAT.

iii) for no n,m, and r ε $\mathbb{N}$ and for no A and B ε CAT: (n,m,A,B) = (r,e) or (n,m,A,B) = (r,t).

iv) for any D, D ε CAT iff D satisfies i) - ii).

Def. 2. (A/B) = (o,o,A,B).

Def. 3. b = (o,t).
a = (o,e).

We have the following rules to combine expressions belonging to the categories.

Let β be an expression of either category (n,e) or of category (n,t) for some n ε $\mathbb{N}$, and let α be an expression of either category (r,s,E,(n,e)) or of category (r,s,E,(n,t)) for some r,s ε $\mathbb{N}$ and E ε CAT. Then the expression α(β) belongs to category E. Let β be an expression of category (r,s,M,N) for some r,s ε $\mathbb{N}$ and M,N ε CAT and let α be an expression of category (n,m,E,(r,t,M,N)) for some n,m,t ε $\mathbb{N}$ and E ε CAT. Then α(β) belongs to category E. Notice that the fact that t might not be identical to s does not matter. However, if α is of, say, category (n,m,E,(t,r,M,N)) and β of category (r,s,M,N), where t ≠ r, then the category of α(β) is not defined. We will use the following abbreviations:

	IV	= (t/e)	Intransitive verb phrases.
	T	= (o,i,(1,t),IV)	for any i ε $\mathbb{N}$. Terms
	T_c	= (o,o,(1,t),IV)	Count terms.
T_m =	T_1	= (o,1,(1,t),IV)	Mass terms
	T_2	= (o,2,(1,t),IV)	
	T_3	= (o,3,(1,t),IV)	

$$T_{ipl} = \begin{cases} T_4 = (0,4,(1,t),IV) \\ T_5 = (0,5,(1,t),IV) \\ T_6 = (0,6,(1,t),IV) \end{cases} \quad \text{Indefinite plurals}$$

TV =	(IV/T)	Transitive verb phrases
CN =	(1,i,b,a)	for any i ε $\mathbb{N}$. Nouns.
CN_c =	(1,0,b,a)	Count nouns.
CN_m =	(1,1,b,a)	Mass nouns.
CN_d =	(1,2,b,a)	Dummy nouns.
ADJ =	((1,i,b,a)/(1,i,b,a))	for any i ε $\mathbb{N}$. Adjectives.

Let B be the function from the set of categories into the set of basic expressions, and P the function from CAT into the set of all expressions of the language as defined by the syntactic rules.

1.1 *Basic expressions.*

B_{T_c} = {*John*}

B_{IV} = {*boil*}

B_{CN_c} = {*gingercake, ant*}

B_{CN_m} = {*water, meat*}

B_{CN_d} = {*entity*$_{\ast}$}

B_{TV} = {*eat, drink, be*}

B_{ADJ} = {*red, indispensable*}

$B_{(T_c/T_3)}$ = {*a pound of*}[3]

$B_{(T_c/T_6)}$ = {*a pound of*}

For any A ε CAT other than those mentioned above, $B_A = \emptyset$.

Def.4 Any member of $D_{(o,t)}$ is a *sentence*.

Def.5 Any expression α of the form $\beta_{\ast}$ is called a *designated expression*.

3 We introduce "a pound of" and not "pound of" for the sake of simplicity.

1.2. *Syntactic Rules*

1.2.1. *Basic Rules*

S1. For any A ε CAT: $B_A \subseteq P_A$.

S2. If $\zeta \ \varepsilon \ P_{CN_c}$, then $F_1(\zeta) \varepsilon \ P_{T_c}$ and $F_1(\zeta) = a\ \zeta$ or $an\ \zeta$, according as the leftmost expression in ζ takes *a* or *an*.
If $\zeta \ \varepsilon \ P_{CN_d}$, then $F_1(\zeta) \varepsilon \ P_{T_c}$ and $F_1(\zeta) = a_{*}\zeta$.

S3. If $\zeta \ \varepsilon \begin{Bmatrix} P_{CN_c} \\ P_{CN_m} \end{Bmatrix}$, then $F_2(\zeta) \varepsilon \ P_{T_1}$ and $F_2(\zeta) = S_{*}\zeta$

S4. If $\zeta \ \varepsilon \begin{Bmatrix} P_{CN_c} \\ P_{CN_m} \end{Bmatrix}$, then $F_3(\zeta) \varepsilon \ P_{T_2}$ and $F_3(\zeta) = \forall_{*}\zeta$

S5. If $\zeta \ \varepsilon \begin{Bmatrix} P_{CN_c} \\ P_{CN_m} \end{Bmatrix}$, then $F_4(\zeta) \varepsilon \ P_{T_3}$ and $F_4(\zeta) = C_{*}\zeta$

S6. If $\zeta \ \varepsilon \ P_{CN_c}$, then $F_5(\zeta) \varepsilon \ P_{T_4}$ and $F_5(\zeta) = S_{*}\zeta'$, where ζ' is the plural form of ζ.

S7. If $\zeta \ \varepsilon \ P_{CN_c}$, then $F_6(\zeta) \varepsilon \ P_{T_5}$ and $F_6(\zeta) = \forall_{*}\zeta'$, where ζ' is the plural form of ζ.

S8. If $\zeta \ \varepsilon \ P_{CN_c}$, then $F_7(\zeta) \varepsilon \ P_{T_6}$ and $F_7(\zeta) = C_{*}\zeta'$, where ζ' is the plural form of ζ.

1.2.2. *Rules of functional application.*

S9. If $\delta \ \varepsilon \begin{Bmatrix} P_{T_c} \\ P_{T_m} \end{Bmatrix}$ and $\zeta \ \varepsilon \ P_{IV}$, then $F_8(\delta,\zeta) \varepsilon \ P_{(1,t)}$ and

$F_8(\delta,\zeta) = \delta\zeta'$, where ζ' is the third person singular simple present of ζ.

S10. If $\delta \ \varepsilon \ T_{ipl}$ and $\zeta \ \varepsilon \ P_{IV}$, then $F_9(\delta,\zeta) \ \varepsilon \ P_{(1,t)}$ and

$F_8(\delta,\zeta) = \delta\zeta'$, where ζ' is the third person plural simple present of ζ.

S11. If $\delta \ \varepsilon \ P_{(IV/T)}$ and $\zeta \ \varepsilon \ P_T$, then $F_8(\delta,\zeta) \varepsilon \ P_{IV}$ and $F_8(\delta,\zeta) = \delta\zeta$.

S12. If $\delta \in P_{(T_c/T_3)}$ and $\zeta \in P_{T_3}$, then $F_8(\delta,\zeta) \in P_{T_c}$ and $F_8(\delta,\zeta) = \delta\zeta$.
If $\delta \in P_{(T_c/T_6)}$ and $\zeta \in P_{T_6}$, then $F_8(\delta,\zeta) \in P_{T_c}$.

S13. If $\delta \in P_{ADJ}$ and $\zeta \in P_{CN}$, then $F_8(\delta,\zeta) \in P_{CN}$.

1.2.3. Editing Rule.

S14. If $\phi \in P_{(1,t)}$, then $F_{10}(\phi) \in P_{(0,t)}$ where $F_{10}(\phi)$ comes from ϕ by deleting all designated expressions.

2. A Few Examples.

Before giving the intensional logic, we present some analysis trees.

(48) *John drinks water.*

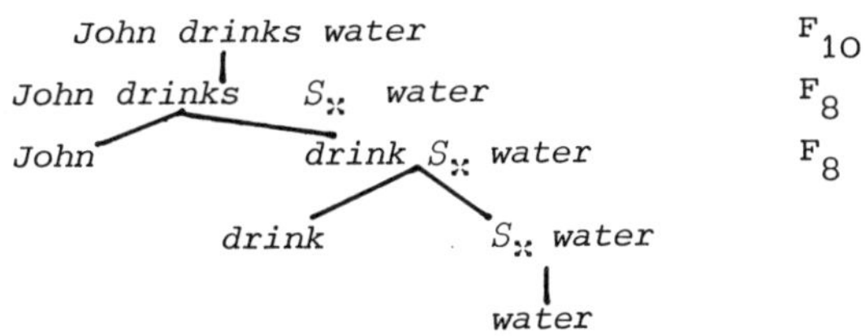

(49) *Water boils.*

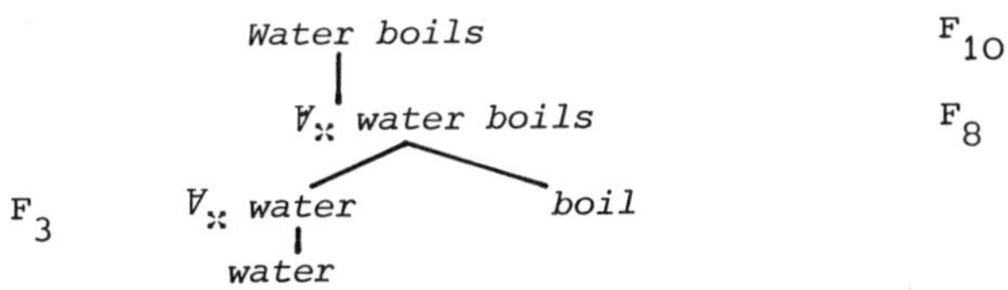

(50) *Water is indispensable.*

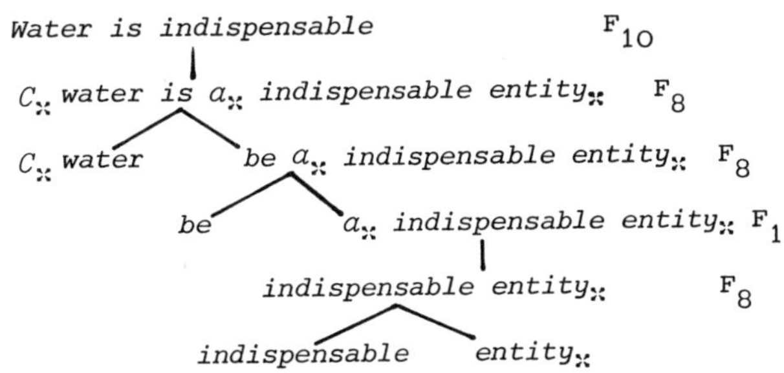

Notice that the expression "water", appearing in (50), (49) and (48) could in each case have been given a different reading than the one actually presented. On the intended interpretations (50), (49) and (48) will then just be false. They will not be ungrammatical.

(51)

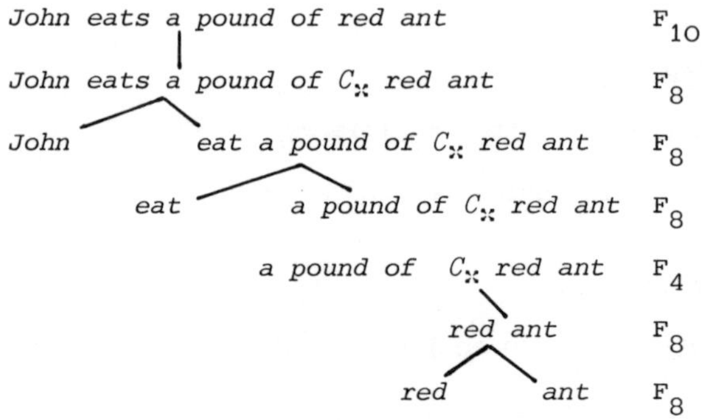

(52)

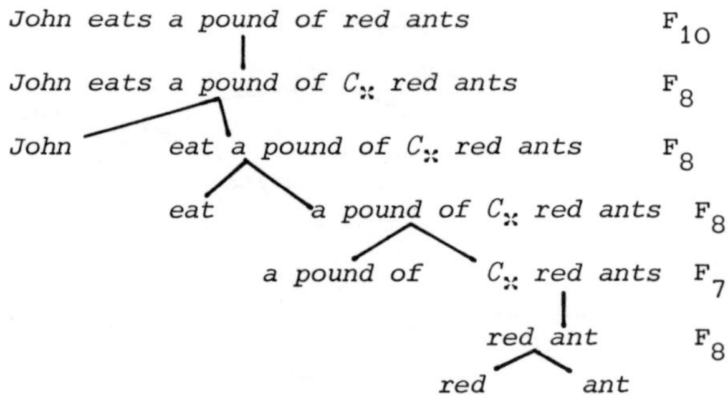

What John eats has a different consistency in (51) and (52). Nevertheless he is eating *ant* both times - which we can prove.

3. *Intensional Logic.*

3.1. *Syntax*

Let s be a fixed object, $s \neq e$, $s \neq t$.

Def.6 TYPE is a set such that

i) e ,t ε TYPE.

ii) if A and B ε TYPE, then (A,B) ε TYPE.

iii) For no (A,B): if (A,B) ε TYPE, then (A,B) =s or (A,B) = e or (A,B) = t.

iv) For any A ε TYPE: (s,a) ε TYPE

v) For any D: D ε TYPE if it satisfies i) - iv).

Let S_i be a set of symbols, B_i a function from TYPE, $R(B_i)$ the range of B_i, $\mathcal{P}(S_i)$ the powerset of S_i, and $R(B_i) \subseteq \mathcal{P}(S_i)$. For any $P,Q \varepsilon R(B_i)$: $P \cap Q = \emptyset$ and for any A ε TYPE : $B_i(A)$ is the set of constants and variables of type A. Me is a function from TYPE such that R(Me) is the set of smallest sets which satisfy:

1) For any A ε TYPE : $B_i(A) \subseteq Me(A)$.

2) If $\alpha \varepsilon$ Me(A,B), $\beta \varepsilon$ Me(A), then $\alpha(\beta) \varepsilon$ Me(B).

3) If $\alpha \varepsilon$ Me(A), u ε Me(B), then $\lambda u \alpha \varepsilon$ Me(B,A).

4) If α, $\beta \varepsilon$ Me(A), then $\alpha = \beta \varepsilon$ Me(t)

5) If ϕ, $\psi \varepsilon$ Me(t) and u is a variable, then $\neg\phi$, $\phi \rightarrow \psi$, $\phi \vee \psi$, $\phi \,\&\, \psi$, $(\forall u)(\phi)$, $(\exists u)(\phi)$ ε Me(t).

6) If $\zeta \varepsilon$ Me(((s,e),t)) then $o\phi \varepsilon$ Me((s,e)) and $+\phi \varepsilon$ Me((s,e)).

7) If u ε Me(A), then smu ε Me((A,t)) and 1pε Me((A,t)).

8) If $\alpha \varepsilon$ Me(A), then $^\wedge\alpha \varepsilon$ Me((s,A)).

9) If $\alpha \varepsilon$ Me((s,A)), then $^\vee\alpha \varepsilon$ Me(A).

3.2. *Semantics*

Let I and J be sets, to be considered as the set of worlds, and the set of moments of time, respectively. We take J to be the set of reals, ordered by the relation <, "smaller than".

Let A_o be a set consisting of non-empty sets. Let U_{A_o} be the universe class of A_o , relative to the axioms of Zermelo-Fraenkel's set theory. Among the members of U_{A_o} we have

M: $M = \{m: \exists b \in A_o \,\&\, m \in b\}$.

A: $A = \mathcal{P}(A_o)$.

$Q : Q = \mathcal{P}(M)$.

$T : T = A_o \cup A \cup Q$.

Intuitively we can say that our ontology is based on the set A_o of "individuals". These individuals, however, are sets themselves, and we also wish to consider as individuals sets consisting of the elements which make up the individuals of a certain kind. Such sets are in Q. Moreover, we wish to consider as individuals certain ensembles of individuals belonging to A_o. These ensembles are the elements of A. The set T contains everything which in our ontology is considered to be an individual, i.e. what can be referred to by expressions of type e.

Let D_a be the total set of denotations of expressions of type a. The sets of denotations of expressions of any type are defined recursively:

$$D_e = T$$
$$D_t = \{0,1\}$$
$$D_{(a,b)} = D_b^{D_a}$$
$$D_{(s,a)} = D_a^{I \times J}$$

The 5-tuple $(T,I,J,<,F)$ is a model $\mathfrak{A}$.

We assume familiarity with the notions of the semantics of the intensional logic of Montague (1974), and with the truth definitions therein. They can be taken over in the present paper. In this fragment we won't need the classical connections anyway. The new clauses which have to be added for the present fragment are the following:

Def. 7 1) If $\delta \varepsilon Me(((s,e),t))$ then $[o\delta]^{\mathfrak{A},i,j,g}$ is that function f, such that for any $(i',j') \varepsilon I \times J : f(i',j') = \{m : (\exists d)([\delta]^{\mathfrak{A},i',j',g}(d) = 1 \;\&\; m \varepsilon d(i',j')\}$.

2) If $\delta \varepsilon Me(((s,e),t))$, then $[+\delta]^{\mathfrak{A},i,j,g}$ is that function h, such that for every (i',j') $h(i',j') = \{d(i',j') : [\delta]^{\mathfrak{A},i',j',g}(d) = 1\}$.

3) If $u \varepsilon Me(A)$, then $[smu]^{\mathfrak{A},i,j,g}$ is $K_{\mathcal{P}([u]^{\mathfrak{A},i,j,g}) - \{\phi\}}$

where for any set $s : K_s$ is the characteristic function of s.

4) If $u \varepsilon Me(e)$, then $[1pu]^{\mathfrak{A},i,j,g}$ is $K_{\mathcal{P}_{1p}([u]^{\mathfrak{A},i,j,g})}$

where $\mathcal{P}_{1p}(A)$ is the set of all subsets y of a set A, such that the weight of y measures one pound.

4. *Translation Rules.*

Let f : CAT → TYPE be a mapping such that f(i,e) = e, f(i,t) = t for any i ε ℕ, and for any m,n ε ℕ and A,B ε CAT:

$$f(n,m,A,B) = f(A/B) = ((s,f(B)),f(A)).$$

We use the conventions of Montague (1974). Let "⟹" abbreviate "translates into". Any basic expression α apart from *John*, *entity*$_{*}$ and *be* translates into the constant of intensional logic α'.

4.1. *Basic Rules.*

T_1 *be* $\Rightarrow \lambda \mathcal{P} \lambda x\, \mathcal{P}\{\hat{y}[{}^{\vee}x = {}^{\vee}y]$

T_2 *John* $\Rightarrow j^{*}$

T_3 *entity*$_{*}$ $\Rightarrow \lambda x(x = x)$

T_4 If $\zeta \varepsilon \begin{Bmatrix} P_{CN_c} \\ P_{CN_d} \end{Bmatrix}$ and $\zeta \Rightarrow \zeta'$, then

$F_7(\zeta) \Rightarrow \lambda P(\exists x)(\zeta'(x)\ \&\ P\{x\})$.

T_5 If $\zeta \varepsilon \begin{Bmatrix} P_{CN_c} \\ P_{CN_m} \end{Bmatrix}$ and $\zeta \Rightarrow \zeta'$, then

$F_2(\zeta) \Rightarrow \lambda P(\exists x)(\forall y)((sm^{\vee}o\zeta({}^{\vee}x)\ \&\ sm^{\vee}x({}^{\vee}y))$

$F_3(\zeta) \Rightarrow \lambda P(\forall x)(sm^{\vee}o\zeta\ ({}^{\vee}x) \rightarrow {}^{\vee}P_{*}\ ({}^{\vee}x))$

$F_4(\zeta) \Rightarrow \lambda P\ P\ \{o\zeta'\}$

T_6 If $\zeta \varepsilon P_{CN_c}$ and $\zeta \Rightarrow \zeta'$, then

$F_5(\zeta) \Rightarrow \lambda P(\exists x)(\forall y)((sm^{\vee}{+}\zeta({}^{\vee}x)\ \&\ sm^{\vee}x({}^{\vee}y)) \rightarrow {}^{\vee}P_{*}\ ({}^{\vee}y))$

$F_6(\zeta) \Rightarrow \lambda P(\forall x)(sm^{\vee}{+}\zeta({}^{\vee}x) \rightarrow {}^{\vee}P_{*}\ ({}^{\vee}x))$

$F_7(\zeta) \Rightarrow \lambda P\ P\{+\zeta'\}$

4.2. *Rules of Functional Application.*

T_7 If $\delta \varepsilon \begin{Bmatrix} P_{T_c} \\ P_{T_m} \end{Bmatrix}$, $\zeta \varepsilon P_{IV}$, $\delta \Rightarrow \delta'$, $\zeta \Rightarrow \zeta'$, then

$F_8(\delta,\zeta) \Rightarrow \delta'({}^{\wedge}\zeta')$.

T_8 If $\delta \varepsilon\ T_{ipl}$, $\zeta \varepsilon P_{IV}$, $\delta \Rightarrow \delta'$, $\zeta \Rightarrow \zeta'$. then

$F_8(\delta,\zeta) \Rightarrow \delta'({}^{\wedge}\zeta')$.

T_9 If $\delta \varepsilon P_{(IV/T)}$, $\zeta \varepsilon P_T$, $\delta \Rightarrow \delta'$, $\zeta \Rightarrow \zeta'$, then

$F_8(\delta,\zeta) \Rightarrow \delta'(\hat{}\zeta')$.

T_{10} If $\delta \varepsilon P_{(T_c/T_3)}$ and $\zeta \varepsilon P_{T_3}$, $\delta \Rightarrow \delta'$, $\zeta \Rightarrow \zeta'$, then

$F_8(\delta,\zeta) \Rightarrow \lambda P(\exists x)(1p^{\vee}o\zeta'(^{\vee}x)\ \&\ ^{\vee}P_{*}\ (^{\vee}x))$.

If $\delta \varepsilon P_{(T_c/T_6)}$ and $\zeta \varepsilon P_{T_6}$, $\delta \Rightarrow \delta'$, $\zeta \Rightarrow \zeta'$, then

$F_8(\delta,\zeta) \Rightarrow \lambda P(\exists x)((1p^{\vee} + \zeta'(^{\vee}x)\ \&\ ^{\vee}P_{*}\ (^{\vee}x))$.

T_{11} If $\delta \varepsilon P_{ADJ}$, and $\zeta \varepsilon P_{CN}$, $\delta \Rightarrow \delta'$, $\zeta \Rightarrow \zeta'$, then

$F_8(\delta,\zeta) \Rightarrow \delta'(\hat{}\zeta')$.

4.3. Editing Rule.

T_{12} If $\phi \varepsilon P_{(1,t)}$ and $\phi \Rightarrow \phi'$, then

$F_{10}(\phi) \Rightarrow \phi'$.

Although we have not presented a formal treatment for locutions like "in an hour", because this would exceed the limits of this paper,[4] we can now explain in intuitive terms why

(6) **John drank water in an hour.*

is unacceptable. The reason for this unacceptability is a semantical one: (6) will be translated by a formula in which a part like (48) occurs.

(48) $(\exists x)(\forall y)(sm^{\vee}o\ water'(^{\vee}x)\ \&\ sm^{\vee}x(^{\vee}y) \rightarrow (in\ an\ hour)'\ ^{\vee}drink^{T}_{*}(^{\vee}y))$.

Here drink^T represents the property connected with the verb "drink" by the function T, which is obtained by anything that has been drunk.[5] (48) expresses that some subpart of all the water (at (i,j)) has obtained this property in an hour, but also that all the subparts of this subpart have obtained it in an hour. It is not difficult to see intuitively, why this cannot be true.

It will be seen that distributivity like in the case of mass expressions can be used to explain at least some of the properties of the imparfait and the explicit negation of such distributivity some of the properties of the passé simple in French.

---

4 Comp. Hoepelman (1976),(1979).

5 It is a so called proto-verb. Comp. Hoepelman (1979).

If you take a man and chop off his little finger, you get one finger and one man, not two men. If you have a piece of cheese and cut it, you get two pieces of cheese. However, if you melt it in a fondue, you no longer have a piece of cheese. On the other hand, if you melt some gold, it will remain some gold. It seems that terms can denote different kinds of individuals. One could build up an ontology corresponding to meaning postulates which could be given for each kind of individuals. We will not attempt this here, but we remark that in the following the formula for the passé simple will resemble the meaning postulate that one could give for strictly non-distributive individuals, i.e. such individuals δ, such that for no proper subpart y of them, it is the case that y is a δ. One could say that in a sense the passé simple defines past intervals as individuals of a certain kind.

We give some more data on the distibution of the imparfait and the passé simple.

IV. Some Data on the use of the Passé Simple and the Imparfait.

a. The imparfait cannot be used with cardinal count[6] adverbs, but the passé simple can.

(49) **Jean dansait trois fois.*

(50) *Jean dansa trois fois.*

(51) *L'année dernière j'allais au cinéma tous les dimanches.*

(52) **L'année dernière j'allais au cinéma tous les dimanches sauf un.*

(53) *L'année dernière j'alla au cinéma tous les dimanches sauf un.*

Adverbs denoting a repetition of an event fall into two classes: cardinal count (meaning that an event took place a number of times) and frequency adverbials (giving an average over a stretch of time).

Examples:

Cardinal count adverbs: une fois, deux fois, ... n-fois, plusieur fois, tous les dimanches sauf un.

Frequency adverbs : Souvent, rarement, toujours, jamais, tous les dimanches.[7]

---

6 We adopt this term from A.Mourelatos (1978).

7 "Tous les dimanches" could be considered to belong to both classes.

Sentences in the imparfait which contain a frequency adverbial often denote habits, dispositions, lawlike behaviour, and the like.[8] This use of the imparfait can be analysed as a generic present pushed into the past.

b. The imparfait cannot be used with adverbs denoting intervals "closed at the right", but the passé simple can. Consider the following examples.

(54) **Jean dansait de deux heures à quatre heures.*

(55) **Jean dansait jusqu'à quatre heures.*

(56) *Jean dansa de deux heures à quatre heures.*

(57) *Jean dansa jusqu'à deux heures.*

c. If the intervals are not "closed to the right", only the imparfait can occur[9].

(58) *Jean dansait depuis deux heures.*

(59) **Jean dansa depuis deux heures.*

d. The passé simple can denote the beginning of an action or state, but the imparfait cannot. Sentence (60) and (61) have a different meaning:

(60) *Maigret fut sous la voûte.*

(61) *Maigret était sous la voûte.*

The same meaning difference clearly comes out in sentences with *quand*. Compare (62) and (63):

(62) *Quand Pierre entra, sa femme téléphonait.*

(63) *Quand Pierre entra, sa femme téléphona.*

In (62) the wife started the phonecall before Pierre came in. In (63) Pierre's action and the phonecall of his wife were simultaneous, or the phonecall began immediately after Pierre entered.

As an illustration of this function of the passé simple, consider the following fragment of "Candide", where - as usual in the case of a series of

---

8 Certain verb classes can be combined with cardinal count adverbs in the imparfait. For instances: "Jean frappais trois fois". Here too, the imparfait has a generic use, meaning that John had the habit of knocking three times.

9 "Comment analyser "depuis"?"in J.David and R.Martin (eds.) "Modèles logiques et niveaux d'analyse linguistique", Paris 1976, p.293 ff.

passé simples in texts - a sequence of beginnings of states or actions is described:[10]

> Elle [scil. Cunégonde] *rencontra* Candide en revenant au château, et *rougit;* Candide *rougit* aussi;elle lui *dit* bonjour d'une voix entre-coupée, et Candide lui *parla* sans savoir ce qu'il disait. Le lendemain après le dîner,comme on sortait de table, Cunégonde et Candide *se trouvèrent* derrière un paravent; Cunégonde *laissa* tomber son mouchoir, Candide le *ramassa*, elle lui *prit* innocemment la main, le jeune homme *baissa* innocemment la main de la jeune demoiselle avec une vivacité, une sensibilité, une grâce toute particulière; leurs bouches *se rencontrèrent*, leurs yeux *s'enflammèrent*, leurs genoux *tremblèrent*,leurs mains *s'égarèrent*. Monsieur le baron de Thunderten-thronckh *passa* auprès du paravent, et, voyant cette cause et cet effet, *chassa* Candide du château à grands coups de pied dans le derrière; Cunégonde *s'évanouit*; elle *fut* souffletée par madame la baronne dès qu'elle *fut revenue* à elle-même; et tout *fut* consterné dans le plus beau et le plus agréable des châteaux possibles.

H.Sten (1952,p.100) supposes that the events of such a sequence are non-overlapping, and gives the following picture

(65)

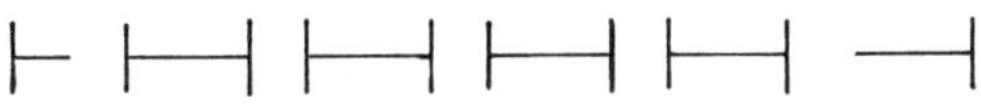

This seems implausible in view of examples like (62), (63) (where Pierre's entrance and the beginning of his wive's phonecall may be simultaneous), and of parts of the preceding fragment of Candide. In "leurs yeux s'enflammèrent, leurs genoux tremblèrent, leurs mains s'égarèrent" it seems probable that the knees were still trembling when their hands began their play (and afterwards).

V. *A Formal Explanation.*

In intuitive terms, our explanation of the phenomena described above will be, that we consider an infinitive, like "*marcher*" to be an individual concept, in the way we suppused "*water*" to be an individual concept. At each point (i,j) the extension of the individual concept "*marcher*" will be the set of people who walk at (i,j). This is in accordance with the syntactic status of the infinitive as a term. To express a sentence like (66) formally

(66) *Jean marchait.*

we will pick out a past part of the concept "marcher", and say that at all subparts of this subpart, it is the case that at each (i,j) belonging to the

---

10 This fragment has been token from H.Weinrich. "Tempus, besprochene und erzählte Welt.",Stuttgart, 1971.

domain of such a subpart, John walks at (i,j).

We will not consider the verb classes in this paper. A classification of verbs, or rather of their "Aktionsarten", can well be made within the framework to be developed.[11] In this paragraph we will just treat the activity verb "*marcher*". Also, we will not develop a full Montague-grammar for the French tenses, as this would lead us beyond the limits of this paper. Instead we will give a few fragments of formalism and semantics.

First let us have a look at the following examples.
We have seen that the imparfait can be used with "depuis" (in contradistinction to the passé simple).

(67) *Jean marchait depuis deux heures.*

Consider the following picture:

(68)
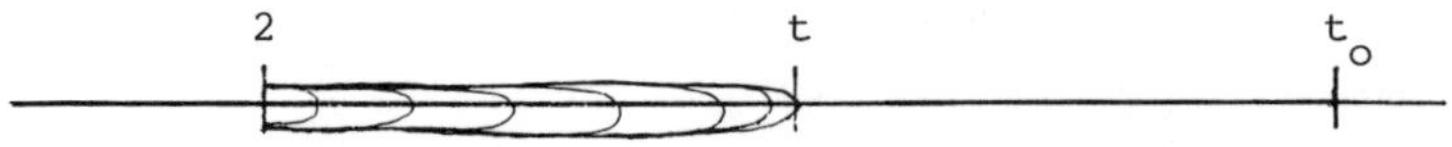

If John walks from two o'clock until a certain point t in the past, then for each interval, stretching from two o'clock it is the case that John walks since two o'clock in that interval. Or, seen in a different way: If the infinitive "*marcher*" denotes the function which at each point (i,j) gives the set of all individuals that walk, then the stretch of time from 2 to t is a subset of the infinitive and for all the subsets of this subset which begin at two o'clock it is the case that John belongs to the set of individuals that walk since two o'clock.

However, the imparfait cannot be used with "jusqu'à (again in contradistinction to the passé simple). Sentence (69) is not acceptable:

(69) **Jean marchait jusqu'à quatre heures.*

Consider picture (70):

(70)
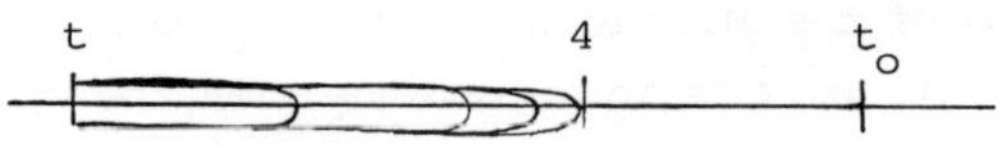

for no interval (t,j) such that $t < j < 4$ it is the case that (t,j) is an interval *until* four o'clock, or that it is an interval of John's walking until four o'clock.

Likewise, the imparfait (but not the passé simple) cannot be used with de ... à ... ; one can have:

11 Comp. Hoepelman (1979).

(71) *Jean marchait de deux à quatre heures.*

In (72)

(72)

we see that no interval (2,j) such that 2 < j < 4 is an interval of John's walking from two to four, although they all are intervals of John's walking *between* two and four. It is interesting to notice that the imparfait *can* be combined with "*entre*". (73) is perfectly correct:

(73) *Jean marchait entre deux et quatre.*

So we see that certain properties distribute over these "accumulative" subintervals, and others do not. Another example. Suppose that during a certain interval x John walks three times.

(73)

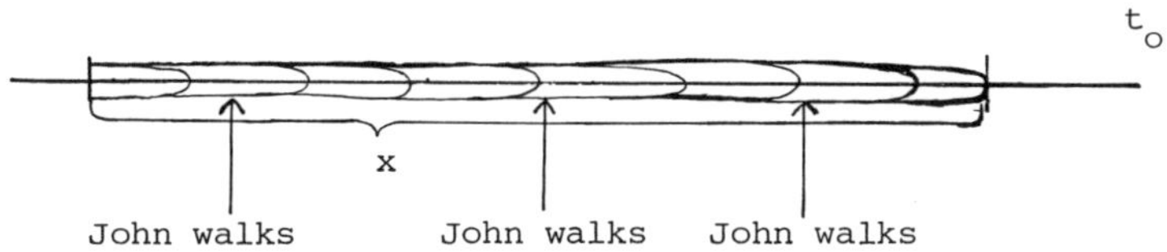

It is easy to see that not in all accumulative subintervals of x, it is the case that John walks three times *in them*.

We will use these peculiarities to define imparfait and passé simple. The main change in the semantics of the Montague grammar to be eventually built, will be, that we no longer consider the denotations of expressions (a,b) and (s,a) (for any a and b) to be *total* functions. For example: $D_{(a,b)}$ is the set of total or partial functions from a to b; $D_{(s,a)}$ is the set of total or partial functions from I x J to a.

In the intensional logic we will have an operator "*inf*"[12] which can be compared with the mass noun operator "o". Semantically: $[\text{inf march}']^{\mathfrak{X},i,j,g}$ is a function f such that at each (i',j'), f(i',j') = $\{d: [\text{march}']^{\mathfrak{X},i',j',g}(d) = 1\}$; that is, at each (i',j'), the value of f is the set of people who walk at (i',j'). The expression "*march*" can be considered as a proto verb (comp. Hoepelman, 1979). It is not to be realized in the surface structure. Suppose we have the following adverbs: "*Depuis*$_2$"[13], "*Jusqu'à*$_4$", "*de*$_2$*à*$_4$", "*tous les dimanches*", "*tous les dimanches sauf un*", "*trois fois*". We propose the following semantics:

12 "*inf*" for "infinitive"

13 For "depuis deux heures", "Jusqu'à quatre heures", "de deux à quatre heures".

$[\mathrm{Depuis}_2\ \phi]^{\mathfrak{A},i,j,g} = 1 \leftrightarrow (\forall t)(\exists t')(2 < t' < t\ \&\ t' \varepsilon\ \mathrm{UUDOM}([{}^\wedge\phi]^{\mathfrak{A},i,j,g}))\ \&$
$(2 < t < j \rightarrow [\phi]^{\mathfrak{A},i,t,g} = 1$

Otherwise $[\mathrm{Depuis}_2\ \phi]^{\mathfrak{A},i,j,g} = 0.$[14]

$[\mathrm{Jusqu'\grave{a}}_4\ \phi]^{\mathfrak{A},i,j,g} = 1 \leftrightarrow$

$(\exists t)(t < 4\ \&\ (\forall t')(t < t' < 4 \rightarrow t' \varepsilon\ \mathrm{UUDOM}[{}^\wedge\phi]^{\mathfrak{A},i,j,g}\ \&\ [\phi]^{\mathfrak{A},i,t',g} = 1))$

Otherwise $[\mathrm{Jusqu'\grave{a}}_4\ \phi]^{\mathfrak{A},i,j,g} = 0.$[15]

$[\mathrm{De}_2\grave{a}_4\ \phi]^{\mathfrak{A},i,j,g} = 1 \leftrightarrow [\mathrm{Depuis}_2\ \phi]^{\mathfrak{A},i,j,g} = 1\ \&\ [\mathrm{Jusqu'\grave{a}}_4\ \phi]^{\mathfrak{A},i,j,g} = 1$

$[\text{tous les dimanches } \phi]^{\mathfrak{A},i,j,g} = 1 \leftrightarrow$

$(\forall z)(\exists t)(t \varepsilon z \rightarrow (\exists t')(t' \varepsilon z\ \&\ [\phi]^{\mathfrak{A},i,t',g} = 1))$.

Otherwise $[\text{tous les dimanches } \phi]^{\mathfrak{A},i,j,g} = 0$, where z ranges over sundays.

$[\text{tous les dimanches sauf un } \phi]^{\mathfrak{A},i,j,g} = 1$

$(\exists z_1)(\exists t_1)((t_1 \varepsilon\ \mathrm{UUDOM}([\phi]^{\mathfrak{A},i,j,g}\ \&\ t_1 \varepsilon z_1 (\forall t_2)(t_2\ \varepsilon\ z_1 \rightarrow [\phi]^{\mathfrak{A},i,t_2,g} = 0)\ \&$

$(\exists z_2)(\exists t_3)(z_2 \neq z_1\ \&\ t_3\ \varepsilon\ z_2\ \&\ t_3\ \varepsilon\ \mathrm{UUDOM}([\phi]^{\mathfrak{A},i,j,g})\ \&$

$(\forall z_3)(z_3 \neq z_1 \rightarrow (\exists t_4)(t_4\ \varepsilon\ z_3\ \&\ [\phi]^{\mathfrak{A},i,t_3 g} = 1)))$, where DOM(f) is the domain of f for any f.

Otherwise $[\text{tous les dimanches sauf un } \phi]^{\mathfrak{A},i,j,g} = 0$

$[\text{trois fois } \phi]^{\mathfrak{A},i,j,g} = 1 \leftrightarrow$

$(\exists t_1)(\exists t_2)(\exists t_3)(t_1 < t_2 < t_3\ \&\ [\phi]^{\mathfrak{A},i,t_1,g} = 1\ \&\ [\phi]^{\mathfrak{A},i,t_2,g} = 1\ \&$
$[\phi]^{\mathfrak{A},i,t_3,g} = 1\ \&\ (\forall t_4)(\neg(t_4 = t_1 \vee t_4 = t_2 \vee t_4 = t_3 \rightarrow [\phi]^{\mathfrak{A},i,t_4,g} = 0)).$

Apart from sm we have the special cases sm_p, sm_v. sm_p gives us past parts of concepts, e.g. of the infinitive:

If $u\ \varepsilon\ \mathrm{Me}(s,A)$, then $\mathrm{sm}_p u\ \varepsilon\ \mathrm{Me}((s,A),t)$ and $[\mathrm{sm}_p u]^{\mathfrak{A},i,j,g} = K_{I(\zeta)-\{\phi\}}$, where ζ is the function such that for every (i',j') with $j' < j$ $\zeta(i',j') = [u]^{\mathfrak{A},i',j',g}(i,j')$ and which is undefined otherwise, and $I(\zeta)$ the smallest

---

14 Of course, if we want to be quite strict, we need additional machinery to handle expressions like "two o'clock" in a precise way.

15 Notice that "*Depuis*" always refers to the past, whereas "*Jusqu'à*" can lead up to past, present and future points.

set of those subsets of ζ such that whenever $i \in I(\zeta)$ and $((i,j),\zeta(i,j))$ and $((i,j'), \zeta(i,j')) \in i$, and $j < j'$, then for any j'', with $j < j'' < j$: $((i,j''), \zeta(i,j'')) \in i$. sm_V gives us our sequences of accumulating subfunctions. We introduce the following notions:

Let X,Y be any sets and f a function from X to Y. Let R be a relation on X. Let for any $x \in X : V(x) = \{a\in X : aRx \ \& \ \neg xRa\}$.
Let $f \restriction_R (x)$ be the restriction of f to $V(x)$.

Let $F^{X}_{V,R} = \{g : (\exists x)(x \in X \ \& \ g = f \restriction_R (x))\}$.
Let $u \in Me(s,A)$, then $sm_V u \in Me((s,A),t)$.

We now define the semantics of sm_V as follows:

$$[sm_V u \ (x)]^{\mathfrak{A},i,j,g} = 1 \longleftrightarrow [x]^{\mathfrak{A},i,j,g} \in F^{[u]^{\mathfrak{A},i,j,g}}_{V,R_<}$$

where for any pairs (i,j) and (i',j') $(i,j) \ R_< \ (i',j')$ iff $j < j'$.

We have an operator "el" which makes properties out of individual concepts."el" is an operator on expressions of type (s,e), such that if $y \in Me(s,e)$, then el $y \in Me(s_1(e,t))$ and $[el \ y]^{\mathfrak{A},i,j,g}$ is a function f, such that for any (i',j') in the domain of y :

$$f(i',j') = K_{[{}^{\vee}y]^{\mathfrak{A},i',j',g}}.$$

Finally we have an operator ⊡ such that ⊡ $\in Me(t,t)$, with the following semantics: If $\phi \in Me(t)$, then $[⊡ \ \phi]^{\mathfrak{A},i,j,g} = 1 \longleftrightarrow (\forall j')(j \in \cup\cup DOM([{}^{\wedge}\phi]^{\mathfrak{A},i,j,g}) \rightarrow [\phi]^{j'} = 1)$.
Otherwise $[⊡ \ \phi]^{\mathfrak{A},i,j,g} = 0$.
Our proposal will be to translate a sentence with the imparfait of "marcher", e.g. "Jean marchait", by a formula of the following form:[16]

16 Actually the presented formulae could (and should) be refined. (75) and (76) express that the extension of "John" as it is *at the moment of evaluation* belonged to the extension of "march" as certain *past* fragments of the infinitive, which is not very natural. It might be interesting to consider past fragments of the individual concept "John" as well and to assign them to the past fragments of the infinitive "marcher". The present perfect (passé composé) might be treated by considering those past fragments of "John" of which the last moment is the moment of evaluation. We will reserve this for another occasion.

(75) $(\exists x)(sm_p\ \text{inf march}'(x)\ \&\ (\exists y)(sm_v\ x(y))\ \&$
$(\forall z)(sm_v x(z) \rightarrow \boxdot\ \text{el}\ z\ \{^\vee j\}))$

The corresponding sentence in the passé simple, "Jean marcha" will be translated by a formula of the following form:

(76) $(\exists x)(sm_p\ \text{inf march}'\ (x)\ \&\ \boxdot\ \text{el}\ x\ \{^\vee j\}\ \&$
$(\forall y)(sm_v x(y) \rightarrow \neg\boxdot\ \text{el}\ y\{^\vee j\})).$

Of course "⊡" is an adverbial temporal operator, like "$Depuis_2$" and "$Jusqu'a_4$", etc. The rules of the grammar will be such that these can be inserted in the same place as "⊡" (there are several ways to achieve this in a Montague grammar). If we do this, we get an explanation for the acceptability or the unacceptability of the sentences containing the connectives which we discussed above.

Let us have a look at a few cases.

(68) *Jean marchait depuis deux heures.*

is acceptable, whereas

(69) ⁎*Jean marchait jusqu'à quatre heures.*

is unacceptable.

Proposed translations:

(77) $(\exists x)(sm_p\ \text{inf march}'(x)\ \&\ (\exists y)(sm_v x(y)\ \&$
$(\forall z)(sm_v x(z) \rightarrow Depuis_2\ \text{el}\ z\ \{^\vee j\}))).$

(78) $(\exists x)(sm_p\ \text{inf march}'(x)\ \&\ (\exists y)(sm_v x(y)\ \&$
$(Vz)(sm_v x(z) \rightarrow Jusqu'à_4\ \text{el}\ z\ \{^\vee j\}))).$

Let us suppose that the truth value of (77) at (i,j) is 1, and let us work out the truth condition for the equation.

(79) $[Depuis_2\ \text{el}\ z\ \{^\vee j\}]^{\mathfrak{A},i,j,g} = 1$ (for all z)

We get

(80) $(\forall t)(2 < t < j \rightarrow [\text{el}\ z\ \{^\vee j\}]^{\mathfrak{A},i,t,g} = 1\ \&\ (\exists t')(2 < t' < t\ \&$
$t' \varepsilon UUDOM([^\wedge \text{el}\ z\{^\vee j\}]^{\mathfrak{A},i,j,g}))).$

So formula (77) expresses that there is a past fragment of the individual concept "marcher", after two o'clock such that for each moment t (between 2 and j) of any of the accumulating subfragments, John[17] belongs to the set of those things which walk at t. It is not difficult to prove that at each such t, $[Depuis_2\ \text{el}\ z\ \{^\vee j\}]^{\mathfrak{A},i,t,g} = 1$. This is of course in accordance with the

17 Or rather the extension of "John".

intuition expressed by picture (68).
Now take formula (78) and suppose that $[\text{Jusqu'à}_4 \text{ el z } \{^{\vee}j\}]^{\mathfrak{A},i,j,g} = 1$, i.e., for all accumulating subfragments z it is the case that

(81) $(\exists t)(t < 4 \ \& \ (\forall t')(t < t' < 4 \rightarrow \text{el z } \{^{\vee}j\}^{\mathfrak{A},i,t',g} = 1 \ \&$
$t' \ \varepsilon \text{UUDOM}([^{\wedge}\text{el z } \{^{\vee}j\}]^{\mathfrak{A},i,j,g}))$.

This cannot be the case. For let x be any subfragment of inf march', then (78) ascertains that there is an accumulative subfragment y of x, so x is not a function which has a domain consisting of one point (i',j'), but a function with a domain consisting at least of all points of an interval (i',j'), (i',j"), where j' < j" (assuming that the set of moments is ordered by < as the set of reals). Take any j''' such that j' < j''' < j". Then there is a subfragment z of x, such that $z = x \upharpoonright_{R_<} V(j''')$. So for any t_o: $t_o \varepsilon \text{UUDOM}([z]^{\mathfrak{A},i,j,g}) \rightarrow t_o < j''' < j"$. It is easy to see, that the moment "quatre heures" has to be at j or before j (and cannot be in the future of j as this would make (78) immediately false)[18]. It is also easy to see (for the same reason) that this moment has to be at j", or after j", i.e. between j" and j (possibly including j). Therefore, for any $t_o \varepsilon \text{UUDOM}([z]^{i,j,g}) \rightarrow$ $t_o < j''' < 4$. Then it cannot be true that $(\forall t')(t < t' < 4 \rightarrow$ $[\text{el z}\{^{\vee}j\}]^{\mathfrak{A},i,t',g} = 1)$, because among these t', there are such for which $j''' < t' < 4$. However, for such t' : $t' \notin \text{DOM}([z]^{\mathfrak{A},i,j,g})$.

So formula (78) cannot be true, explaining why (69) is unacceptable.

We have seen that in contradistinction to the imparfait, the passé simple plus depuis_2 is unacceptable, whereas it is acceptable with "jusqu'à":

(82) *⁎Jean marcha depuis deux heures.*

(83) *Jean marcha jusqu'à quatre heures.*

We propose the following translations for (82) and (83) resp.:

(84) $(\exists x)(sm_p \text{ inf march}'(x) \ \& \ \text{Depuis}_2 \text{ el } x\{^{\vee}j\}))$
$(\forall y)(sm_v x(y) \rightarrow \neg \text{Depuis}_2 \text{ el } y\{^{\vee}j\}))$

(85) $(\exists x)(sm_p \text{ inf march}'(x) \ \& \ \text{Jusqu'à}_4 \text{ el } x\{^{\vee}j\} \ \&$
$(\forall y)(sm_v x(y) \rightarrow \neg \text{Jusqu'à}_4 \text{ el } y \{^{\vee}j\}))$.

---

18 The restriction of "Jusqu'à" to the past in case of a main verb in a past tense has always been a major problem. It is seen to be solved in a natural way in the presented formalizm.

(84) gives us at (i,j) a past subfragment of inf march, such that at each (i,t), where 2 < t < j, if $t \varepsilon UUDOM([x]^{\mathfrak{A},i,j,g})$ then John walks at t. But then, this has to be the case for any subfragment of x to, a fortiori for all accumulative subfragments. So $[\neg Depuis_2 el\ y\ \{^{\vee}j\}]^{\mathfrak{A},i,j,g} = 0$ (for all subfragments y of x). (84) cannot be true, explaining the unacceptability of (82). On the other hand, taking (85), it is possible that there be a past subfragment of inf march such that for a certain t < 4 it is the case that for all t' such that t < t' < 4 : $[el\ x\{^{\vee}j\}]^{\mathfrak{A},i,t',g} = 1$, and $t' \varepsilon UUDOM([x]^{\mathfrak{A},i,j,g})$. $[^{\vee}j]^{\mathfrak{A},i,t',g} \varepsilon [^{\vee}x]^{\mathfrak{A},i,t',g}$. In other words x is a subfragment stretching up to 4 o'clock, and at each (i,t) of x, after some t, it is the case that John walks at t. But then for none of the accumulative subfragments y of x it is e the case that after a certain t *all* $t' \varepsilon UUDOM([y]^{\mathfrak{A},i,t',g})$. Therefore $[\neg Jusqu'\grave{a}_4 el\ y\{^{\vee}j\}]^{\mathfrak{A},i,j,g} = 1$ (for all accumulative subfragments y of x), making (85) acceptable.

Let us go back for a moment to formula (76), proposed as the translation for the passé simple without any surface temporal adverb, like "Jean marcha". We have seen that sentences with activity or state verbs with the passé simple used in this way express the beginning of an action. This is explained by the formula for the adverbless passé simple as in "Jean marcha".
Look at (76):

$$(\exists x)(sm_p\ inf\ march'(x)\ \&\ \boxdot\ el\ x\{^{\vee}j\}\ \&$$
$$(\forall y)(sm_v x(y) \rightarrow \neg \boxdot\ el\ y\{^{\vee}j\})).$$

In (76) it is stated that there exists a past fragment x of inf march such that for any moment t, if(i,t) is in the domain of x, then John walks at (i,t). Moreover it is stated that for no accumulative subfragment y of x it is the case that for any moment t, if (i,t) is in the domain of y, then John walks at (i,t). It is not difficult to see that this is impossible when the domain of x stretches over an interval. The only way for (76) to be true, is when x is a function, consisting of only one point (i,j'), with j' < j , and a set of individuals who walk at (i,j').

As a last case consider sentences with the adverb "trois fois".

(86) **Jean marchait trois fois.*

(87) *Jean marcha trois fois.*

Translation:

$$(88)\quad (\exists x)(sm_p\ inf\ march'(x)\ \&\ (\exists y)(sm_v x(y))\ \&$$
$$(\forall z)(sm_v x(z) \rightarrow trois\ fois'\ el\ z\{^{\vee}j\})).$$

(89) $(\exists x)(sm_p \text{ inf march}'(x) \;\&\; \text{trois fois}' \text{ el } x\{^{\vee}j\} \;\&$
$(\forall y)(sm_v x(y) \rightarrow \neg\text{trois fois el } y\{^{\vee}j\}))$.

Suppose that (88) is true. Then *for each accumulative subfragment z* of a certain past part x of inf march it is the case that $[\text{trois fois}' \text{ el}z\{^{\vee}j\}]^{\mathfrak{A},i,j,g}$, i.e. $(\exists t_1)(\exists t_2)(\exists t_3)(t_1 < t_3 < t_3 \;\&\; [\text{el}z\{^{\vee}j\}]^{\mathfrak{A},i,t_1,g} = 1 \;\&\; [\text{el}z\{^{\vee}j\}]^{\mathfrak{A},i,t_2,g} = 1 \;\&\; \text{el}z\{^{\vee}j\}^{\mathfrak{A},i,t_3,g} = 1 \;\&\; (\forall t_4)(\neg(t_4 = t_1 \vee t_4 = t_2 \vee t_4 = t_3) \rightarrow [\phi]^{\mathfrak{A},i,t_4,g} = 0$.
It is again not difficult to see, that if even smaller and smaller accumulative subfragments z have three distinct points at which John walks, that there must be more than three points in the bigger subfragments, at which John walks. So (88) cannot be true.
Now take (89). It states that there is a subfragment x of inf march, such that at three different points (and no others) John walks at those points, and that for no accumulative subfragments of x, it was the case that John walked at three different points (and no others in them). This means in the first place that the duration of the subfragment x has to be restricted to the last point at which John walks. If it would pass beyond this point, there would be accumulative subintervals of x, which also contained just three points at which John walked. But if x is limited to the third point, then for any accumulative subinterval y, y contains at most *two* points at which John walks. So for each accumulative subfragment y of x $[\neg \text{ trois fois el } y\{^{\vee}j\}]^{\mathfrak{A},i,j,g} = 1$, making (89) true.
We leave it to the reader to compute why "tous les dimanches" goes with the imparfait, whereas "tous les dimanches sauf un" does not.

A final word on adverbs like "souvent". Although "Jean marchait trois fois" is not acceptable,

(90) *Jean marchait trois fois par jour*

is completely normal.
The expression "trois fois par jour" gives a kind of average frequency (over a certain period), and such an average is of course the same at each moment. This might explain why "trois fois par jour ϕ "can distribute over all the accumulative subfragments of a past fragment. One could suppose that "souvent" and other adverbs like it behave in the same way., they express the measure of a certain average frequency (quite high, in the case of souvent), which of course is the same at each moment of a certain interval.

Bibliography

Bennett, M., 1974, "Some Extensions of a Montague Fragment of English", Ann Arbor, Michigan.
Dowty, D., 1972, "Studies in the Logic of Verb Aspect and Time Reference in English", University of Texas at Austin.
Hoepelman, J.Ph., 1976, Mass-nouns and Aspects, or: "Why we can't eat gingercake in an hour." In J.Groenendijk, M.Stokhof (eds.), Amsterdam, *Papers in Formal Grammar*, Vol.1.
Hoepelman, J.Ph., 1979, Verb Classification and the Russian Verbal Aspect: A Formal Treatment (to appear).
Imbs, P., 1968, "L'emploi des temps verbaux en français moderne", Paris.
Martin, R., 1971, "Temps et aspect", Paris.
Mourelatos, A., 1978, "Events, Processes and States", *Linguistics and Philosophy*, 2, pp.415-434.
Montague, R., 1974, "The Proper Treatment of Quantification in Ordinary English", in R.H. Thomason (ed.) *Formal Philosophy, Selected Papers of Richard Montague*, New Haven and London.
Sten, H., 1952, "Les temps du verbe fini en français moderne", København.

ON PROBLEMS CONCERNING THE QUANTIFICATION RULES IN MONTAGUE GRAMMAR

Theo M.V. Janssen

Abstract

In Montague grammar variables such as he_1, he_2 ... give rise to two kinds of problems. They can occur in cases where they should not, and they can be lacking in cases where they are wanted. Several proposals for dealing with these problems are discussed and a principle is proposed which guarantees that variables occur just in case they are required. In order to give this principle the form of a restriction on grammars, some new syntactic tools are used, which are borrowed from van Wijngaarden grammars. These tools are also useful for the treatment of verb tenses.

1. *Problems with variables*

MONTAGUE 1973, henceforth PTQ, contains in its vocabulary he_1, he_2 ... as expressions of the category of terms. Such words which bear indices, I call syntactic variables (or shortly: variables). On the one hand these variables are not words of English; on the other hand they occur in intermediate stages of PTQ productions of English. Occurrences of these variables are removed by rules like S_3 (relative clause formation) and S_{14} (term substitution); I call them quantification rules. When using these rules two kinds of problems arise.

The first problem is that variables may arise in situations where they are not desired. According to the rules of PTQ, the expression

(1) He_7 *runs*

is a well-formed expression of the category of sentences. It translates into an expression containing a free variable:

(2) $run_*(x_7)$.

The word he_7, however, is not an English word; therefore (1) cannot be a correct English sentence. So the PTQ grammar is not adequate in this respect.

The second kind of problems concerns situations in which specific syntactic variables are wanted, but where there are no such occurrences. According to the rules of PTQ, one may apply $S_{14,1}$ to the termphrase *a unicorn,* and the sentence *John loves Mary*. Syntactically there are no problems: one has to substitute *a unicorn* for the first occurrence of he_1, and this is quickly done since there are no such occurrences. Thus the sentence

(3) John loves Mary

is produced with translation:

(4) $\lambda P[\exists u[unicorn_*(u) \wedge P(^\wedge u)]](^\wedge\lambda x_1\ love_*(j,m))$

Notice that this translation contains a vacuous λ-abstraction. Formula (4) reduces to (5):

(5) $\exists u[unicorn_*(u) \wedge love_*(j,m)]$

So (3) obtains a reading which implies the existence of unicorns. This is, however, not a possible reading of (3). A related problem arises for relative clauses. A relative clause has to contain a pronoun as is demonstrated by the incorrectness of

(6) John loves the man such that Mary walks.

Consider now the following example (GROENENDIJK & STOKHOF 1976). According to PTQ, the sentence

(7) John loves the man such that he walks

can be formed, using $S_{14,1}$, from the termphrase *John* and the sentence

(8) He_1 loves the man such that he_1 walks.

The subphrase *man such that* he_1 *walks* is formed, using $S_{3,2}$, from *man* and he_1 *walks*. With this production, the translation of (7) says that the person who walks, is John. The only possible reading of (7) is, however, that the man who is loved by John is the one who is walking. This demonstrates that a relative clause has to contain a pronoun which refers to the head noun. In other words: if one uses $S_{3,n}$, then the sentence has to contain an occurrence of he_n.

2. *Proposals from the literature*

Although it is not the main theme of his paper, RODMAN 1976 makes some interesting remarks on the signalized problems concerning syntactic variables. His approach is of syntactic nature. Concerning sentences containing unbound syntactic variables he states that they are 'thrown out' because of 'a convention that is implicit in Montague's work, viz, that an expression containing a free variable is ill-formed' (p.167). Concerning wanted variables, his opinion can be concluded from the following. His rule for relative clause formation is, just as in PTQ, formulated as a rule working for all combinations of nouns and sentences. When he considers a combination of a common noun with a sentence which does not contain a relativizable pronoun, he writes 'derivation blocks' (p.170). So there seems to be a general convention as well for the other kinds of problems. The idea that the derivation should stop in such a case, can also be found with other authors, but they present it as explicit conditions on the input sentence (GROENENDIJK & STOKHOF 1976, KARTTUNEN 1977).

This syntactic approach is attractive, since it has as a consequence that, after a small change of the grammar, the correct structures are produced. In fact, certain aspects of this approach will figure in the solution proposed in section 3. But in the form presented by Rodman, it is a dangerous approach since it amounts to a filter; and a filter is a powerful tool which can also be used for unintended applications. I will demonstrate that Rodman's throw-out convention is such a dangerous tool. A related argumentation applies to the solutions to the other problem (blocking-convention as well as explicit restrictions).

Consider the situation that one wishes to incorporate in the grammar an obligatory rule which transforms sentences into sentences by performing some action which we call swopping. Let us assume that we have a formulation F_{sw} of the swopping action (it is irrelevant for our discussion to know what swopping really is). The obligatory character of the rule can be obtained by means of the throw-out filter and some changes in the rules. The rules which produce a sentence from expressions which are not sentences themselves, have to be changed. After having formed the sentence, two occurrences of he_7 are written in front of the sentence (below I will explain why 'two'). So S_4 becomes:

S_4: If $\alpha \in P_T$ and $\beta \in P_{IV}$ then $F'_4(\alpha,\beta) \in P_t$,
where $F'_4(\alpha,\beta) = he_7\ he_7\ F_4(\alpha,\beta)$
and where F_4 is as in PTQ.

The swopping rule becomes:

Swop: If $\alpha \in P_t$ then $F_{sw}(\alpha) \in P_t$, where $F_{sw}(\alpha) = \underline{\alpha}$, and $\underline{\alpha}$ is obtained from α by first deleting a sequence of two consecutive occurrences of he_7 and next performing the swopping action.

An expression of the category sentence to which the swopping rule is not applied, will contain the sequence $he_7\ he_7$ and thus be filtered out according to the convention. A sentence which has undergone swopping does not contain such a sequence and will pass the filter. So we used the sequence $he_7\ he_7$ as a mark for the swopping rule. (Conjoined sentences will have two such marks and both sentences will be swooped.) One has to be careful that the marks are not disturbed by other rules. An application of $S_{14,7}$ which substitutes *John* for the first he_7 would demolish the mark. Therefore $S_{14,n}$ has to be changed into

$S_{14,n}$: If $\alpha \in P_T$ and $\phi \in P_t$, then $F'_{10,n}(\alpha,\phi) \in P_t$, where $F'_{10,n}$ is as $F_{10,n}$ in PTQ with the difference that $F'_{10,n}$ does not change sequences of two consecutive he_n's.

Here one sees the use of having added in S_4 two, rather than one occurrence. A variable used for the usual purposes cannot occur in a sequence of two consecutive variables, so the ones used for rule ordering can now be recognized. If we would have used just one, this would not have been the case.

The above example concerns a single obligatory rule. Related methods can be used to encode any rule ordering. I have objections against this use of variables. It is not in the style of Montague grammar to have obligatory rules or any other restriction on the ordering of rules. This is expressed by Partee in her work concerning the wellformedness constraint (PARTEE 1979). If one accepts this view, then one should not allow to have arbitrary rule ordering encoded by the he_n's. And if one does not accept this view and one wishes to have a prescribed rule-ordering, then one should do so in an explicit component for the rule ordering (as such is done in transformational grammars). Variables are introduced for pronouns and scope treatment, not for rule ordering. That is an abuse of variables which should be made impossible.

COOPER 1975, explicitly discusses the problems concerning variables and he presents approaches of a semantic nature. His ideas are especially interesting since they form the fundament for the approach of BACH & COOPER 1978, in which it is argued that it is possible to have in Montague grammar a term-phrase-sentence analysis of relative clause constructions instead of the usual common noun-sentence analysis.

We have seen that constructions, in which variables are wanted but are not there, give rise to vacuous λ-abstraction. Cooper proposes to assign no truth value to such an expression containing vacuous abstraction. He obtains this effect by a restriction on the usual definition of interpretation of $\lambda u\alpha$. '... the function denoted by the abstraction expression $\lambda u\alpha$ is only defined for entities within its domain if a different assignment to the variable u will yield a different denotation for α' (p.246). As he signalizes, this has a consequence 'that $\lambda u\alpha$ is undefined not only if α does not contain a free occurrence of u, but also if α is a tautology. Thus, for instance, according to this definition, $\lambda u[u=u]$ represents a function which is undefined for any entity. However, the technique of supervaluation ... will show these expressions to be defined but not those where α is not a tautology'. A complicated approach (probably incorrect for $\lambda u[x=x]$).

One of Coopers proposals concerning free variables is to 'disign the semantics in such a way that any formula containing a free occurrence of a variable is shown to be truth-valueless' (p.257). As consequence of this proposal

(9) *walk* (x)

gets for no point of reference a truth value; it represents the nowhere defined intension. As a subexpression in the formula:

(10) $\exists x[man(x) \wedge walk(x)]$

it does have another meaning. So we lost the Fregean principle of compositionality: the meaning of (10) is not built up from the meaning of (9). A related objection arises if one observes that also

(11) *talk*(x)

gets for no point of reference a truth value. So (9) and (11) are intensionally equivalent and according to a rule of intensional logic interchangeable in all contexts. Now (10) constitutes a context in which this is not the case. So acceptance of this proposal would turn the so basic substitution law unvalid.

Another approach finds its origin in Universal Grammar (MONTAGUE 1970). It is practised by GROENENDIJK & STOKHOF 1976 and it is one of the proposals of

COOPER 1975. In syntax one has a rule which deletes indices, thus delivering a correct expression. In the semantics, free variables are interpreted by the current variable assignment. Notice that this leads us away from PTQ since there a formula is defined to be true for a point of reference if and only if it is true for every variable assignment (p.250). So *run*(x) means the same as its universal closure: $\forall x[run(x)]$. Therefore this definition has to be dropped, when following this approach to free variables.

There are more problems with this approach. Consider:

(12) He runs

with the translation

(13) $run(x_7)$.

For each variable assignment (13) obtains an interpretation. One of the possible assignments is that x_7 refers to the utterer of the sentence under consideration; so (12) is interpreted as *I run*. It may also be the case that x_7 is the person spoken to, thus meaning *You run*, or it may handle about a female person, meaning *She runs*. Beside these syntactic inadequacies, there are also semantic ones; depending on the linguistic and situational context, the pronoun *he* can refer only to certain individuals (e.g. the person mentioned in the last utterance or the person pointed at by the speaker). None of the authors mentioned above does explicitly signalize these complications. They can be dealt with by means of a device from Universal Grammar (MONTAGUE 1970): considering not all variable assignment functions for evaluating a complete sentence, but only a subset thereof. This approach, however, gives rise to a strange situation. In the light of the above arguments, just a few assignments for occurring free variables are possible in any particular situation. The free variables cannot vary and are more like constants. Moreover, the variable assignment function g is by its nature independent of context or point of reference, whereas the function for interpreting constants depends on information concerning the time and possible worlds. Why not then translate such pronouns by constants as is done in BENNETT 1978?

3. *The variable principle*

In the previous section, we have considered several proposals dealing with the problems concerning variables. The syntactic approach aims at avoiding the problematic cases, but, in doing so, unattractive tools are introduced. The semantic approaches try to do something useful with the arising problematic cases. These semantic proposals do not give me the impression that the authors are quite satisfied by such cases, but that they rather try to find an escape from a situation they would prefer not to be confrontated with. I will put forward a proposal which combines syntactic and semantic aspects and which has as consequence that variables arise just in case they are required.

In my opinion the existing proposals all ignore the special character of variables. This special character is, in an informal way, expressed by *the variable principle:*

Syntactic variables correspond closely with semantic variables.

Several arguments for this correspondence can be found in PTQ and related works. The most immediate is that syntactic variables translate into logical variables (The term variable he_n is translated as $\lambda PP\{x_n\}$, and the CN-variable one_n, which is discussed in HAUSSER 1979, is translated as P_n). Another correspondence is that if a syntactic rule removes all occurrences of a syntactic variable, the corresponding translation rule binds the corresponding logical variable ($S_{3,n}$ and $S_{14,n}$). And rules not binding logical variables do not remove syntactic variables. The consequence of these correspondences is that in PTQ an expression contains an indexed syntactic variable if and only if its translation contains a free logical variable. I consider the correspondences among syntactic and logical variables not as accidental, but as fundamental, and propose to follow in Montague grammar the variable principle. The above formulation of the principle is, of course, rather informal; a more exact one is given below.

Let us start by recalling that by a syntactic variable is understood a word which bears an index, and that by a syntactic quantification rule is understood a rule that has the effect that all occurrences of a certain variable are in some way removed. By the *variable principle* we understand the following requirements:

1a) The translation of a syntactic variable contains a free occurrence of a logical variable.

1b) This is the only way to introduce a free occurrence of a logical variable.

2a) If a syntactic rule removes occurrences of a variable, the related translation rule binds the corresponding logical variable.

2b) If a translation rule introduces a binder for a free logical variable, the related syntactic rule removes the corresponding syntactic variables.

3a) In the complete analysis of a sentence, each syntactic variable has to be removed by a syntactic quantification rule.

3b) If a syntactic quantification rule is used, it has actually removed the occurrences of a syntactic variable.

Clause 3 is formulated for syntactic rules. Since the clauses 1 and 2 force a correspondence of syntactic and semantic rules, clause 3 has an equivalent logical counterpart. This semantic version of clause 3 cannot be formulated as easy as the syntactic one, since it demands for a logical counterpart of the notion 'complete analysis of a sentence'. Therefore I prefer the syntactic formulation.

As the reader has learned from the previous discussion, the principle is intended for the usual syntactic and logical variables. It does not apply to the so-called context variables which are especially introduced for dealing with several aspects of contextual influence (examples are Γ of HAUSSER 1979 and v_c of GROENENDIJK & STOKHOF 1979). Such variables are not interpreted by the variable assignment function g and they might be called context constants as well.

Let us go back to the principle and see what it says about the several proposals we considered. As mentioned before, requirement 1a is obeyed by all existing proposals. This requirement is also proposed by Partee as a constraint on indexed notations (PARTEE 1979). Requirements 1b and 2b are not fulfilled by the proposals of COOPER 1975 and BACH & COOPER 1978.
My abuse-example concerning Rodman's 'throw-out' convention does not obey 1b nor 2a. Clause 3 makes that the original problems concerning PTQ we discussed before, do not arise. Restrictions with the same effect as clause 3 are used in JANSSEN 1976. The conventions of Rodman express the same aims as 3a and 3b, but now the abuse of variables is prevented by clauses 1 and 2. So the variables arise just in case they are wanted.

The clauses 1 and 2 define restrictions on possible grammars. In order to make sure that these restrictions are decidable, I should define a format for rules from which it is possible to check whether these restrictions are satisfied. Since in all concrete proposals it is easily checked whether these clauses are satisfied, I will not define such a format. It seems that clause 3 has to be taken as a global constraint on possible derivations. If one has no objections against such a constraint, then the variable-principle has thus solved the problems concerning variables. I do have objections against global constraints. They are even stronger than filters since filters impose constraints on the final outcome of a production process, whereas global constraints take the whole process into account. As we have seen before, filters (and thus global constraints) are dangerous since they are so powerful. Global constraints allow for making the information contained in other parts of the grammar worthless. In Montague grammar, the syntactic rules provide for the information which kinds of expressions may be combined to form new expressions of a certain kind, but global constraints would make it possible that the rules combine rubbish to rubbish, leaving it to the constraint to say which combinations make sense. Because of such arguments I would not like to incorporate global constraints in general. But accepting clause 3 as the only global one would be rather ad hoc. Therefore I will describe in the next section tools which make it possible to consider clause 3 as a restriction on the possible rules in a Montague grammar instead of considering it as a global one. After having done so, the variable principle can be taken as a decidable restriction on possible Montague grammars.

4. *Hyperrules and variables*

In this section I describe a syntax which is in some aspect a generalization of the PTQ syntax. I start with a description of the framework, some simple examples follow thereafter.

The categories may be compound symbols and it is allowed that there are infinitely many of them. We can handle an infinite number by means of a finite number of schemes of rules. These schemes are called hyperrules and they constitute the hypergrammar. Such hyperrules are alike PTQ rules, but they contain variables. By an appropriate substitution for these variables a hyperrule is transformed into a rule which may actually be used in a production. The information what is to be substituted for a certain variable is described by grammatical tools as well. Besides the hyperrules, we have another set of rules, called metarules. These metarules constitute a grammar (say context free) in which the variables from the hyperrules occur as auxiliary symbols. The set of strings of terminal symbols of the metagrammar which can be produced from a variable, constitutes the set of possible substitutions for this variable. For each metavariable M we have the variants $M_1,\ldots M_9$. For these variants we have the same metarules as for M. If a hyperrule contains occurrences of different variants of a metavariable, these may be replaced by different terminal productions. The same variants are (convention) to be replaced by the same terminal productions. For convenience, we will present the metarules only for one variant (say M) and not for the others $(M_1,\ldots M_9)$. So our syntactic framework consists of two levels: the metagrammar and the hypergrammar. This conception of a grammar (and the related terminology) is due to van Wijngaarden who developed this kind of grammar for the formal description of the programming language Algol 68 (VAN WIJNGAARDEN 1975). In certain details our framework deviates from the standard form of a Van Wijngaarden grammar (see appendix). A linguistically oriented example of such a grammar can be found in VAN WIJNGAARDEN 1970. The main use of the organization of the syntax in two levels is that it thus provides for a technical handsome, readable, and understandable way for dealing with a large set of rules.

Compound category symbols and metarules are new in Montague grammar, but hyperrules are not. Consider the rule for relative clause formation:

$S_{3,N}$: If $\zeta \in P_{CN}$ and $\phi \in P_t$ then $F_{3,N}(\zeta,\phi) \in P_{CN}$ where $F_{3,N}(\zeta,\phi) = \zeta$ such that ϕ'.. and ϕ' comes from ϕ by replacing each occurrence of he_N... .

This rule cannot be considered as involving the literal expression he_N since the fragment has no such an expression.
The rule rather is a scheme which, for each choice of a number for N, gives a rule. One might add the following eleven metarules to the grammar (a bar: | separates alternatives):

$$N \rightarrow 1 \mid 2 \mid 3 \mid 4 \mid 5 \mid 6 \mid 7 \mid 8 \mid 9 \mid NO \mid NN.$$

Note that due to the convention, all occurrences of the metavariable N in $S_{3,N}$ have to be replaced by the same terminal production of N, so by the same representation of a number. The above demonstrates that one does not have to understand numbers in order to handle $S_{3,N}$; one can express everything with pure grammatical tools. Note that we are using the convention to write metavariables in capital (N).

Another example of a PTQ hyperrole is S_1. It reads:

$B_A \subset P_A$ for every category A.

Here A is variable over categories. Its range might be made explicit by means of metarules:

A → Term | Sent | Intr Verb| ... .

A first new example of the framework is the use of compound category symbols for subcategorization. There are arguments for distinguishing among the nouns the mass nouns (e.g. *gold*) and the count nouns (e.g. *ring*). In some respects they behave the same (relative clause formation), and in other respects they behave differently (term formation). Using two separate categories would have as consequence that we have to duplicate the rules for which they behave the same. This is unattractive and can easily be avoided by the use of hyperrules. Let CM be a metavariable which can be replaced by 'Count' as well as by 'Mass'. Then we may have the hyperrules

S_2 if $\alpha \in$ Count Noun then $F_0(\alpha)$, $F_1(\alpha)$, $F_2(\alpha) \in T$ where F_0, F_1 and F_2 are as in PTQ;

$S_{3,N}$ if $\alpha \in$ CM Noun and $\beta \in S$ then $F_{3,N}(\alpha,\beta) \in$ CM Noun, where F_3 is as in PTQ.

Note that S_2 is a hyperrule without a metavariable, so it is a rule in its present form. The rule says that the phrase *a ring* can be produced, but *a gold* cannot. Hyperrule S_3 expresses that a count noun with a relative clause (*ring such that it is in Amsterdam*) remains a count noun and a mass noun with a relative clause (*gold such that it is in Amsterdam*) remains a mass noun.

The use of a metagrammar is especially clear in case there are more levels of subcategorization and possibly crosslinks. I present a metagrammar for the subcategorization which is mentioned in CHOMSKY 1965 (p.85). It is striking that Chomsky uses rewriting rules for the formal representation of subcategorization as well.

COMMON → COUNT | NONCOUNT.
COUNT → ANIMATE | -animate cn.
NONCOUNT → SGN abstract cn.
ANIMATE → SGN human cn.
SGN → + | - .

Again, the auxiliary symbols of the metagrammar are written in capitals. This metagrammar implicates, according to the substitution convention, that a hyperrule containing ANIMATE as metavariable, represents 2 actual rules (for the subcategories +human cn and -human cn) and that a hyperrule containing COMMON represents 5 actual rules.

Note that up till now, we used the terminology category/subcategory without defining their difference. On the level of actual rules there is no distinction between them: they both are indices of sets of expressions. The above kind of grammar makes it possible to make a formal distinction. Categories are names from the metagrammar which do not occur at the right hand side of a metaproduction rule. Subcategories of a certain category are all names that can be produced from that category using the metarules. In the sequel I will use in most cases 'category' in the PTQ sense: as an index of a set of expressions. So it may in many cases be replaced by 'subcategory'.

Below I will formulate clause 3 of the variable principle within the syntactic rules. For this purpose I consider a category name as a compound symbol consisting of a category name as the ones in PTQ, and a bag of integers. A bag (or multiset) is a set in which an element may occur more than once. This bag indicates which indices occur unremoved in the expression of that category. So he_2 or he_3 is an expression of the category (T,{2,3}) or equivalently (T,{3,2}). We write this as he_2 or he_3 ε (T,{2,3}). Other examples are he_4 or he_4 ε (T,{4,4}) and *John* ε (T,{ }). The language generated by the grammar is the set of English sentences which contain no indexed terms, so it is the set of expressions of the category (S,{ }) or in PTQ terminology (t,{ }). The hyperrules we need, contain variables for integers (N) and for bags (BAG_1,BAG_2,...). Let us adopt the following notations:

$BAG_1 \cup BAG_2$	denotes the bag consisting of the bag-union of the bags BAG_1 and BAG_2. So {1,2} ∪ {2,3} = {1,2,2,3}.
BAG <u>w</u> N	Is a compound expression indicating that BAG contains element N (w=with).
BAG - N	Idem for the bag obtained from BAG by removing all occurrences of N. So {1,2} - 1 = {2}.

The hyperrule corresponding with S_4 reads:

HS_4: If α ε (T,BAG_1) and β ε (IV,BAG_2) then $F_4(\alpha,\beta)$ ε $(S,BAG_1 \cup BAG_2)$

This hyperrule states that the syntactic variables in the sentences are the union of syntactic variables in the T-phrase and the IV-phrase. An instance of a rule obtained from HS_4 is:

if α ε (T,{1,2}) and β ε (IV,{ })
then $F_4(\alpha,\beta)$ ε (S,{1,2}).

This rule may be used in the production of *He*$_1$ or *he*$_2$ *runs*. Corresponding with S_2, $S_5..S_{13}$, and S_{17} we have analogous hyperrules. With $S_{14,N}$ corresponds:

$HS_{14,N}$: if $\alpha \in (T, BAG_1)$ and $\beta \in (S, BAG_2 \ \underline{w}\ N)$ then $F_{10,N}(\alpha,\beta) \in$
$\in (S, BAG_1 \cup [BAG_2 - N])$.

An instance of $HS_{14,N}$ is:

if $\alpha \in (T,\{\ \})$ and $\beta \in (S,\{2,3\})$ then $F_{14,2}\ (\alpha,\beta) \in (S,\{3\})$.

This rule may be used in the production of *John loves him*$_3$ from *John* and *he*$_2$ *loves him*$_3$. The hyperrules corresponding with S_3, S_{14}, and S_{15} are analogous.

One observes that the expression $BAG_2\ \underline{w}\ N$ guarantees that $HS_{14,N}$ is only applied to sentences containing occurrences of *he*$_N$. So the clause 3b of the variable principle that a quantification rule must remove an actual occurrence of a variable, can be expressed within the syntactic rules: the rule only applies to the subcategory consisting of expressions containing such a variable. So clause 3b can be checked by looking to the rules: does a rule which removes occurrences of a variable *he*$_N$, indeed operate on the related subcategory of expressions containing *he*$_N$, and does it deliver an expression of a subcategory from which the index in question is indeed removed? So 3b is now a restriction on the rules of the grammar. Clause 3a that each variable is removed, is guaranteed by defining the generated languages as being the set of expressions of category (S,{ }), (or in PTQ-like terminology of category (t,{ })).

I have implicitly assumed that the reader has some non grammatical knowledge since what bags are, and what ∪, w and - mean is not formulated within the grammar. This can be provided for in the following way. One adds metarules which produce from BAG actual bags and one adds hyperrules which produce from any instance of BAG-N the bag in question (so from {1,2}-1 they produce {2}). It requires some technics to formulate such hyperrules; therefore they are presented in an appendix.

5. *An application to verb tense*

Hyperrules are extremely useful in cases where a lot of closely related rules are needed. Such a case is constituted by the rules for tense and sign. In order to demonstrate this, I will present a grammar for the treatment of the sign and the PTQ tenses of verbs. The grammar deals with the following phenomena which are not treated in PTQ.

1. Negated iv-phrases are possible:

(14) John tries not to run.

Notice that *try to* cannot be combined with a tensed iv-phrase. This sentence is also incorporated in the fragment of BENNETT 1976. His solution is improved in the sense that the iv-phrase

(15) not to run slowly

can, in our fragment, not obtain the reading in which *slowly* has wider scope than the negation.

2. As in PTQ, tensed variants of conjunctions and disjunctions are possible. Syntactic details are changed in order to produce

(16) John has walked and talked

instead of the PTQ version

(17) John has walked and talk.

3. Distinct from PTQ, we allow for conjunction and disjunction of arbitrarily tensed sentences and iv-phrases such as

(18) A president has walked and will talk.

The translation of this sentence will express that both actions concern a present president.

4. The PTQ grammar provides for two readings of the sentence

(19) Every president will not walk.

These are

(20) $\forall x[\mathit{president}(x) \rightarrow \neg W[\mathit{walk}(x)]]$

(21) $\neg W[\forall x[\mathit{president}(x) \rightarrow \mathit{walk}(x)]]$

If (21) indeed is a possible meaning of this sentence, then, in my opinion, also (22) is:

(22) $\neg[\forall x[\mathit{president}(x) \rightarrow W\ \mathit{walk}(x)]]$

The grammar presented below produces all these three readings. If one judges that (21) and (22) cannot be readings of the sentence, a single hyperrule has to be removed from the grammar.

The main prupose of presenting a grammar for the above phenomena is to demonstrate the use in syntax of hyperrules. I do not intend to say something new about semantical concepts concerning verb tenses: the same model and the same semantical operators for tense are used as in PTQ. This means that we only deal with the reportive use of tenses (see BENNETT 1977). The semantical problems raised by the new constructions mentioned above, all have to do with the scope of the operators W, H and $\neg$. Also the treatment of morphology is as simple as in PTQ: I do not use features. In the formulation of the hyperrules I assume that it is always possible to identify the mainverbs of iv-phrases. The auxiliary verbs *will*, *have* and *do* may function as mainverbs as well. The possibility of identification of mainverbs can be obtained by using recursive definitions as in FRIEDMAN 1979 (or the #-mark of BENNETT 1976 or the subfunctions of PARTEE 1979). Moreover we assume that by related means the iv-phrase of a sentence can be identified. The syntactic formation rules do not treat (as is done in PTQ), the action of 'replacing a verb by its negative future form' as a single syntactic action, but rather as the composition of 'replacing by its future form' and 'negating' which are performed by two independent syntactic rules.

The PTQ category t (in our terminology S) will have several subcategories. For instance, there is a subcategory for each tense. This relationship subcategory - tense is indicated by the following pairs: sent - present tense, past sent - simple past, fut sent - future tense. Each negated tense constitutes a subcategory and there is also a subcategory of unspecified tense (for conjunctions and disjunctions). All subcategories translate into expressions of the same logical type: the type of sentences. For IV-phrases there is an analogous subcategory system. The intransitive verbs themselves are in the subcategory iv. The subcategory system is formally described by means of the following metarules (the symbol λ denotes the empty string):

S	$\rightarrow$	TENSED sent.
IV	$\rightarrow$	TENSED iv.
TENSED	$\rightarrow$	SGN TNS $\mid$ tns.
SGN	$\rightarrow$	neg $\mid$ λ.
TNS	$\rightarrow$	past $\mid$ fut $\mid$ λ.

The hyperrules presented below replace the PTQ-rules S_4(IV+T), S_8(IV/IV+IV), S_{10}(IV//IV+IV) and S_{17}(tensed variants of S_4). The description of an hyperrule has the following parts:

H: Information about the categories involved in the hyperrule. We write e.g. "$\phi \in$ neg sent" for "ϕ is a phrase of the subcategory neg sent".

T: The translation rule is defined by presenting its output. In this, by ϕ', ψ', α' and β' are understood the translations of ϕ, ψ, α and β mentioned in the H-part.

F: A specification of the string manipulation function used in the H-part. For the ease of reference, the function F will bear the same index as the hyperrule in which it is used, even where the same operation is used in more than one rule. (This convention and the one for T is also practised in DOWTY 1978).

E: -Optionally- an example.

R: -Optionally- some remarks.

H_{101}: If $\alpha \in T$ and $\beta \in$ TENSED iv then $F_{101}(\alpha,\beta) \in$ TENSED sent.

T_{101}: $\alpha'(^\wedge\beta')$

F_{101}: $F_{101}(\alpha,\beta) = \alpha\underline{\beta}$, where $\underline{\beta}$ is obtained from β by replacing the main verbs in β by their third person singular present form, provided that they are not yet in some third person singular form.

E_{101}: F_{101}(*John, walk*) = *John walks*

F_{101}(*John, will walk*) = *John will walk.*

R_{101}: Since there are, according to the metarules, six possible substitutions for TENSED, one can make six actual rules out of H_{101}.

H_{102}: If $\alpha \in$ iv then $F_{102}(\alpha) \in$ past iv.

T_{102}: $\lambda x\ H[\alpha'(x)]$

F_{102}: Remove all occurrences of the mainverb *do* in α and next replace all mainverbs in α by their past participle and write *have* in front of the thus changed α.

E_{102}: F_{102}(*walk and talk*) = *have walked and talked*

F_{102}(*walk and do not talk*) = *have walked and not talked.*

R_{102}: The above iv yields the reading in which the subject is outside the scope of the tense operator.

H_{103}: If $\phi \in$ sent then $F_{103}(\phi) \in$ past sent.

T_{103}: $H(\phi')$.

F_{103}: Apply F_{102} to the iv-phrase of ϕ, next replace mainverb *have* by *has*.

R_{103}: The subject is inside the scope of the tense operator (neglecting H_{14}).

H_{104}: If $\alpha \in$ iv then $F_{104}(\alpha) \in$ fut iv.

T_{104}: $\lambda x\ \ W[\alpha'(x)]$.

F_{104}: Remove all occurrences of the mainverb *do* in α and write *will* in front of the thus changed α.

E_{104}: F_{104}(*walk and do not talk*) = *will walk and not talk*.

R_{104}: See R_{102}.

H_{105}: If $\phi \in$ sent then $F_{105}(\phi) \in$ fut sent.

T_{105}: $W(\phi')$.

F_{105}: Apply F_{104} to the iv-phrase of ϕ.

R_{105}: See R_{103}.

H_{106}: If $\alpha \in$ TNS iv then $F_{106}(\alpha) \in$ neg TNS iv.

T_{106}: $\lambda x \neg[\alpha'(x)]$.

F_{106}: If α has just one mainverb and this is *will, have, do,* or *be,* then insert *not* behind it; in all other cases write *do not* in front of α.

R_{106}: Conjuncted and disjuncted tensed iv-phrases are of the subcategory tns iv. Therefore no instance of H_{106} applies to them.

H_{107}: If $\phi \in$ TNS sent then $F_{107}(\phi) \in$ neg TNS sent.

T_{107}: $\neg[\phi']$.

F_{107}: Apply F_{106} to the iv-phrase of ϕ, next replace mainverb *do* by *does*.

R_{107}: Conjuncted and disjuncted sentences are of the subcategory tns sent. Therefore no instance of H_{107} applies to them. We have discussed the ambiguity of sentence (19). If one wishes to have only one reading for such sentences, then H_{107} has to be removed from the grammar.

H_{108}: If $\alpha \in$ IV//IV and $\beta \in$ SGN iv then $F_{108}(\alpha,\beta) \in$ iv.

T_{108}: $\alpha'(\hat{}\beta')$.

F_{108}: If β starts with *do not* then delete this and insert *not* before the first *to* in the IV//IV-phrase. Otherwise concatenate α and β.

E_{108}: F_{108}(*try to, run*) = *try to run*

F_{108}(*try to, do not run*) = *try not to run*.

H_{109}: If $\alpha \in$ IV//IV and $\beta \in$ iv then $F_{109}(\alpha,\beta) \in$ iv.

T_{109}: $\alpha'(\hat{}\beta')$

F_{109}: $F_{109}(\alpha,\beta) = \beta\alpha$.

E_{109}: F_{109}(*slowly, run*) = *run slowly*.

R_{109}: Since adverbs can only be combined with an untensed and unsigned iv-phrase, a phrase like *do not run slowly* can only have the reading with slowly inside the scope of negation (see remark on (15)).

H_{110}: If $\alpha \in$ iv and $\beta \in$ SGN iv then $F_{110}(\alpha,\beta) \in$ iv.

T_{110}: $\lambda x[\alpha'(x) \wedge \beta'(x)]$.

F_{110}: $F_{110}(\alpha,\beta) = \alpha$ *and* β.

R_{110}: The asymmetry in this hyperrule makes that the negation in:

(23) *John will not walk and talk*

applies to *will walk and talk,* and not just to *walk* (cf. FRIEDMAN 1979).

H_{111}: Analogously to H_{110} for disjunction.

H_{112}: If $\alpha \in$ TENSED$_1$iv and $\beta \in$ TENSED$_2$iv then $F_{112}(\alpha,\beta) \in$ tns iv.

T_{112}: see T_{110}.

F_{112}: see F_{110}.

R_{112}: This rule is used in the production of

(24) *John will not walk and talks.*

(cf. R_{110}).

H_{113}: Analogously to H_{112} for disjunction.

H_{114}: If $\phi \in$ TENSED$_1$sent and $\psi \in$ TENSED$_2$sent then $F_{114}(\phi,\psi) \in$ tns sent.

T_{114}: $\phi' \wedge \psi'$.

F_{114}: $F_{114}(\phi,\psi) = \phi$ *and* ψ.

H_{115}: Analogously to H_{114} for disjunction.

These 15 hyperrules describe in a handsome way 164 actual rules, so the use of hyperrules reduced the length of the grammar by a factor 10. If I would also incorporate other tenses such as the past perfect, the benefit would still be more. Note that there is no syntactic problem in replacing H_{112} - H_{115} by two rules using one metavariable for both iv and sent, but then we would not be able to find two corresponding semantic rules.
Below I present three examples. All expressions are preceded by the number of the hyperrule used, and the subcategory they belong to.

1)

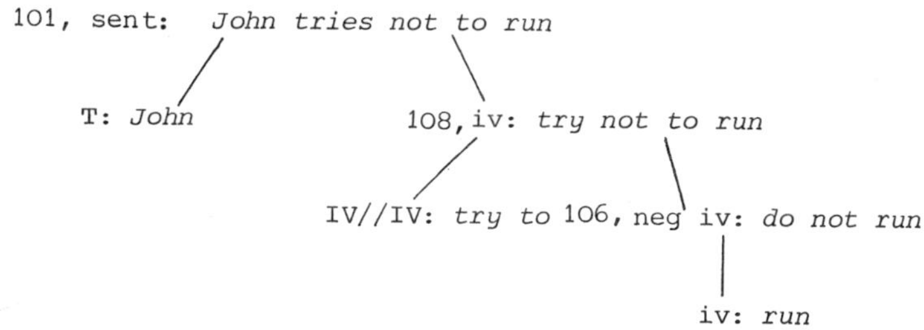

the corresponding translation yields:

$try\ to_{*}\ (j, \lambda u \neg run_{*}\ (u))$

2)

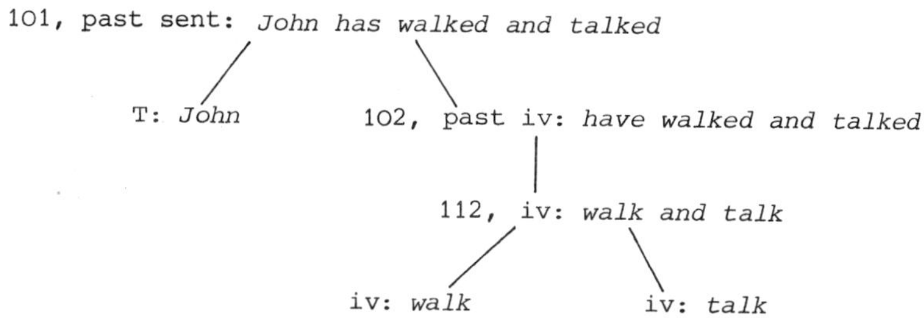

the corresponding translation yields:

$H(walk(j) \wedge talk(j))$

An alternative analysis (with the same semantic result) would apply H_{103} to:

101, sent: *John walks and talks.*

3)

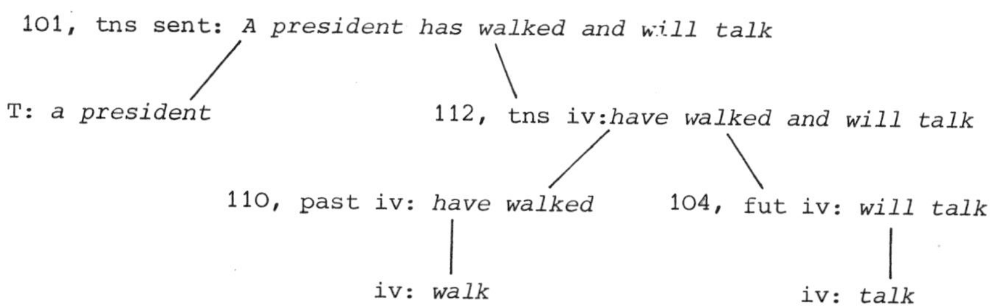

translation:

$$\exists u[president_{*}(u) \wedge \mathrm{H}\ walk_{*}(u) \wedge w\ talk_{*}(u)]$$

6. *Conclusions*

Two kinds of problems arise in connection with syntactic variables in Montague grammar. The existing proposals for treating these problems are not free of objections, whereas the *variable principle* is able to cope with them. The application of hyperrules, metarules and compound category symbols permits the incorporation of the variable principle as a restriction on possible grammars. These tools are also useful for other purposes (subcategorization, verb tense). The variable principle excludes the complicated NP-S analysis of relative clauses as proposed by Bach & Cooper. Concerning the demonstrative use of pronouns and related phenomena, the principle guides us towards the use of constants as practised by Bennett.

Acknowledgements

I thank Arnim von Stechow and Thomas Zimmermann for their remarks concerning the treatment of variables in Universal Grammar, and Peter van Emde Boas, Jeroen Groenendijk and Martin Stokhof for their comments on several topics treated in this paper.

7. *Appendix*

Rules for bags.

The aim of this appendix is to demonstrate that is possible to describe the bag manipulations used in section 4 by means of pure grammatical tools. Before going so, some remarks for experts in Van Wijngaarden grammars. The hyperrules presented in this paper do not have the format of hyperrules according to the literature, but they can be transformed to such a format. The hyperrule $S_{14,N}$ then obtains the shape

$$X(S,BAG_1 \cup [BAG_2 - N])Y ::=$$
$$X\ F_{10,N}((T,BAG_1),(S,BAG_2\underline{w}\ N))Y.$$

The metanotions X and Y stand for arbitrary contexts (cf. VAN WIJNGAARDEN 1970). The syntactic operation $F_{10,N}$ has to be described by means of other hyperrules. The phrase '*unless* N_1 *is* N_2' (used below) is to be considered as a separate hypernotation. The rules below use the rotation technique of VAN WIJNGAARDEN 1974 and the blind alley technique Sintzoff as practised in VAN WIJNGAARDEN 1975.

The metarules are (λ denotes the empty string)

$$BAG \rightarrow \{SEQ\} \mid \{\ \}$$
$$SEQ \rightarrow N \mid N,SEQ$$
$$N \rightarrow 1 \mid 2 \mid 3 \mid 4 \mid 5 \mid 6 \mid 7 \mid 8 \mid 9 \mid NN \mid NO$$
$$LESQ \rightarrow SEQ, \mid \lambda.$$
$$RSEQ \rightarrow ,SEQ \mid \lambda.$$

The hyperrules for union are

$$\{SEQ_1\} \cup \{SEQ_2\} \rightarrow \{SEQ_1,SEQ_2\}$$
$$\{SEQ_1\} \cup \{\ \} \rightarrow \{SEQ_1\}$$
$$\{\ \} \cup \{SEQ_2\} \rightarrow \{SEQ_2\}\ .$$

For with

$$\{LSEQ\ N\ RSEQ\}\ \underline{w}\ N \rightarrow \{LSEQ\ N\ RSEQ\}$$

For minus

$$\{SEQ\} - N \rightarrow \{\ast,SEQ\} - N$$
$$\{\ \} - N \rightarrow \{\ \}$$
$$\{LSEQ\ \ast\ RSEQ,N_1\} - N_2 \rightarrow \{N_1,LSEQ\ \ast\ RSEQ\} - N_2\ \textit{unless}\ N_1\ \textit{is}\ N_2$$
$$\{LSEQ\ \ast\} - N_2 \rightarrow \{LSEQ\}.$$

Hyperrules concerning the *unless* phrase are

$$\textit{unless false} \rightarrow \lambda$$
$$\textit{0 is 0} \rightarrow \textit{true};\quad \textit{0 is 1} \rightarrow \textit{false};\quad \textit{0 is 2} \rightarrow \textit{false} \ldots$$
$$N_1N_2\ \textit{is}\ N_3N_2 \rightarrow N_1\ \textit{is}\ N_3;\ N_1N_2\ \textit{is}\ N_1N_3 \rightarrow N_2\ \textit{is}\ N_3.$$

So the phrase '*unless* N_1 *is* N_2' reduces to the empty string just in case N_1 is not equal to N_2, otherwise one cannot get rid of the phrase and thus the rule cannot be used in an actual derivation.

References

Bach, E. & Cooper, R.H., 1978, The NP-S analysis of relative clauses and compositional semantics, *Ling. and Phil.* 2, pp.145-150.
Bennett, M., 1976, Variation and extension of a Montague fragment, in: *Partee* 1976, pp.119-163.
Bennett, M., 1977, A guide to the logic of tense and aspect in English, *Logique et Analyse* 80, pp.491-518.
Bennett, M., 1978, Demonstratives and indexicals in Montague grammar, *Synthese* 39, pp.1-80.
Chomsky, N., 1965, *Aspects of the theory of syntax,* The MIT Press, Cambridge (Mass.).
Cooper, R.H., 1975, *Montague's semantic theory and transformational syntax,* unpublished dissertation, Univ. of Massachusetts, Amherst.
Dowty, D.R., 1978, Governed transformations as lexical rules in a Montague grammar, *Ling. Inq.* 9, pp.393-426.
Friedman, J., 1979, An unlabelled bracketing solution to the problem of conjoined phrases in Montague's PTQ, *Jour. of Phil. Log.* 8, pp.151-169.
Groenendijk, J. & Stokhof, M., 1976, Some notes on personal pronouns, reflexives and sloppy identity in a Montague grammar, in: K.Braunmüller & W.Kürschner (eds.), *Akten des 10. linguistischen Kolloquiums, Tübingen 1975,* Max Niemeyer Verlag, Tübingen, pp.301-315.
Groenendijk, J. & Stokhof, M., 1979, Infinitives and context in Montague grammar, in: S.Davis & M.Mithun (eds.), *Proc. of the conf. on Montague grammar at SUNY Albany,* University of Texas Press (to appear).
Hausser, R.R., 1979, How do pronouns denote?, in: *Heny & Schnelle 1979.*
Heny, F.W. & Schnelle,H.S. (eds.), 1979, Selections from the third Groningen round table, *Syntax and Semantics 10,* Academic Press, New York.
Janssen, T., 1976, A computer program for Montague grammar: Theoretical aspects and proofs for the reduction rules, in: J.Groenendijk & M.Stokhof (eds.), *Proc. of the Amsterdam coll. on Montague grammar and related topics,* Amsterdam papers in formal grammar I, Centrale Interfaculteit, University of Amsterdam, pp.154-176.
Karttunen, L., 1977, Syntax and semantics of questions, *Ling. and Phil.* 1, pp.3-44.
Montague, R., 1970, Universal grammar, *Theoria* 36, pp.373-398. Reprinted in: *Thomason 1974,* pp. 222-246.
Montague, R., 1973, The proper treatment of quantification in ordinary English, in: J.Hintikka, J.Moravcsik and P.Suppes (eds.), 1973, *Approaches to natural language,* Synthese library, Reidel, Dordrecht, pp.221-242, reprinted in: *Thomason,1974,* pp.247-270.
Partee, B., 1976, *Montague grammar,* Academic Press, New York.
Partee, B., 1979, Montague grammar and the well-formedness constraint, in: *Heny & Schnelle 1979.*
Rodman, R., 1976, Scope phenomena, "Movement transformations", and relative clauses, in: *Partee 1976,* pp.165-176.
Thomason, R.h., 1974, *Formal philosophy, selected papers of Richard Montague,* Yale University Press, New Haven.
Van Wijngaarden, A., 1970, On the boundary between natural and artificial languages, in: B.Visentini et al., *Linguaggi nella società e nella tecnica,* Edizioni di comunita, Milan, pp.165-176.
Van Wijngaarden, A., 1974, The generative power of two-level grammars, in: J.Loeckx (ed.), *Automata, languages and programming, 2nd coll. (Saarbrücken),* Lect. Notes in Comp. Sc. 14, Springer, Berlin, pp.9-16.
Van Wijngaarden, A., et al., 1975, Revised report on the algorithmic language Algo 168, *Acta Informatica* 5, pp.1-236.

SOME REMARKS ON THE LOGIC OF CHANGE, PART I

Hans Kamp, Bedford College, London

I. I knock the egg against the rim of the frying pan. This produces a change in the egg: a moment ago it was in one piece; now it is kaput. Whenever there is change there is a succession of incompatible states. And whenever there is such a succession there is a change.

Suppose a given state p is followed by an incompatible state q.* Then a change occurs. But when? Answer: Not before p has ended and not after q has

---

* The terminology of this paper will in certain respects be quite loose. Thus I will often speak of states, or conditions, and of the sentences which characterize these states or conditions without distinguishing between them. I will also speak of predicates as characterizing conditions or events, meaning thereby that the condition or event is characterized by a sentence which consists of the predicate followed by the name(s) of individual(s) whose identity is clear from the context. It will become explicit in section 3 that I am only concerned with conditions, states and events which are characterizable by means of sentences of a given language; so this kind of sloppiness does not affect anything of substance that is discussed in this paper.

In view of the intimate relationship between linguistic expressions on the one, and such entities as states, conditions etc., on the other hand it is also harmless to talk, as I shall sometimes do, of two conditions or states as being incompatible, or contradictories, or of one state or condition implying another. I also sometimes speak of a predicate Π as implying, being incompatible with, or being the contradictory of, another predicate R . Π implies R iff any sentence of the form $\Pi(a)$ implies any sentence of the form R(a), and analogously for the other two relations mentioned. So this way of speaking should not lead to confusion either.

Truly problematic is my use of the terms 'incompatible', 'contradictory' and 'implies' (irrespective of the nature of the terms that they are made to stand between). My basic intention has been to use these expressions in accordance with what I take to be their accepted meanings in logic; roughly: sentences s_1 and s_2 are incompatible iff of necessity s_1 is false whenever s_2 is true; they are contradictories iff it is logically necessary that s_1 is true iff s_2 is false; and s_1 implies s_2 iff by logical necessity s_2 is true whenever s_1 is. In the course of our investigations we shall have to abandon some of the familiar connections between truth, falsehood and negation, however, and this means that the concepts of incompatibility and contradictoriness cannot retain the significance that they have in classical logic. Contradictoriness in particular is a notion that will have to be reassessed and redefined after some other logical issues have been discussed - this in fact touches upon the most difficult issues to which our investigations will lead us.

begun. If p and q are incompatible without being contradictories this may leave a lot of room for the time of change. But if p and q are contradictories the situation is different. Let us consider this case first and refer, while discussing it, to q as not-p. When does the change from p to not-p occur? What answer we are prepared to give to this question is tied up with various logical issues. In fact there are two logical principles, one of long standing and the other plausible enough, which in conjunction leave no room for the change at all. The first of these is the *Principle of Bivalence* (PB), which says that at any time t it must be either that the condition p obtains or else that the contradictory condition not-p obtains.

The second, which I shall call the *Principle of Incompatibility* (PI), asserts that at the time of change from p to q neither p nor q obtains. Clearly these two principles exclude the possibility of there being any time of change from p to not-p at all. For according to PB either p or not-p should hold at such a time; and according to PI neither of them should hold.

Since there are no times of change from a given state to its contradictory, a present tense report of such a change cannot possibly be true. It does not prohibit the truth of reports of such changes in the past, however, nor of predictions that such changes will happen in the future.

What seems to me the natural setting for the study of the logical implications of this theory of change is provided by recent (and as yet unpublished) work of J. Stavi. Stavi has shown that four binary temporal connectives S,U,S',U' suffice for the characterization of all temporal properties of and relations between or among propositions that can be specified in the first order theory of the earlier later relation (assuming only that this relation is a linear order). The four are semantically defined by the following clauses (where χ is any formula I write: $[\chi]_t = 1$ for "χ is true at t", and $[\chi]_t = 0$ for "χ is false at t"; $t < t'$ means "t is earlier than t' ").

$$[S(\phi,\psi)]_t = 1 \text{ iff } (\exists t')(t' < t \ \& \ [\phi]_{t'} = 1 \ \& \ (\forall t'')(t' < t'' < t \rightarrow [\psi]_{t''} = 1))$$

$$[U(\phi,\psi)]_t = 1 \text{ iff } (\exists t')(t < t' \ \& \ [\phi]_{t'} = 1 \ \& \ (\forall t'')(t < t'' < t' \rightarrow [\psi]_{t''} = 1));$$

$$\begin{aligned}[S(\phi,\psi)]_t = 1 \text{ iff } & (\exists t')(t' < t \ \& \ (\forall t'')(t' < t'' < t \rightarrow [\phi]_{t''} = 1)) \ \& \\ & \neg(\exists t')(t' < t \ \& \ (\forall t'')(t' < t'' < t \rightarrow [\phi]_{t''} = 1) \ \& \\ & (\forall t'')(t'' < t' \rightarrow (\exists t''')(t'' < t''' < t' \ \& \\ & [\phi]_{t'''} = 0))) \ \& \ (\exists t')([\phi]_{t'} = 0 \ \& \\ & (\forall t'')((t' < t'' < t \rightarrow ((\exists t''')(t'' < t''' < t \ \& \\ & [\phi]_{t'''} = 0) \rightarrow [\psi]_{t''} = 1)));\end{aligned}$$

$$[U'(\phi,\psi)]_t = 1 \text{ iff } (\exists t')(t < t' \& (\forall t'')(t < t'' < t' \rightarrow [\phi]_{t''} = 1)) \& (\exists t')(t < t' \& (\forall t'')(t < t'' < t' \rightarrow [\phi]_{t''} = 1) \& (\forall t'')(t' < t'' \rightarrow (\exists t''')(t' < t''' < t'' \& [\phi]_{t'''} = 0))) \& (\exists t')([\phi]_{t'} = 0 \& ((\forall t'')(t < t'' < t' \rightarrow ((\exists t''')(t < t''' < t'' \& [\phi]_{t'''} = 0) \rightarrow [\psi]_{t''} = 1))).$$

Thus $S(\phi,\psi)$ is true at t provided ϕ is true at some earlier time t', since which time ψ has been true uninterruptedly until t, and $U(\phi,\psi)$ is true at t provided ψ is true from t uninterruptedly until some later time, at which ψ is true. $S'(\phi,\psi)$ says, at t, that there is a *gap* before t between which and t ϕ is uninterruptedly true, while there are times arbitrarily close before the gap at which ϕ is false; moreover there is a stretch of time of which the gap is the least upper bound throughout which ψ is true. $U'(\phi,\psi)$ says the same as $S'(\phi,\psi)$, but with past and future reversed (just as $U(\phi,\psi)$ says the same as $S(\phi,\psi)$ with reversal of past and future).

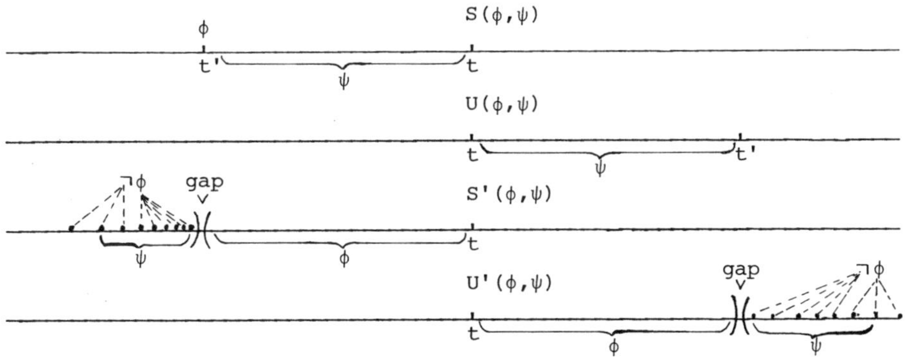

With the help of these operators we can express "there was a change from p to not-p" as

$$P(S'(\neg p,p) \vee S(\neg p \; \& \; S(p,p),\neg p) \vee S(p \; \& \; S(p,p),\neg p)$$

(where $P\phi$ ("it was the case that ϕ") is short for $S(\phi,\phi \rightarrow \phi)$).
"There will be a change from p to not-p" and "there has been a change from p to not-p" (where this last phrase is to be understood as implying that the condition not-p into which the change resulted has continued until the present) can be expressed similarly. And the same is true of more complicated temporal relations that involve change from some condition into its contradictory. Methods related to those used in Stavi's proof that S,U.S' and U' are capable of expressing all first order topological relations between propositions, establish the deductive completeness, relative to a variety of linear order-

ings and classes of linear orderings of appropriate axiomatizations of the language containing these four connectives. All these results are quite complicated, however, and this is not the place to present them.

We often say such things as: "It is now becoming dark", "The tree is changing its colour", "He is turning pale", etc. Moreover there are many verbs, such as e.g. 'fade', 'mature', 'die', 'appear', 'mellow', which convey change of a particular sort in their subjects and which are nevertheless used perfectly naturally in the present progressive. And an even larger class of transitive verbs (sometimes classified as 'accomplishment verbs'), which imply a change in their direct objects, also occur quite naturally in the present progressive. How could our present theory of change account for these uses? The reply to this must be that in all these cases there is a transition from one state p into another state q which is incompatible with p but not its contradictory. These are separated by a number of intermediate states. Each such intermediate state begins and subsequently ceases, and thus is associated with two changes of the kind that our theory deals with directly.
Such an account of protracted change raises a number of questions: What changes between contradictory states must be contained within a period of time in order that a certain protracted change can be said to be going on during that period? Is it possible to spell out in a systematic way, for any expressions of protracted change conditions on sequences of changes between contradictions that are necessary and sufficient for the applicability of the expression? Assuming such a systematic correlation could be given, should we consider such expressions as shorthands for the statements of the corresponding conditions on sequences? There are complex and difficult problems. Before tackling any of them *, however, we must, I think, consider a more fundamental issue: Is it really correct to hold that changes between contradictories do not, strictly speaking, ever occur?

II. It is common to represent time as a linearly ordered structure of instants, e.g. as a structure isomorphic to that of the real numbers, the representation employed in theoretical physics. Such a representation of time is also presupposed by the vast majority of results in what has become known as 'tense logic', (and in particular in the results of Stavi which I mentioned in the previous section). Such representations are quite abstract. For anything that can be *recognized* as standing in some temporal relation to some-

* In fact none of them could be dealt with in the first part of the paper published here.

thing else must, unlike the elements which these representations postulate, have some finite duration - this is so whether the relation is a relation between experiences, or between events which need not be experiences themselves but which must be experienced in order to be recognized as standing in the relation.

In the early years of this century Norbert Wiener showed how from such items of finite duration as experiences or other events, and the temporal relations by which they can be recognized as being connected, one can construct a linear ordering of instants;[1] the original items reemerge as convex sets[2] of such instants, and the original temporal relations between them are easily defined in terms of the single relation which linearly orders the instants. Formally Wiener's construction comes to this. We start from a certain set E, the members of which (think of them as events of some sort of another) enter into two relations, that of complete precedence $\propto$, and that of overlap, O. Intuitively $e_1 \propto e_2$ if e_1 has come to an end by the time e_2 begins; and $e_1 O e_2$ iff e_1 and e_2 are (at least in part) simultaneous.

Suppose we want to construct a linear structure of instants such that each member of E can be made to correspond to a convex set of these instants - the instants at which one might say that the event 'goes on'. This requires:

R1. that each event goes on at at least one instant;

R2. that whenever $e_1 O e_2$ then there is an instant at which they both go on; and, more generally, if E' is any finite subset of E such that for any two $e_1, e_2 \in E'$ $e_1 O e_2$ then there is an instant at which all members of E' go on; and

R3. that whenever $e_1 \propto e_2$ then if t_1 and t_2 are any instants at which, respectively, e_1 and e_2 go on then t_1 is earlier than t_2.

It is intuitively plausible that one way to satisfy these three requirements is to define instants as maximal sets of pairwise overlapping elements of E:

---

1 One finds references to and descriptions of this in many works of Russell's, (who may well have been the one to have conceived the idea on which Wiener's paper is based.)

2 By a *convex xet of* an ordering $\langle X,<\rangle$ I understand any subset Y of X such that for all $x,y,z \in X$ if $x < y < z$ and $x,z \in Y$ then $y \in Y$. An *interval of* $\langle X,<\rangle$ is a convex subset of a special kind, vz. a subset Y such that there are $x,z \in X$ such that Y is the set of elements of X bounded below by x and above by z. In this last definition 'bounded' is intentionally ambiguous between < and <. Thus with the particular boundaries x and z are associated four distinct intervals, vz. $\{y: x < y < z\}$, or $[x,z]$; $\{y: x < y < z\}$,or $(x,z]$; $\{y: x \leq y < z\}$, or $[x,z)$; and $\{y:x < y < z\}$, or (x,z). In my related [ ] I unwisely used the term 'interval' to refer to what I have here, consistently with common mathematical terminology, termed convex sets.

t is an *instant* (of the structure $\mathcal{E} = \langle E, \propto, O\rangle$) iff $t \subseteq E$; for all $e_1, e_2 \in t$ $e_1 O e_2$; there is no t' such that $t \subset t' \subseteq E$ which also satisfies the condition that for any $e_1, e_2 \in t'$ $e_1 O e_2$;

while putting moreover, for any two such instants t_1, t_2:

$t_1 < t_2$ iff there are $e_1 \in t_1$ and $e_2 \in t_2$ such that $e_1 \propto e_2$.

Clearly these definitions meet R1. - R3. It is not difficult, moreover, as Wiener was the first to point out, to show that < linearly orders the set of instants of $\mathcal{E}$ provided that $\mathcal{E}$ satisfies the following postulates:

A1. $e_1 \propto e_2 \rightarrow e_2 \propto e_1$;

A2. $e_1 \propto e_2$ & $e_2 \propto e_3 \rightarrow e_1 \propto e_3$;

A3. $e_1 O e_2 \rightarrow e_2 O e_1$;

A4. $e_1 O e_1$;

A5. $e_1 \propto e_2 \rightarrow \neg e_1 O e_2$;

A6. $e_1 \propto e_2$ & $e_2 O e_3$ & $e_3 \propto e_4 \rightarrow e_1 \propto e_4$;

A7. $e_1 \propto e_2 \vee e_1 O e_2 \vee e_2 \propto e_1$.

Of these postulates A1. - A6. are readily justified in terms of the intended meanings of < and O. We shall presently discuss the acceptability of A7.

We shall denote the set of instants of $\mathcal{E}$ as $T(\mathcal{E})$, the order between these instants as $<(\mathcal{E})$ and the pair $\langle T(\mathcal{E}), <(\mathcal{E})\rangle$ as $\mathcal{T}(\mathcal{E})$ is the minimal structure which satisfies the requirements R1. - R3.[4]

---

3 Actually A2. is redundant as it follows from A7. and A4.

4 There is a difficulty here which relates to certain infinite sets of pair-wise overlapping events. Suppose e.g. that E contains for each natural number n two elements e_n, f_n such that

i) for any n, m e_n and e_m overlap; e_n and f_m overlap iff $m > n$, and if $m \leq n$ then $e_n \propto f_m$. An example of this situation is provided by a set E whose members are intervals of the real number line and where e_n is the interval $(0, \frac{1}{2n})$ and f_n the interval $(\frac{1}{2n}, 1)$. In this situation there will be an instant t which contains all the e_n, and at which therefore each e_n 'goes on'. If moreover E contains no other members than e_n and f_n then there is no need of t for the satisfaction of R1. - R3., so that $\mathcal{T}(\mathcal{E})$ is not exactly the minimal structure to satisfy these requirements. I will ignore the complication in what follows, and continue to speak as if $\mathcal{T}(\mathcal{E})$ were in fact the smaller structure satisfying R1. - R3. There is an additional problem with my earlier stipulation that if $e \in t$ then e can be said to 'go on' at t. Intuitively it would seem more appropriate to say, in the case we are contemplating, that t is the starting point of each of the e_n than that e_n goes on at t. Thus in general the membership relation $e \in t$ is less restricted than what is normally understood by 'e goes on at t'. For this latter phrase to be applicable we should presumably demand not only that $e \in t$ but in addition that t be internal to some interval such that $e \in t$ for all t belonging to that interval.

However, there may be considerations other than those incorporated in R1.-R3. which favour a richer instant structure than $\mathcal{T}(\mathcal{E})$. In physics for instance one normally insists that time is like the real numbers, rather than e.g. the rationals. The only rationale for this is, it seems, provided by the consideration that the application of physical laws, such as e.g. those of classical or relativistic mechanics, becomes highly artificial (though not strictly impossible) if time is assumed to have a structure other than that of the real numbers. Of course nothing that has been said indicates whether $\mathcal{T}(\mathcal{E})$ might not itself have the structure of the reals; but the commitment that is made in physics does not seem to depend on this. The question of the structure of temporal order is thus a fairly involved one to which that of $\mathcal{T}(\mathcal{E})$ might provide a partial, but can hardly be expected to give a complete answer.

In any case, however, little if anything can be ascertained about the structural properties of $\mathcal{T}(\mathcal{E})$ before we clarify the precise nature of E. What I have said about $\langle E, \propto, O\rangle$ so far still leaves room for a wide range of interpretations; and different interpretations are bound to lead to very different structures of instants. For instance, it is compatible with my earlier remarks about the nature of E to interpret E as consisting of the experiences of a given individual X at some particular time t together with those past experiences of his of which he is still aware at t. On this interpretation of E $\mathcal{T}(\mathcal{E})$ will be absurdly limited. It will contain no times in the future of (its times corresponding to) t; and its times in the past of t are unlikely to allow him to incorporate adequately new information about events preceding t which he might acquire subsequently. Without being told more about X's experiences it is impossible to say much about the instant structure derived from this set E. Not that this matters much. For the features of this instant structure will have only the remotest bearing on what would be regarded an adequate representation of time. What X himself will accept as an adequate interpretation, in particular, will depend more on certain general inductively based beliefs, e.g. that he has had more experiences than he can now remember; that he will have experiences in the future; that others have experiences too, many of which are not matched by his own experiences; or that of the vast variety of the world's events only some are experiences, and only some of the remaining ones are experienced.

Like the considerations which lead to real-number-like time in physics these hypotheses lead to richer instant structures. But unlike those physical considerations they can be interpreted, more directly, as assumptions of the existence of richer *event* structures - which will lead to richer instant structures when subjected to Wiener's construction. In fact each of the hypotheses I listed corresponds to a different way of interpreting E:

as the set of X's present as well as all of his past experiences; as the set of all of X's experiences whether past, present or future; as the set of all experiences ever had by anyone; or as the set of all (observer-independent) events.

Saying that any of these interpretations is possible would seem to imply that it induces, or at least can be combined with, interpretations of < and O so that the resulting structure <E, ∝ , O> satisfies A1. - A7. I already remarked that the verification of A1. - A6. is unproblematic. Indeed, each of the interpretations mentioned, in conjunction with any interpretations of ∝ and O that are consistent with the informal descriptions of their meanings on p. verifies these postulates. A7. on the other hand is not so easily seen to be valid, and here it might make a difference which particular interpretations of E, ∝ and O we consider. If for instance we take E to consist of X's experiences up to t, and ∝ and O to be such that $e_1 \propto e_2$, or e_1Oe_2, is to hold only if X is aware at t that e_1 precedes, or overlaps, e_2, then there is no reason whatever why A7. should be true: X may be perfectly well aware of e_1 and e_2 separately without being aware of the temporal relation in which they stand to each other. If on the other hand we stipulate that, say $e_1 \propto e_2$ iff e_1 *was* entirely before e_2, and e_1Oe_2 iff they did overlap, whether X is aware of these facts or not, then A7. would seem a good deal more plausible. Similarly if E is the set of all observer-independent events and ∝ and O are the relations of actual precedence an overlap A7. might appear to be satisfied. (I am consciously ignoring the difficulties that might be thought to arise here in connection with the theory of relativity.). I believe, however, that even these last two interpretations of E, and O do not render A7. true. The grounds for this belief derive from the following two convictions:

1. The concept of an event is not absolute. What can be singled out as a separate event depends on the tools available for singling it out. There are no good reasons to suppose these tools, - the components of some particular conceptual scheme, which enables its users to distinguish one thing from another, to see it as different from others in various ways, and to see it as different now from what it was before, - to be completely determined by the intrinsic features of the phenomena for whose structural perception they are employed. Nor if we restrict attention to human users, are there grounds for supposing them to be fully determined by a combination of the intrinsic features of nature on the one hand and biologically fixed characteristics of homo sapiens on the other. Nor, finally, are we at present entitled to the

belief, that there is one optimal conceptual scheme to which any other could be reduced. It appears unwarranted therefore to speak of 'the class of events' as if this were a totality that was fixed absolutely. All we can meaningfully refer to is the class of events relative to a given conceptual scheme $\mathfrak{C}$; and where $\mathfrak{C}$ is not mentioned explicitly it is the conceptual scheme of the speaker which must be understood as presupposed.

2. The concepts which enable us to discern difference and change in the empirical world are typically vague: there always may be, and generally will be, cases where the criteria associated with the concept fail to determine whether it applies. I suspect that vagueness is inevitable - that it follows from certain general premisses concerning the ways in which concepts are used and can be learned by creatures such as ourselves. I shall not attempt to work out such an argument here, however. For my present purpose it is enough that many of the concepts we do employ are in fact vague; and that there exists at present no viable prospect of replacing them by concepts that are fully precise. We often treat, it appears, vague predicates as if they were 'potentially sharp', i.e. as if they had sharp boundaries which could be in any one of a certain range of possible locations. In doing so we manage to hold on to the principles of classical logic, - such as, in particular, the law of excluded middle - in spite of the circumstance that certain sentences (e.g. some sentences of the form Q(a).) are neither true nor false.[5]
Vagueness in the concepts used to individuate events may lead to certain indeterminacies in the relations of temporal overlap and precedence. This may happen in particular where the events are changes involving vague concepts. Suppose a screen shows two light spots, one, s_1, changing from red to orange, and the other, s_2, changing from green to blue. Are the changes simultaneous or does one of them precede the other? There are at least two ways in which we may approach this question:
a) We treat the colour predicates involved as potentially sharp. Suppose first, just for the sake of discussion, that events other than the two in

---

5 That ordinary discourse does treat vagueness in this way is suggested for instance by our tendency to treat such statements as "a is red and a is not red" as contradictory, and also, though perhaps not quite as readily, statements of the form "a is red or a is not red" as tautologous, even where we admit the possibility that a lies in the truthvalue gap of 'red'. [5] contains a formal analysis of this use of vague predicates, a variant of which will play a role in later parts of this paper. See also [3] .

question already determine a time structure $\langle T,<\rangle$ and that we are asking with respect to this time structure at what instant, or during which interval, each of these changes took place. Then we can reason as follows: Each of the possible ways in which we can resolve the ambiguity between orange and red will determine the exact time t_1 where s_1 ceases to be red and starts being orange. Similarly each of the ways of resolving the ambiguity between green and blue determines a precise instant t_2 for the transition of s_2 from green to blue. It is possible that no matter how the ambiguities are resolved t_1 is always before t_2. In that case it is reasonable to maintain that the change in s_1 *is* indeed before that in s_2. By the same token, if no matter how we resolve the ambiguities in the four colour predicates t_1 and t_2 coincide then this justifies the claim that the changes are simultaneous; and if t_2 is in all cases before t_1 then we may say that the second change precedes the first. It is perfectly possible, however, that on some resolutions t_1 precede t_2 while on others t_2 precedes t_1 while on yet others they coincide. In that case the temporal order of the two changes must be regarded as undetermined. As this last possibility cannot be excluded we have no reason for believing A7. to be satisfied.[6]

To assume that the structure of time has already been determined by events whose mutual temporal relations are completely fixed is, within the present context, to put the cart before the horse. For it is precisely the question whether the structure of events can be assumed to yield a suitable - in particular linear - instant structure which is at stake. We cannot permit ourselves, therefore, to describe the situation by referring, in the way we have just done, to the 'times' of the two changes. It ought to be evident, however, that this does not affect the essence of our conclusion: the temporal relations between events need not be fully determined.

---

6 There is a possible source of misunderstanding here. According to the theory of supervaluations that is discussed in [3] and [5] A7. *would* be satisfied, in as much as it comes out true in each of the structures that result from resolving the ambiguities. Our present concern, however, is whether $\langle E, \propto, O\rangle$, taken as a model in the usual sense satisfies A7. directly, i.e. if the union of $\propto$, O and $\breve{\propto}$ equals E^2. Only this will guarantee that Wiener's construction yields a linear, and not only a partial, ordering.

Moreover, the intuition that each way of resolving the ambiguities results in an instant of change can still be expressed, but now rather as the proposition resolutions yields an event structure in which the change, call it e_0, is instantaneous in the precise sense of not being divided by other events:

$$(1)\quad \neg(\exists e_1, e_2)(e_1 O e_0 \; \& \; e_2 O e_0 \; \& \; e_1 \propto e_2)$$

b) It is tempting to try to avoid indeterminacy of temporal relation by analyzing the changes as lasting for as long as s_1 and s_2 belong to the truthvalue gaps of the relevant predicates. But to think that temporal determinacy could thus be restored is an illusion. Talk of truthvalue gaps easily leads us to think of a vague predicate Q as dividing its range of application into three, rather than two, parts: positive extension, negative extension and truthvalue gap. But we must be careful here. An object a belongs to the truthvalue gap of Q if it is undetermined whether the criteria associated with Q are satisfied by a or not. To treat the gap as on a par with positive and negative extension is to suggest that there is an other set of criteria which decide whether (i.e. are satisfied iff) satisfaction of the former criteria is undetermined. In general, however, there are no grounds for supposing that there is such a second set of criteria. If there isn't then no object a can be regarded as *definitely* belonging to the truthvalue gap of Q; and, by the same token, if a changes in the respects relevant to Q, there is no time which can be counted as definitely belonging to the period of transition. The following parallel with intuitionistic mathematics may be of some help to get this point across.

Where a is in the truthvalue gap of Q the proposition Q(a) is in a position similar to that of a proposition p of intuitionistic mathematics which is neither provable nor refutable. No proof may be possible that p is undecidable by mathematical argument, and thus the proposition that p is undecidable would, even if it could be expressed, never be assertable.[7]

This is not to say that it is never possible to ascertain that an object belongs to the truth value gap of a predicate. In fact it is precisely in the context of change that we find the best evidence that this possibility is sometimes realized. There are occasions, for instance, where it seems right to say: "I was getting angry" to report a gradual transition from a state of non-anger to a state of anger, passing through a number of intermediate states which are recognizable as being neither states of anger nor states of non-anger. Here there is a period of some finite duration during which I was in the truthvalue gap of 'angry'. But note that this does not help us at all to establish that the temporal relations between events are determinate. For let us suppose that the account I just gave of "I was getting angry" also applies to the changes in the spots s_1 and s_2 which we discussed above; i.e. there is a period during which s_1 is definitely in the truthvalue gaps of 'red' and 'orange' and also a period during which s_2 is definitely in the truthvalue gaps of 'green' and 'blue'. In order that it be

---

7 I cannot say that I find this account of the nature of truthvalue gaps a great deal more satisfactory than others with which I have been aquainted (including earlier attempts of my own). Such accounts run into an apparently fundamental difficulty. Truthvalue gaps constitute a phenomenon for which there is no room in scientific languages as we have come to conceive of them, largely under the influence of the development of modern logic during the past century. As it is only of such languages that we understand the logical properties reasonably well it is a sound methodological canon to insist on their use in all scientific exposition. It seems, however, that when we describe truthvalue gaps in such a language we cannot help distorting the phenomenon that we are trying to capture. This is particularly true of model-theoretic analyses of vagueness, such as can be found e.g. in [3] and [5] , which are formulated in a language which builds the theory of sets on a basis of classical predicate logic. The only proper way to give an account of vagueness is to formulate the logic of languages which contain vague predicates *directly* (i.e. without reducing it, as one does in modeltheoretic treatments, to that of some other type of language). This is a problem that is still largely unsolved. (It might be thought that the supervaluation theory of vagueness has resolved this problem by showing that the logic of languages with vague predicates is the same as that of languages all of whose predicates are sharp. But note that all the supervaluation theory asserts is that the laws governing those operators on which modern logic has primarily concentrated (vz. truthfunctions and quantifiers) are not affected by vagueness. So if the supervaluation account is at all correct, this only means that to show the distinctive logical features of vagueness we would have to formulate the logic of languages which contain additional logical notions besides these familiar operators.

determined whether the first change completely preceded the second, it must be determined where the first of these periods ends and where the second of them begins. But there is no reason why the boundary between the truth value gap of, say, 'orange' and its positive extension should be sharply defined and thus no reason why it should be determined exactly when s_1 leaves the truthvalue gap of 'orange' and enters the positive extension. Similarly the time at which s_2 moves from the positive extension of 'green' into its truth-value gap need not be precisely fixed. Thus temporal indeterminacy can no more be excluded on this second account of change than it could on the first.[8]

---

8 The same point might also be made in a slightly different way: As soon as there is the possibility of ascertaining that a belongs to the truth-value gap of Q there arises therewith the possibility of replacing the single predicate Q by three incompatible predicates (we might call them 'definitely Q', 'non-Q' and 'neither Q nor non-Q') whose extensions are, respectively, the positive extension of Q, its negative extension, and its truthvalue gap. In fact we achieve precisely this effect by introducing, as we often do in ordinary discourse, two distinct negation-operators, roughly those which are familiar from the theory of three-valued logic as "strong", or "inner", and "weak", or "outer", negation. I did make this very distinction just now when I used both the "strong" 'non-' and the "weak" 'neither ... nor'. These predicates are ideally both mutually exclusive and jointly exhaustive. This means that e.g. the negative extension of 'definitely Q' coincides with the union of the (positive) extensions of non-Q and 'neither Q nor non-Q'. For each of these predicates, however, exactly the same question arises that we earlier felt obliged to ask about Q: What reasons could we have for believing that, say, 'definitely Q' doesn't have any truthvalue gap; or, in other words, that the boundary between the positive extension of Q and its truthvalue gap is sharp? In general there are no such reasons.

It should be possible in principle to adopt criteria that decide whether an object lies in the truthvalue gap of some such predicate as 'definitely Q'. Then the same procedure could be repeated, now introducing three new predicates to replace the single 'definitely Q'. [3] suggests a number of ways in which we might analyse languages in which this process is repeatable ad libitum. Whether all vagueness could be eliminated through an infinite number of iterations though not through any finite segment thereof is a matter of speculation.

III. The theory of change that I wish to explore starts from the idea that it is the changes themselves which are constitutive of time. It seems to me that we generally take for granted that changes are observer-independent events; and I shall assume from now on that they are. In the sequel of this paper therefore, it is only event structures whose members are such observer-independent events that will be of importance. Henceforth event structures shall be understood to be of this sort. We have seen, however, that even such event structures cannot be assumed to satisfy A7. Only in ideal modifications of such structures, where all vagueness has been removed, will A7.be valid.

For what is to follow it will be helpful to have at hand a more formal representation of the ideas that have been discussed in the last section. Such a representation will be somewhat simplified if we assume, as I will do from now on, that an event-structure is determined by a language L (of which we should think in this connection as embodying the conceptual scheme used in individuating the events). Such a language will contain, in particular, for each of the concepts used in the individuation of events a simple or complex predicate which expresses that concept. I shall make further assumptions about L as we go along.

The event structure determined by L must incorporate, besides the three components E, $\propto$, O which we have already discussed at some length, also a component to which I shall refer as $\mathcal{A}_1$, which conveys what sort of event each of the members of E is. That is, it must specify for each e ε E the conditions, as expressed in L, in terms of which e is singled out. It is natural to represent this component as a function from E to appropriate expressions of L. Which expressions are appropriate? Let us assume that L has at least the structure of a language of 1st order predicate logic, i.e. contains predicates, names for and variables ranging over individuals, truth functions and quantifiers; and further that L is *tensed*, in the sense that its expressions may take different semantic values at different times. In particular, the sentences of L may be true at one time while false at another. The sentence Q(*a*), e.g., will be true at those times where a satisfies Q and false at those times where it does not satisfy Q. In such a tensed language the changing conditions that serve to individuate events are naturally expressed by complete sentences. So we shall assume that the fourth component is a function from E to sentences of L. There are, however, several ways in which a condition can be used to individuate an event. The event might be the presence of that condition during some limited period; or it might be the event of that condition's absence; or it might be the event of that condition coming to obtain; or of its ceasing to do so. (Especially these last two types of events are important here, of course). It is therefore not enough for $\mathcal{R}_1$ to associate with an event e from E just a sentence of L, it must also make clear in which of the four mentioned ways that sentence is related to e. I shall assume therefore that $\mathcal{R}_1$ maps each e onto a pair consisting of a sentence and one of the symbols P (for Present), F (for Failing), B (for Becoming) and C (for Ceasing).

The next question we must ask is: Which sentences of L serve for the individuation of events? One is tempted to answer to this: Any meaningful sentence of L. For let s be any such sentence and suppose for instance that s holds at one time and fails at some later time. Then there ought to be somewhere between these two times an event which is the ceasing of the condition expressed by s. There are two difficulties with this answer. The first is that until we know more of the language L we cannot be quite so sure that each of its meaningful sentences expresses a condition change in which determines an event. This is a question to which I shall return later.

There is, however, another, and more fundamental,difficulty. Suppose s is a grammatically complex sentence. Then the conditions for its obtaining or failing at any particular time are determined by the semantic values, at various times, of its constituents - this is what grammatical complexity means! It is therefore not possible that the complex sentence alters its truthvalue without any accompanying change in the value of at least one of its atomic components.

I shall assume that such changes in the values of atomic constituents can always be reduced to changes in the satisfaction of atomic *predicates*. If one also assumes, as I shall do, that L contains a name for each of the individuals it talks about[9], then it follows that all events are determined by atomic sentences (i.e. sentences consisting of an atomic predicate followed by an appropriate number of names.[10]

---

9 The assumption that every individual has a name, which will also be convenient later, can be eliminated by methods familiar from standard developments of symbolic logic. I will not dwell on this any more.

10 Where L is assumed to have the structure of a language of first order predicate logic (possibly with some additional apparatus) the notion of an atomic sentence is unproblematic. When we apply the present analysis to any natural language, however, we shall have to consider very carefully which expressions are to be regarded as playing the role I have here attributed to atomic sentences.

In addition to the information, provided by $\mathcal{A}_1$, as to how the members of E are individuated we shall also want further facts concerning what was the case when a given event occurred. Some such information could be extracted from the quadruple $\langle E, \propto, O, \mathcal{A}_1\rangle$ as it stands. E.g. we may infer that s_2 remained green while s_1 turned orange from the fact that the event e_1 of s_1's turning orange is included in some event e_2 which is identified as the event of s_2's being green - where inclusion of e_1 in e_2 can be expressed as

(2) $(\forall e)(e \propto e_2 \rightarrow e \propto e_1) \ \& \ (\forall e)(e_2 \propto e \rightarrow e_1 \propto e) \ \& \ (\forall e)(eOe_1 \rightarrow eOe_2)$

To assume that all such information can be so extracted is to put a heavy strain on the notion of an event, however. It would require that even conditions which, although contingent, nevertheless happen to hold forever would determine "events" characterized by their presence; and such unbounded "events" would be very different from what normally counts as an event. As I do not wish to commit myself to such requirements I shall assume that event structures come with yet a fifth component, $\mathcal{A}_2$, which supplies the information about what was the case during each of the members of E. This may still not encompass all information about the members of E that we shall need. For it may happen that a given event e is individuated as the change from, say, not-Q(a) to Q(a) but that it is also the change in some other condition, e.g. from R(b) to not-R(b). It is therefore desirable to let $\mathcal{A}_2$(e) map atomic sentences of L to one of the four values P,F,B,C. $\mathcal{A}_2$(e)(s) = P will mean that s is true throughout e; $\mathcal{A}_2$(e)(s) = F that it is false throughout e; $\mathcal{A}_2$(e)(s) = B that e is a change from not-s to s; and $\mathcal{A}_2$(e)(s) = C that e is a change from s to not-s.

The sets $\mathcal{A}_2$(e) will have to satisfy certain consistency requirements. In the first place there must be consistency between the assignments to any two overlapping events: If e_1 and e_2 overlap and the atomic sentence s belongs to the domains of both $\mathcal{A}_2$(e) and $\mathcal{A}_2(e_2)$ then it must yield the same value for these two functions. Furthermore the information provided by $\mathcal{A}_2$ must not contradict that provided by $\mathcal{A}_1$. The simplest, and most natural, way to ensure this is to stipulate that the first member of $\mathcal{A}_1$(e) always belong to the domain of $\mathcal{A}_2$(e), and that the values agree (i.e. if $\mathcal{A}_1$(e) = $\langle$s,B$\rangle$ then $\mathcal{A}_2$(e)(s) = B etc.). This renders the component $\mathcal{A}$, strictly speaking superfluous for the semantic considerations that follow. Yet conceptually it seemed better to include it as a separate component into our representations.

I shall refer to quintuples $\langle E, \propto, O, \mathcal{A}_1, \mathcal{A}_2\rangle$ which satisfy the mentioned conditions (i.e. $\langle E, \propto, O\rangle$ satisfies A1. - A6., and the functions $\mathcal{A}_1$ and $\mathcal{A}_2$

satisfy the consistency requirements just listed) as *event models*.

I suggested earlier that event structures $\langle E, \propto O\rangle$ are typically underdetermined in that A7. may fail, and that this indeterminacy is due to vagueness in the predicates used in individuating the events in question. Let us assume that such vagueness is the only source of indeterminacy of temporal relations. Then in each (ideal) situation where all gaps have been removed the temporal structure will be fully determinate (i.e. A7. will hold). One might also expect that the model $\mathcal{M}' = \langle E' \propto ', O' \mathcal{A}_1', \mathcal{A}_2'\rangle$ corresponding to such an ideal situation is determinate in yet another respect, vz. that at each of the instants which result when Wiener's construction is applied to the structure $\mathcal{E}' = \langle E', \propto', O'\rangle$ every atomic sentence of L has a definite value. Suppose, however, that the instant contains an event e which is a change from not-s to s. Then, according to the Incompatibility Principle, s can be neither true nor false at t.

It seems reasonable, in view of the assumptions we have made already, to assume in addition that these are the only cases where an atomic sentence lacks a value at an instant:

(3) unless t contains an e such that $\mathcal{A}_2(e)(s) = B$ or $\mathcal{A}_2(e)(s) = C$
$[s]_t = 1$ or $[s]_t = 0$.

Let us make the intuitively obvious stipulation that the truthvalue of any atomic sentence s at an instant t is determined by $\mathcal{E}'$ via the condition:

(4) $[s]_t = {1 \atop 0}$ iff $(\exists e)(e \in t \ \& \ \mathcal{A}_2(e)(s) = {P \atop F})$

Thus any structure $\mathcal{M}'$ corresponding to a situation where vagueness has been resolved must satisfy the further condition that, given definition (4) of "$[s]_t$ ", (3) holds. This leaves us with the pairs $\langle t,s\rangle$ where t contains an event e that is a change with respect to s. The dilemma we meet here is the one already signaled in section 1: if there are genuine times of change (from a given condition into its contradictory), then either Bivalence of Incompatibility must be given up.

There are various options which we can pursue at this point. The simplest of these, in the sense that it allows us to stay as closely as possible to familiar metamathematical terrain, succeeds in maintaining in a fashion, both PB and PI.

The means of expressing change will be quite limited; this can be done only by applying to a primitive predicate of L either the operator B ("-- is becoming ..." or the operator C ("-- is ceasing to be ..."). (BQ) and

(CQ) are again predicates[11] and (BQ)(a), for instance, is to be read "a is becoming Q", or "a is changing from non-Q to Q" - similarly for (CQ)(a). As always where we deal with truthvalue gaps the real difficulty arises over negation. Here it presents itself in the following guise. We have already admitted that a's becoming, or ceasing to be, Q is incompatible both with its being Q and with its 'failing to be' Q. This may sound incoherent. For if a's becoming Q, say, is incompatible with a's being Q does that not imply that if a is becoming Q then a fails to be Q? In an informal setting it is difficult to escape this appearance of incoherence; a consistent description requires two distinct concepts of negation, whereas prevailing logical doctrine has made us prone to overlooking what differences in meaning between various expressions for negation we might otherwise have been able to discern. What words we use in our attempt to capture the difference between these two negation concepts may not matter all that much. What does matter is that our present commitments force a bifurcation of the concept of negation upon us - into what corresponds, roughly, to the "strong", or "inner", and the "weak", or "outer", negation that one finds in the theory of three-valued logic, and especially in the theory of presupposition.
Here, as in certain systems of three-valued logic, one of these negations becomes definable as soon as the other is introduced as a primitive. Thus if we represent the "strong" concept of negation, at which I gestured above when using the expression 'fails to be Q', by a predicate operator F, then the disjunction

(FQ)(a) v (BQ)(a) v (CQ)(a)

exhausts the conditions under which it is not the case that a is Q, and thus represents the negation of 'Q(a)' in a second ("weak") sense. I shall represent this second negation as a sentential operator, $\neg$.

So far we have only defined $\neg$ as applying to sentences of the form Q(a). However, the same thesis that led to its definition in those contexts, vz. the view that Q,BQ,CQ, and FQ are pairwise incompatible and jointly exhaustive, suggests similar definitions for it where it stands in front of sentences of the forms (BQ)(a), (CQ)(a), and (FQ)(a):

---

11 Strictly speaking the operators B and C, as well as the operator F that is introduced below, are systematically ambiguous, in that they transform 1-place predicates into 1-place predicates, 2-place predicates into 2-place predicates, etc. In what follows I shall only deal with 1-place predicates; but what I have to say will invariably apply also to predicates of more places.

$$\neg(BQ)(a) \equiv_{df} (Q(a) \vee (CQ)(a) \vee (FQ)(a),$$

etc.

In all these contexts $\neg$ has the properties of classical negation, and, in particular, obeys the law of excluded middle. Note, however, that we have only accounted for the operator as applying to atomic sentences. The truth definition below will make it behave like classical negation also in the context of complex formulae. If we introduce $\neg$ as a primitive then F can be contextually defined by:

$$(FQ)(a) \equiv_{df} \neg Q(a) \ \& \ \neg(BQ)(a) \ \& \ \neg(CQ)(a)$$

We shall adopt both F and $\neg$ as primitives, identifying $\neg$ with the negation operator which L already contains (in as much as it encompasses first order predicate logic). Let L' be the extention of L which we obtain by adding the predicate operators B, C, and F. (We shall later differentiate between a number of languages L_i; for each of these we will then have a corresponding extention L_i').
Sentences of the form Q(a) can be negated in two ways. One of these - that involving F - behaves in accordance with the principle of Incompatibility, the other, involving $\neg$, preserves Bivalence. Thus, in a sense both principles have been saved, but at the cost of giving up a unified account of negation.

In conjunction with any event model $\mathcal{M} = \langle E, \propto, O, \mathcal{A}_1, \mathcal{A}_2 \rangle$ we must consider the various models which describe the ideal situations in which all indeterminacies of $\mathcal{M}$ have been resolved. We shall refer to such models as *completions* of $\mathcal{M}$, and usually refer to them as '$\mathcal{M}'$' or $\langle E', \propto', O', \mathcal{A}_1', \mathcal{A}_2' \rangle$. Each such $\mathcal{M}'$ leads in turn to a model $\mathcal{N} = \langle T', <', \mathcal{F}' \rangle$ where $\langle T', <' \rangle$ is the instant structure derived by Wiener's construction from $\mathcal{E}' = \langle E', \propto', O' \rangle$. We shall refer to this last model as the *instant model derived from* $\mathcal{M}'$. $\mathcal{F}'$ is to convey the truthvalues of the atomic sentences of L at the instants $t \in T'$. In the context of the option we are now pursuing this is best done in the format familiar from ordinary model theory, i.e. by specifying the extensions of the primitive predicates of L, as well as of the predicates that can be formed out of these by application of the operators B,C and F. Thus $\mathcal{F}'$ will map each pair consisting of a $t \in T'$ and one of these predicates onto the set of individuals which satisfy the predicate at that time. More specifically:

$$\begin{array}{lll}
(5)\ \text{i)} & \mathcal{F}'(t,Q) & = \{a: (\exists e)(e \in t \ \& \ \mathcal{A}_2'(e)(Q(a)) = P\}; \\
\text{ii)} & \mathcal{F}'(t,FQ) & = \{a: (\exists e)(e \in t \ \& \ \mathcal{A}_2'(e)(Q(a)) = F\}; \\
\text{iii)} & \mathcal{F}'(t,BQ) & = \{a: (\exists e)(e \in t \ \& \ \mathcal{A}_2'(e)(Q(a)) = B\}; \\
\text{iv)} & \mathcal{F}'(t,CQ) & = \{a: (\exists e)(e \in t \ \& \ \mathcal{A}_2'(e)(Q(a))) = C\};
\end{array}$$

On the basis of this information we can then give a (two-valued) truth definition for the sentences of L, as follows. The truthvalues of the atomic sentences of L' are given by the obvious clauses:

$$\begin{array}{lll}
(6)\ \text{i)} & [Q(a)]_{\mathcal{N}',t} & = 1 \text{ iff } \mathcal{F}'(t, Q(a)) = P \\
\text{ii)} & [(BQ)(a)]_{\mathcal{N}',t} & = 1 \text{ iff } \mathcal{F}'(t, Q(a)) = B \\
\text{iii)} & [(CQ)(a)]_{\mathcal{N}',t} & = 1 \text{ iff } \mathcal{F}'(t, Q(a)) = C \\
\text{iv)} & [(FQ)(a)]_{\mathcal{N}',t} & = 1 \text{ iff } \mathcal{F}'(t, Q(a)) = F
\end{array}$$

For the truthvalues of complex sentences we shall adopt clauses familiar from standard developments of the model theory of tense logics.[12]
In particular we have for sentence negation simply:

$$(7) \qquad [\neg\phi]_{\mathcal{N}',t} = 1 \text{ iff } [\phi]_{\mathcal{N}',t} = 0$$

(Precisely what further clauses are required depends on the resources of L. We shall come back to this). The supervaluation theory (as it is developed e.g. in [3] and [5]) suggests that the values which the sentences of L receive in the instant models derived from the completions of $\mathcal{M}$ be used to characterize truth and falsehood relative to $\mathcal{M}$ itself. There are at least two ways in which we might go about this. First we might stipulate, for any t belonging to the (not necessarily linear) instant structure $\langle T,<\rangle$ derived from $\mathcal{M}$, that a sentence s is true (or false) at t iff s is true at t in each model $\mathcal{N}'$ derived from a completion $\mathcal{M}'$ of $\mathcal{M}$. This cannot be the correct specification, however, for in general the "instants" of T need not occur in the instant structures derived from completions of $\mathcal{M}$; nor need they correspond to the instants of such a derived structure in a one-to-one fashion. It is possible for instance that two sets t_1 and t_2 of pairwise overlapping events which are both maximal relative to the temporal relations of $\mathcal{M}$ will collapse into a single instant t' of $\mathcal{N}'$. This happens, to be precise, if for any $e_1 \in t_1$ and $e_2 \in t_2$ $\mathcal{M}$ does not specify the temporal relation between them, while in $\mathcal{M}'$ any two such events overlap. In this way several "instants" of $\mathcal{M}$ may correspond to a single instant of $\mathcal{N}'$.

---

12 see e.g. [2],[6]

Formally this difficulty could be overcome by the stipulation that s be true, or false, at t iff it is true, or false, in each $\mathfrak{N}'$ at each t' which corresponds to (i.e. settheoretically includes) t. But conceptually this remains somewhat unsatisfactory in as much as it is dubious whether the so-called "instants" of T can always be regarded as genuine instants at all.

The second alternative is to define truth and falsity directly with respect to events of $\mathfrak{M}$:

(8) s is true (false) in $\mathfrak{M}$ at e iff for each $\mathfrak{N}'$ and each instant t' of $\mathfrak{N}'$ such that e ε t' s is true (false) at t' in $\mathfrak{N}'$.

This definition is actually a quite natural explication of the central insight of tense logic - and more generally of the theory of indexicals - that the central task of semantics is to characterize what it is for a sentence (which contains context-sensitive elements) to be true at the time at which it is uttered. Utterances are events of usually relatively short, but nevertheless finite, duration. It is thus possible that a sentence lacks a truth-value at the time of its utterance in virtue of the fact that that time can be subdivided into smaller portions during some of which the sentence is true while during others it is false. (8) succeeds in capturing that possibility.[13]

---

13 In actual discourse indeterminacy of this sort tends to be resolved by a form of the principle of charity: where X utters s and s is true only during some part of the time that it takes X to make his utterance there is a certain conversational presumption that one should take, as the time of evaluation for s precisely that part of the duration of the utterance during which s *is* true. It might be noted in this context that the notion 'time of utterance' is not as clear-cut as some discussions in which it plays a role may seem to imply. There is a problem of interpretation, in particular, with written utterances, such as one finds e.g. in letters. What is the time of utterance of a particular sentence s occuring somewhere in a letter of some length? is it the period that it took to write the entire letter; or just the time it took to write that particular sentence; or some period of intermediate duration, during which the passage was written in which the sentence figures? Usually these questions do not matter as the sentence does not fluctuate in truthvalue even over the longest period that could qualify as the time of utterance. There is, however, enough indeterminacy here to provide the principle of charity with scope for action. Evidently these problems are intimately connected with those, usually discussed in a different context, relating to the so-called "specious present". Indeed, once one accepts that "the present" can only be a certain *event* (which is recognized as being simultaneous with its recognition), the interminacy of its extent becomes, if not inescapable, at least very plausible.

Precisely how we solve the problem of characterizing truth and falsehood relative to $\mathfrak{M}$ should not affect the characterization of validity (of sentences) and logical consequence. These notions seem to require unversal quantification both over possible completions and over the instants of the models derived from these completions. Thus we could define

(9) s is valid in $\mathfrak{M}$ iff s is true at each instant of each instant model derived from a completion of $\mathfrak{M}$.

(a similar definition could be given for the notion of logical consequence. I shall ignore this second concept in the discussions that will follow. It will in all cases relevant here be connected with the concept of validity in the manner familiar from classical logic - i.e. via the Deduction Theorem). This is admittedly not a very natural notion. We are not so much interested in what has to come out true at all times of a particular model - which after all represents a particular way in which the world is or could have been - but rather in what has to come out true always in *any* such model. Earlier work in tense logic has taught us, however, that there is an important source of complications here. As soon as the object language contains even the simplest tense operators (such as Prior's P ('it was the case that') or F ('it will be the case that')) the notions of validity and logical consequence become sensitive to the properties of temporal order. For instance the set of valid sentences will be different depending on whether time is supposed to be discrete or dense; and even depending on whether we take its ordering to be like that of the rationals or that of the reals. This leads naturally to the relativization of these notions to classes of models - where it is natural to think of any such class as containing precisely those models whose time structures possess all those features of temporal ordering which, in the context of the characterization of validity in question, are thought of as logically necessary aspects of time.

In the work on tense logic to which I referred above it was possible, and natural, to specify various intended classes of models by simply listing the conditions (such as density, having no beginning and no end point, etc.) that the temporal orderings of their models must satisfy. But here that doesn't seem to be the most satisfactory way to proceed. For the members of such a model class should now be event models and not instant models. They should, moreover, be allowed to be partially indeterminate. It is by no means clear that the conditions with which tense logic has been traditionally concerned can be captured as conditions on such partial event models. How e.g. is one to state a condition on $\mathfrak{M}$ which is necessary and sufficient therefor that each instant model derived from a completion of $\mathfrak{M}$ is discretely, or densely, ordered?[14] I should add immediately at this point that even where such conditions on partially determinate event models can be found, it is not necessarily desirable that we insure that the instant models have a feature such as density or discreteness by imposing this condition on the underlying event models. For, as I already observed in section 2, our reason for believing time to have a certain property need not be our conviction that the structure of actual events (under some scheme of individuation) yields such an instant structure under Wiener's construction; but that it may be of a quite different sort. If it were, then it would be wrong to try and insure that the instant structures have the property by imposing on the underlying event models a condition that warrants it. Rather the intended instant models should result from embedding the instant models that Wiener's procedure produces from the completions of the event models in the class $\mathcal{K}$ into certain time structures that do have the desired property. We shall not pursue. this issue any further here, however.

Besides conditions which concern the structure of temporal order there are others which also could, and in some cases should, be imposed on the models of our present theory. These conditions relate to the temporal connections between the predicates Q, BQ, CQ, FQ. The first set consists of different versions of what I shall call the *Principle of Separation* (PS). According to PS any two events e_1, e_2 which are such that $e_1 \propto e_2$, and during which two contradictory conditions obtain, are separated by an event which is a change from the first condition to the second. We shall formulate the principle as applying to atomic sentences only:

PS1: Suppose $e_1 \propto e_2$, $\mathcal{A}_2(e_1)(Q(a)) = F$, and $\mathcal{A}_2(e_2)(Q(a)) = P$; then there is an e such that $e_1 \propto e \propto e_2$ and $\mathcal{A}_2(e)(Q(a)) = B$.

PS2: Suppose $e_1 \propto e_2$, $\mathcal{A}_2(e_1)(Q(a)) = P$ and $\mathcal{A}_2(e_2)(Q(a)) = F$; then there is an e such that $e_1 \propto e \propto e_2$ and $\mathcal{A}_2(e)(Q(a)) = C$.

14 Some such problems have commanded a certain interest from those who were involved with the reduction of instants to events which I have here primarily associated with Wiener. See e.g. Russell's [8] , where a number of such questions are discussed. (Of course Russell's problem was different from the present one in as much as he only considered temporally determinate event structures.).

Equally plausible, though, as we shall see, reducible to these conditions in the presence of certain other principles to be stated below, are the following.

PS3: Suppose $e_1 \propto e_2$, $\mathcal{A}_2(e_1)(Q(a)) = C$, $\mathcal{A}(e_2)(Q(a)) = P$; then there is an e such that $e_1 \propto e \propto e_2$ and $\mathcal{A}_2(e)(Q(a)) = B$.

PS4: Suppose $e_1 \propto e_2$, $\mathcal{A}_2(e_1)(Q(a)) = F$ and $_2(e_2)(Q(a)) = C$; then there is an e such that $e_1 \propto e \propto e_2$ and $\mathcal{A}_2(e)(Q(a)) = B$.

PS5: Suppose $e_1 \propto e_2$, $\mathcal{A}_2(e_1)(Q(a)) = B$ and $\mathcal{A}_2(e_2)(Q(a)) = F$; then there is an e such that $e_1 \propto e \propto e_2$ and $\mathcal{A}_2(e)(Q(a)) = C$.

PS6: Suppose $e_1 \propto e_2$, $\mathcal{A}_2(e_1)(Q(a)) = P$ and $\mathcal{A}_2(e_2)(Q(a)) = B$; then there is an e such that $e_1 \propto e \propto e_2$ and $\mathcal{A}_2(e)(Q(a)) = C$.

PS7: Suppose $e_1 \propto e_2$, and $\mathcal{A}_2(e_1)(Q(a)) = \mathcal{A}_2(e_2)(Q(a)) = C$; then either there is an e such that $\mathcal{A}_2(e)(Q(a)) = C$ and which temporally includes both e_1 and e_2,[15] or there is an e such that $e_1 \propto e \propto e_2$ and $\mathcal{A}_2(e)(Q(a)) = B$.

PS8: Suppose $e_1 \propto e_2$ and $\mathcal{A}_2(e_1)(Q(a)) = \mathcal{A}_2(e)(Q(a)) = B$; then either there is an e such that $\mathcal{A}_2(e)(Q(a)) = B$ and which temporally includes both e_1 and e_2,[15] or there is an e such that $e_1 \propto e \propto e_2$ and $\mathcal{A}_2(e)(Q(a)) = C$.

The Principle of Separation asserts the existence of various times of change. As such it directly contradicts the theory we proposed in section 1, but it is in keeping with the central idea of the theory we are developing in this section.

It is natural to impose PS1 - PS8 not only on the event models $\mathfrak{M}$ themselves, but also on their completions. It is readily seen, however, that the fact that $\mathfrak{M}$ itself satisfies these conditions does not warrant that they also hold in its completions. We must therefore stipulate separately that only such completions as agree with PS1 - PS8 are permitted.

For this, and other, similar, reasons it is desirable to introduce, besides the semantic units already discussed, in addition what I shall here call *interpretations*: pairs consisting of an event model $\mathfrak{M}$ and a set $\mathcal{S}$ of completions of $\mathfrak{M}$. It is then possible to define classes of interpretations by imposing restrictions both on $\mathfrak{M}$ and $\mathcal{S}$. Let me state, once more, the

---

15 By e *temporally includes* e' I understand the condition (2) on p.150: $(\forall e_1)(e_1 \propto e \rightarrow e_1, \propto e') \,\&\, (\forall e_1)(e \propto e_1 \rightarrow e' \propto e_1) \,\&\, (\forall e_1)(e_1 O e' \rightarrow e_1 O e)$.

operators S, U, S', U'. Let the resulting languages be L_2 (without B,C,F) and L'_2 (with these operators). I have already remarked that the set of valid sentences of such a language is sensitive to the structure of time. Thus we are bound to find different sets of valid sentences relative to model classes $\mathcal{K}'$ which derive, in the way just indicated, from different classes $\mathcal{K}$ of linear orderings. Moreover the conditions PS1 - PS8 are now reflected by the validity of certain sentences. Thus PS1 warrants the validity relative to any class $\mathcal{K}'$ consisting exclusively of interpretations that satisfy it of

(S1) $Q(a) \mathbin{\&} P((FQ)(a) \mathbin{\&} \phi) \rightarrow P((BQ)(a) \mathbin{\&} P((FQ)(a) \mathbin{\&} \phi))$

and

(S2) $(FQ)(a) \mathbin{\&} P(Q(a) \mathbin{\&} \phi) \rightarrow P((CQ)(a) \mathbin{\&} P(Q(a) \mathbin{\&} \phi))$

Similarly PS3-PS6 warrant the validity of, respectively,

(S3) $Q(a) \mathbin{\&} P((CQ)(a) \mathbin{\&} \phi) \rightarrow P((BQ)(a) \mathbin{\&} P((CQ)(a) \mathbin{\&} \phi))$,

(S4) $(CQ)(a) \mathbin{\&} P((FQ)(a) \mathbin{\&} \phi) \rightarrow P((BQ)(a) \mathbin{\&} P((FQ)(a) \mathbin{\&} \phi))$,

(S5) $(FQ)(a) \mathbin{\&} P((BQ)(a) \mathbin{\&} \phi) \rightarrow P((CQ)(a) \mathbin{\&} P((BQ)(a) \mathbin{\&} \phi))$, and

(S6) $(BQ)(a) \mathbin{\&} P(Q(a) \mathbin{\&} \phi) \rightarrow P((CQ)(a) \mathbin{\&} P(Q(a) \mathbin{\&} \phi))$.

PS7 moreover warrants the validity of

(S7) $(CQ)(a) \mathbin{\&} P(\psi \mathbin{\&} P((CQ)(a) \mathbin{\&} \phi)) \rightarrow [P((CQ)(a) \mathbin{\&} \psi \mathbin{\&} P((CQ)(a) \mathbin{\&} \phi)) \vee P((BQ)(a) \mathbin{\&} P((CQ)(a) \mathbin{\&} \phi))]$

and PS8 that of

(S8) $(BQ)(a) \mathbin{\&} P(\psi \mathbin{\&} P((BQ)(a) \mathbin{\&} \phi)) \rightarrow [P((BQ)(a) \mathbin{\&} \psi \mathbin{\&} P((BQ)(a) \mathbin{\&} \phi)) \vee P((CQ)(a) \mathbin{\&} P((BQ)(a) \mathbin{\&} \phi))]$

For the weaker language L'_1, which contains besides B,C,F and the devices of first order logic only the Priorean operators P and F (not to be confused with our predicate operator!) these schemata summarize, I believe, the effect of imposing PS1 - PS8 on interpretations. To be precise, I conjecture that they are *complete* in the following sense: Let $\mathcal{K}$ be a class of linear orderings and let $\mathcal{K}''$ be the class of interpretations determined by it as indicated in (2). Let $\mathfrak{A}$ be an axiomatization for the language L_1 (which lacks B,C,F but contains the Priorean operators), which is complete relative to $\mathcal{K}$. Then we obtain a complete axiomatization for L'_1 relative to $\mathcal{K}''$ by adding to $\mathfrak{A}$ the schemata (11) i - iv) and S1 - S8. S1 - S8 do not suffice to generate, in the presence of appropriate axioms for L_2, all sentences of L'_2 for whose

validity PS1 - PS8 are responsible. It is not difficult, however, to find axioms which are complete for L'_2 in the same sense in which I conjectured S1 - S8 to be complete for L'_1. Indeed it follows from the functional completeness of S,U,S',U' that we can express with the help of these operators a three place connective PS, such that PS(ϕ ,ψ ,χ) is true at t iff either ϕ is false at t or else for all t' before t at which ψ holds there is a t" between t' and t so that χ holds at t". The principle PS1 warrants the validity of each formula of the form PS(Q(a),(FQ)(a),(BQ)(a)). To each of PS2 - PS6 corresponds another set of instances of PS(ϕ,ψ,χ). In very much the same way we can also find formulae corresponding to PS7 and PS8. The resulting 8 schemata will be complete for L'_2 in the sense explained.
Before concluding this section I must mention a few other principles which interpretations might be asked to obey.

The first of these is the principle, already mentioned in section 2, that all changes from one state to its opposite are what I termed 'ideally instantaneous'. Let us call this principle the P(rinciple of) I(nstantaneity of) C(hange). We already saw how the condition that e is instantaneous can be expressed, vz. as (1) on p 145. What we need here is a condition which makes explicit that the event in question is indeed a change. We thus get:

PIC $(\exists s)(\mathcal{A}_2(e)(s) = B \vee \mathcal{A}_2(e)(s) = C) \rightarrow \neg(\exists e',e'')(eOe' \,\&\, eOe'' \,\&\, e' \propto e'')$

PIC also has its effect on validity. Thus if all completions of interpretations in $\mathcal{K}''$ satisfy PIC then the following formulae of L'_2 are all valid relative to $\mathcal{K}'$:

(13) $S(\phi,(BQ)(a)) \rightarrow S(\phi,\perp)$,
$S(\phi,(CQ)(a)) \rightarrow S(\phi,\perp)$,
$U(\phi,(BQ)(a)) \rightarrow R(\phi,\perp)$,
$U(\phi,(CQ)(a)) \rightarrow U(\phi,\perp)$. [15]

Moreover in the presence of PIC PS7 and PS8 can be simplified. Thus PS7 reduces to the condition that any two successive ceasings of Q(a) are separated by at least one becoming of Q(a), and PS8 to the condition that any two becomings are separated by a ceasing. These new versions of PS7 and PS8 warrant the validity of corresponding strengthenings of S7 and S8, so that PIC also has an effect on validity for L'_1. I suspect that to get complete axiomatizations for classes $\mathcal{K}'$ which are restricted to interpretations whose completions satisfy PIC the only additional axioms required are, in the case of L'_2, the formulae (13), and that in the case of L'_1 the strengthened

15 By $\perp$ I understand any explicit contradiction of ordinary propositional calculus, e.g. the formula Q(a) & $\neg$Q(a).

versions of S7 and S8 will suffice.

The Principle of Separation has a counterpart in what I shall call (by lack of a more poignant name) the *Second Principle of Separation*, or PS_2. This principle says that between two opposite changes - e.g. a change from not-s to s followed by a change from s to not-s there is an intervening time when the end-state of the first, which is also the initial state of the second change, obtains. Formally:

$PS_2 1$: Suppose $e_1 \propto e_2$, $\mathcal{A}_2(e_1)(Q(a)) = B$ and $\mathcal{A}_2(e_2)(Q(a)) = C$; then there is an e such that $e_1 \propto e \propto e_2$ & $\mathcal{A}_2(a)(Q(a)) = P$.

Suppose $e_1 \propto e_2$, $\mathcal{A}_2(e_1)(Q(a)) = C$ and $\mathcal{A}_2(e_2(Q(a)) = B$; then there is an e such that $e_1 \propto e \propto e_2$ and $\mathcal{A}_2(e)(Q(a)) = F$.

It appears that PS_2 is, if possible, even less controversial than PS. For how could an event e be a change from not-s to s if it was not preceded by a time where s failed and followed by a time where s holds? And how could e be followed by a reverse change if a situation in which s obtained had not been reached first? Both these questions invite negative answers. In fact the negative answer to the first embodies yet another uncontroversial principle, to which I shall refer as the *P*(*rinciple of*) *C*(*ompleted*) *C*(*hange*):

PCC1 Suppose $\mathcal{A}_2(e)(Q(a)) = B$; then there is an e_1 such that $e_1 \propto e$ and $\mathcal{A}_2(e_1)(Q(a)) = F$ and there is an e_2 such that $e \propto e_2$ and $\mathcal{A}_2(e_2)(Q(a)) = P$.

PCC2 Suppose $\mathcal{A}_2(e)(Q(a)) = C$; then there is an e_1 such that $e_1 \propto e$ and $\mathcal{A}_2(e_1)(Q(a)) = P$ and there is an e_2 such that $e \propto e_2$ and $\mathcal{A}_2(e_2)(Q(a)) = F$.

Note that there is a certain asymmetry between PS and PS_2. PS expresses what might be called a 'realist', as opposed to a 'virtual', conception of change. When a time where s fails is followed by a time where s holds then there has been a change. This much is uncontroversial. But the claim that in such cases there always was a time *at* which the change occurred may be regarded as controversial, and this is what PS asserts. PS_2 on the other hand seems to do no more than unpack what is implicit in the characterization of an event as a change from one condition to its opposite - and the same goes of course for PCC.

Axioms corresponding to these principles can readily be formulated, both for L_1' and for L_2'. But, after all that has already been said on this subject in connection with other principles, it would only be tedious to do yet another exercise of this sort.

Although PS_2 and PCC derive from the same consideration they are formally independent from each other (an event model could satisfy either of them without satisfying the other). There is, however, a stronger formal principle which implies both of them and which itself would also seem to follow from the informal consideration that justifies their adoption. I shall refer to this principle - my baptismal powers are deteriorating - as the *Principle of Discrete Change*, or PDC. PDC asserts that a change is *flanked* by its initial and end conditions with nothing intervening between it and the events flanking it. The principle admits of two formulations, a strong and a weak version. I shall give the weak version first:

PDC_w1 Suppose $\mathcal{A}_2(e)(Q(a)) = B$; then there is an e_1 such that $e_1 \propto e$, $\mathcal{A}_2(e_1)(Q(a)) = F$ and $\neg(\exists e_3)(e_1 \propto e_3 \propto e)$; and there is an e_2 such that $e \propto e_2$, $\mathcal{A}_2(e_2)(Q(a)) = P$ and $\neg(\exists e_3)(e \propto e_3 \propto e_2)$.

Suppose $\mathcal{A}_2(e)(Q(a) = C$; then there is an e_1 such that $e_1 \propto e$, $\mathcal{A}_2(e_1)(Q(a) = P$ and $\neg(\exists e_3)(e_1 \propto e_3 \propto e)$; and there is an e_2 such that $e \propto e_2$, $\mathcal{A}_2(e_2)(Q(a)) = F$ and $\neg(\exists e_3)(e \propto e_3 \propto e_2)$.

This weak version says that there is no event intervening between the change and either of the events which flank it. This does not exclude the possibility, however, that in an instant model derived from an event model in which PDC_w1 and PDC_w2 hold an instant intervenes between that or those of the change and those at which the initial or end condition obtains. To exclude this eventuality we must strengthen the first half of the conclusion of PDC_w1 to:

there is an e_1 such that $e_1 \propto e$, $\mathcal{A}_2(e_1)(Q(a)) = F$ and $\neg(\exists e_3,e_4)(e_3 \propto e \;\&\; e_1 \propto e_4 \;\&\; e_3 O e_4)$

The second half of the conclusion, as well as the two halves of the conclusion of PDC_w2, need to be strengthened analogously. I shall refer to the resulting principles PDC_s1 and PDC_s2.

Although PDC may seem to be as well motivated as its consequences PS_2 and PCC we shall find in the next section that there are reasons that might induce us to give it up, or at least modify it, but which do not affect the validity of the two weaker principles.

IV. Consider once more an event e which is a change from a situation e_1 where Q(a) fails to a situation e_2 where Q(a) holds. During e BQ is true of a; during e_2 it cannot be true of a, for Q is true of a there, and what is the case cannot possibly at the same time be going to be the case. There is thus a transition from a situation where (BQ)(a) holds to one where it fails. It might seem therefore that the very same considerations which led us, in the previous section, to the view that e_1 and e_2 must be separated by an event such as e, now force us to postulate an event e' to mediate the transition from (BQ)(a) to its oposite.

The very same question can be raised from the vantage point of the predicate Q. Having stipulated that e is an event during which ¬Q(a) is true should not the very same reasoning that led to the assumption of e now also commit us to the existence of an event e' which mediates between it and the following e_2 ?

It would be a mistake to suppose that our commitment to such events as e necessarily carries with it the further commitment to such events as e'. One could defend the first commitment while rejecting the second by drawing a distinction between predicates such as Q on the one and predicates such as BQ and CQ on the other hand: the former would express conditions which cannot be fulfilled at one time after having not been fulfilled at some earlier time without an intervening change; the second type of predicate, however, would itself already express that a change w.r.t. Q is happening, so that there is no reason for supposing that an event where such a predicate is satisfied must be separated from one where it is not by yet another event; similarly the contrast between sentences Q(a) and ¬Q(a) is not the sort of contrast that requires for any pair of successive events where these respective sentences are true, an additional event that marks the transition; for, on the interpretation that we gave to ¬ in section 3, ¬Q(a) is compatible with the occurrence if a transition to, or from, a situation where Q(a) holds. Although we might justify our acceptance of the first and rejection of the second commitment, and thus of the theory presented in the preceding section, by an appeal to this distinction, I have given as yet no compelling grounds why we should not make the second commitment as well as the first. In this section I wish to explore, albeit in a preliminary fashion, what is implied by making it.

It is one question whether we must assume that there is always an event which marks the transition from (BQ)(a) to Q(a). It is another question how a change from (BQ)(a) to Q(a) might be expressed. Clearly we cannot use the sentence (BQ)(a) for this purpose for e marks a transition *from* the condition expressed by that sentence to some other condition (expressed by Q(a)). We could introduce an entirely new predicate operator (say B^2) - so that $(B^2Q)(a)$ would express the condition in question. But it seems also possible to represent the required predicate as CBQ, using the operator C which is already present in the languages L_i' , and that is the option I shall pursue. Similarly we shall express the condition obtaining at an event which marks the transition from a time where we have (FQ(a) to a time where (BQ)(a) holds as (BBQ)(a); and BCQ and CCQ shall serve to express transitions to and from events described by (CQ)(a). The considerations that earlier led us to stipulate that (BQ)(a) is false when Q(a) or (FQ)(a) are true now yield the similar conclusion that (BBQ)(a) is not true when either Q(a) or (BQ)(a) is true. It appears intuitively clear, moreover, that (BBQ)(a) must be false in all these cases, and the same should hold for (BCQ)(a) and (CCQ)(a). It ought to be equally plausible that the new predicates BBQ, CBQ, BCQ, and CCQ are mutually exclusive. We thus have a set of eight pairwise incompatible predicates.

Having thus extended the application range of the operators B and C we cannot avoid the question whether a similar extention would not be apposite for F also. Previously we distinguished between events where Q(a) fails and events which separate these from events where Q(a) holds, and we agreed to express this difference as that between the truth of (FQ)(a) and that of both $\neg$Q(a) and $\neg$(FQ)(a).

In the same way we might now, taking our departure from the predicate BQ instead of Q, distinguish between events such as e_2, at which (BQ)(a) definitely fails, and events such as e' which mark the transition either from such a condition to one where (BQ)(a) holds or to such a condition from one where (BQ)(a) holds; and stipulate that at the former (FBQ)(a) holds while at the latter both $\neg$(BQ)(a) and $\neg$(FBQ)(a) are true. By this principle (FBQ)(a) will be implied by Q(a) as well as by (FQ)(a), while on the other hand (FBQ)(a) will be false when either (BBQ)(a) or (CBQ)(a) is true. (CQ)(a) expresses the occurrence of a transition from one condition, expressed by Q(a), to another, expressed by (FQ)(a), both of which have already been recognized as implying (FBQ)(a). I shall take this as a sufficient reason for stipulating that (CQ)(a) itself also implies (FBQ)(a). This is in fact a particular case of a general principle, to which I shall presently have to refer again. I have called it the P(rinciple of) C(ontinuity); the

The exact formulation of it will be given below.

By symmetric considerations one is led to the stipulation that FCQ is implied by Q, FQ, and BQ, and excluded by BCQ and CCQ.

In section 3 I stipulated that (FQ)(a) is, like Q(a), false when either (BQ)(a) or (CQ)(a) is true. The motivation for that stipulation was that these latter sentences express the occurrence of a transition to or from the condition described by (FQ)(a). We must now consider whether (FQ)(a) is true or false when a satisfies any of the new predicates that we have allowed ourselves to form on the basis of Q.

First let us look at the predicate BBQ. This predicate expresses a transition from FQ to BQ, and our reason for stipulating that (FQ)(a) is false when (BQ)(a) is true now leads us to postulate that (FQ)(a) must be false when (BBC(a) is true. Similarly, FQ must be excluded by CCQ.
CBQ and BCQ on the other hand express transitions between the condition expressed by Q and those expressed by BQ and CQ. We agreed earlier that each of these excludes FQ in the weak sense that (FQ)(a) is false when Q(a), (BQ)(a) or (CQ)(a) is true. Yet another application of the Principle of Continuity leads to the conclusion that CBQ and BCQ also exclude FQ.

It would not be natural, now that we have extended the syntax of the languages $L_i^!$ this far, to prohibit the application of B, C and F to predicates of the form FQ. In defining the extensions of the resulting predicates I shall be guided by the idea, implicit in all that has been said about F so far, that this operator forms out of any predicate its complementary predicate: If a satisfies Π then a fails to satisfy FΠ , if a satisfies FΠ then a fails to satisfy Π, and it is not determined whether a satisfies Π iff it is not determined whether a satisfies FΠ . Thus FFQ is equivalent to Q; and by the same token BFQ is equivalent to CQ and CFQ is equivalent to BQ.

Once we have made the second commitment - that there are event such as e', which marks the transition from (BQ)(a) to Q(a) - and moreover accepted that the conditions which distinguish such events are expressible - e.g. by the sentence (CBQ)(a) - it is hard to see why we should stop here: should there not be supposed to be an event e", for instance, which marks the transition from the condition expressed by (CBQ)(a), which prevails at e', to the incompatible condition Q(a); and could we not express what distinguishes this event from the two between which it mediates as (CCBQ)(a)? Again there are no conclusive reasons why we should take this additional step. Still, it is worth while to reflect on the implications of making this third commitment also, as well as all those which successively arise out of each other in the same way as the second arose from the first and the third from the second.

This infinite sequence of commitments leads to languages L_i'' in which the operators B, C and F can be iterated arbitrarily often. That is, L_i'' has the same vocabulary as L_i', but its syntax differs in that it has a wider set of predicates, recursively defined by:

(14) i) each predicate of L is a predicate of L_i'' ;

ii) if Π is a predicate of L_i'' then so are B Π , C Π and F Π .

Let me state, in conjunction with this definition, that of a related notion which will be useful in what is to come:

(15) a) the set S(Π) of predicates of L" *positively generated by* the predicate Π is defined recursively by:

i) $\Pi \in S(\Pi)$

ii) if $Q \in S(\Pi)$ then B Π and C Π ε S(Π).

The event models for a language L_i'' must also provide information about the extensions of the new predicates which do not occur in L_i' . In view of this it turns out to be convenient to change the format of the functions $\mathcal{A}_1$ and $\mathcal{A}_2$. We make $\mathcal{A}_1$ into a function that maps events onto atomic sentences (thus instead of $\mathcal{A}_1(e_1) = \langle Q(a),B\rangle$ we shall now have: $\mathcal{A}_1(e) = (BQ)(a))$ and $\mathcal{A}_2$ into a function from events to functions from sets of atomic sentences of L_i'' to {0,1} - so that instead of $\mathcal{A}_2(e)(Q(a)) = P$ we now have $\mathcal{A}_2(e)(Q(a)) = 1$, while the cases in which a belongs to the extension of BQ, CQ or FQ are conveyed thereby that $\mathcal{A}_2(e)$ assigns the value 1 to, respectively, (BQ)(a), (CQ)(a) or (FQ)(a).

Besides the consistency requirements already noted in section 3 (see p. 150) we must now state constraints which insure matching between the assignments to atomic sentences that involve the same individuals and structurally related predicates. Some of these constraints have already been stated informally, and it is straightforward to translate these into conditions on models for L_i'' . Thus we get, among others,

(16) i) if $\mathcal{A}_2(e)$ assigns 1 to one of Π(a), (FΠ)(a), (BΠ)(a) and (CΠ)(a) then it assigns 0 to the remaining three.

ii) if $\mathcal{A}_2(e)$ assigns 1 to Π(a), (C Π)(a) or (F Π)(a) then it assigns 1 to (FB Π)(a).

iii) if $\mathcal{A}_2(e)$ assigns 1 to one of Π(a), (B Π)(a) and (C Π)(a) then it assigns 1 to (FC Π)(a).

iv) $\mathcal{A}_2(e)(FF\Pi)(a) = {1 \atop 0}$ iff $\mathcal{A}_2(e)(\Pi(a)) = {1 \atop 0}$, $\mathcal{A}_2(e)(BF\Pi)(a)) = {1 \atop 0}$ iff $\mathcal{A}_2(e)((C\Pi)(a)) = {1 \atop 0}$ and $\mathcal{A}_2(e)((CF\Pi)(a)) = {1 \atop 0}$ iff $\mathcal{A}_2(e)((B\Pi)(a)) = {1 \atop 0}$.

The other connections between the extensions of structurally related predicates which I stated earlier can similarly be converted into conditions on event models. There still remain a number of such connections, however, which our discussion has so far left untouched. It appears that all of these can be settled by appeal to two principles. The first of these is the Incompatibility Principle, in the form in which we used it in section 3, according to which a predicate Π which expresses the occurrence of a transition between two conditions expressed by respectively the predicates R and Σ, excludes each of these. The second is the Principle of Continuity, which we shall now state in its intended generality:

PC Suppose that the predicate R expresses the occurrence of a transition between conditions expressed by Σ and T.

a) Suppose that Π is a predicate which is implied by both Σ and T. Then Π is implied by R.

b) Suppose Π is excluded by both Σ and T. Then Π is excluded by R.

c) Suppose Π implies both $F\Sigma$ and FT. Then Π implies FR.

It is easy to verify that if R and Σ are distinct predicates which are positively generated by the same predicate Π then they are mutually exclusive. Moreover, if R is positively generated by $C\Pi$ then Π as well as all predicates positively generated by $B\Pi$ imply FR, and if R is positively generated by $B\Pi$ then Π as well as all predicates positively generated by $C\Pi$ imply FR. Clearly the condition which we could impose upon completions of models for L_i' - that in the derived instant model at each instant at least one of Q, BQ, CQ, and FQ must be true - must now be dropped. We can, however, replace it by a similar condition, to the effect that

(17) for each instant t and each primitive predicate Q at least one sentence $\Pi(a)$ is true at t, where Π is either FQ, or else a predicate that is positively generated by Q.

All these conditions are also easily translated into restrictions on the function $\mathcal{A}_2$, but there seems to be no point in carrying this out explicitly. We have already committed ourselves informally to a generalization of the Principle of Separation, to apply to arbitrary predicates of L_i''. In view of the way I have chosen to treat the operator F where it applies to composite predicates this generalized principle is correctly expressed by the conditions PS'1 - PS'8, which we obtain through replacing in PS1 - PS8 the letter Q, which ranges over predicates of L, by the letter Π whose intended range is the set of all predicates of the L_i''.

It is clear that once we impose PS'1 - PS'8 we cannot also impose the Principle of Discrete Change. For any event characterized by, say, $(B\Pi)(a)$ will be separated from any other that is characterized by $\Pi(a)$ or by $(F\Pi)(a)$ by infinitely many other events. PS'1 - PS'8 are compatible, however, with a weaker condition than PDC, to the effect, roughly, that a change is flanked by two events, characterizable as the presence of the initial and the end condition of the change, respectively, between which there are just the events which PS'1 - PS'8 require. Unfortunately, although the intuitions behind this condition are quite straightforward, it is nonetheless a little difficult to state.

Let $\mathfrak{M}$ be an event model, e_1 and e_2 members of E such that $e_1 \propto e_2$, $\mathcal{A}_2(e_1)(F\Pi)(a)) = 1$ and $\mathcal{A}_2(e_2)(\Pi(a)) = 1$. We say that a subset E' is *generated by* Π *and* e_1 *and* e_2 iff

> i) $e_1, e_2 \in E'$; ii) if $e, e' \in E'$, $e \propto e'$ and for some predicate R generated by Π $\mathcal{A}_2(e)((F\,R)(a)) = 1$ and $\mathcal{A}_2(e)(R(a)) = 1$ then there is an $e'' \in E'$ such that $e \propto e'' \propto e'$ and $\mathcal{A}_2(e'')((B\,R)(a)) = 1$; iii) if $e, e' \in E'$, $e \propto e'$ and for some predicate R generated by Π $\mathcal{A}_2(e)(R(a)) = 1$ and $\mathcal{A}_2(e')((F\,R)(a)) = 1$ then there is an $e'' \in E'$ such that $e \propto e'' \propto e'$ and $\mathcal{A}_2(e'')((C\,R)(a)) = 1$; iv) if $E'' \subseteq E'$ and E" satisfies i) - iii) then E" = E'.

In a similar way we define what it is for E' to be generated by Π and a from e_1 and e_2 for the case where $\mathcal{A}_2(e_1)(\Pi(a)) = 1$ and $\mathcal{A}_2(e_2)(F\Pi)(a)) = 1$. We can now formulate the condition referred to above as follows:

(18) i) Suppose $\mathcal{A}_2(e)(B\Pi)(a)) = 1$; then there are e_1 and e_2 such that a) $e_1 \propto e \propto e_2$; b) $\mathcal{A}_2(e_1)((F\Pi)(a)) = 1$; c) $\mathcal{A}_2(e_2)(\Pi(a)) = 1$; d) there is a unique event set E' generated by Π and a from e_1 and e_2; and e) for each set E" of pairwise overlapping events such that for each $e' \in E''$ either $e_1 \propto e'$ or $e' \propto e_2$ there is an $e' \in E'$ such that e' overlaps each of the events in E".

ii) Suppose that $\mathcal{A}_2(e)(C\Pi)(a)) = q$. Then there are e_1 and e_2 such that $e_1 \propto e \propto e_2$, $\mathcal{A}_2(e_1)(\Pi(a)) = q$, $\mathcal{A}_2(e_2)((F\Pi)(a)) = 1$, and the remaining conditions are as i) a and i) c.

The meaning of (18) should be clear: the events e_1 and e_2 which "flank" a change are so to speak *only* separated from e by those events which PS'1 - PS'8 demand - this in the precise sense that each of the predicates generated by Π is represented by at most one event between e_1 and e_2, and that these representatives moreover fill the space between e_1 and e_2 in the sense that any instant of the instant structure derived from the model in question which lies between e_1 and e_2 [16] will contain one such representative.

Clearly it is a consequence of (18) that each change lies in a dense interval, i.e. let e be a change, e_1 and e_2 the events flanking it whose existence is ensured by (18), and t_1 and t_2 two elements of the corresponding instant structure containing e_1 and e_2 respectively. Then the interval (t_1, t_2) of the instant structure will be dense.

Consider moreover the following condition:

(19) Let E' and E" be two sets of pairwise overlapping events, containing events e' and e" such that $e' \propto e''$. Then either there is a change (i.e. an event which verifies at least one sentence of the form $(B\Pi)(a)$ or the form $(C\Pi)(a)$) which overlaps all the events in E', or there is such an event overlapping all events in E", or else there is a change e and members e' and e" of E' and E" respectively such that $e' \propto e \propto e''$.

(19) implies that if t_1 and t_2 are distinct instants then either at least one of them is an instant of change - in the sense that at least one change is occurring at it - or else they are separated by a change. It thus embodies the idea that only change can produce temporal succession.

If the model satisfies (19) as well as (18) then the derived instant structure is dense.

The languages L_i'' suggest many questions of the sort we considered in section 3 in connection with their counterparts L_i'. I shall not try to formulate, let alone answer, these here, however. In fact, before occupying himself seriously with the metamathematics of these systems, one might wish for more evidence that they are conceptually adequate. There have been a number of places in the development of the syntax and semantics of the L_i'' at which we might have taken out a different option from the one I chose, and some of these should be considered with some care, if only to strengthen our conviction that the choices which I did make were the right ones. This task I shall also leave largely unaccomplished. In the next section I will address myself to what I think is the most serious difficulty with the present

---

16 t *lies between* e_1 and e_2 iff $(\exists e_3 \varepsilon t)(e_1 \propto e_3)$ and $(\exists e_4 \varepsilon t)(e_4 \propto e_2)$

languages. I wish to conclude this section by drawing attention to a peculiarity by which the reader may already have been struck, and perhaps mystified.

This peculiarity is related to the particular way in which I have dealt with the operator F. As I have set up the semantics FBQ is implied by Q while on the other hand FQ is excluded by BQ. F thus fails to obey one of the typical properties of negation, vz. a form of the Law of Contraposition, according to which, if Π implies the negation of R then R implies the negation of Π. The failure of this principle is indicative of an asymmetry between certain predicates, e.g. between Q and BQ, which is not only central to the ideas that led to the formulation of the L_i' but which has at least been partly retained by the L_i'' : Whereas BQ is treated as indicating change from FQ to Q - which entails that it excludes FQ as well as Q - Q is not treated as marking a transition to or from a situation where BQ is satisfied; rather, BQ is introduced as a predicate which can never be satisfied when Q is: Q(a) is taken as conclusive evidence that a fails to satisfy BQ, in the strong sense expressed by (FBQ)(a). A similar asymmetry is found between, e.g., BQ and BBQ: BQ implies FBBQ, but FBQ is incompatible with BBQ; and so on.

In the preceding paragraph I use the phrase 'Bϕ is *introduced* as a predicate ...'. This locution suggests a way of looking at the languages L_i'' which the presentation I gave did not make fully explicit, vz. as the limits of sequences of ever stronger languages: $L_i' \subset L_i^2 \subset \ldots \subset L_i^n \subset \ldots$, where L_i^n contains only those predicates of L_i'' which consist of a primitive predicate preceded by at most n operators. The predicates which contain exactly n operators are those which express true changes in L_i^n. In L_i^{n+1} these same predicates come to function as expressions for conditions which cannot turn into their contraries without an intervening change; these intervening changes are then expressed with the help of the new predicates of L_i^{n+1}.

Thus conceived the transition from L_i^n to L_i^{n+1} might be expected to involve not only a change in status of these predicates but also a modification of the way in which they interact with negation. Thus we might expect that in, say, L_i^2 the predicate BQ should be taken to exclude Q in a stronger sense than that in which it excludes that predicate in L_i' and in which for instance BBQ excludes BQ in L_i^2 . We might contemplate the possibility of using F to express this stronger sense in which BQ excludes Q in L_i^2. This policy would lead in the limit represented by L_i'' to the situation that whenever a predicate Π excludes a predicate R in the weak sense (i.e. for all a if Π(a) then $\neg$R(a)) it also excludes R in the strong sense (i.e. Π implies FR).

Thus F has, where applicable, the same force as $\neg$, and would thereby become redundant. The resulting languages are simpler than the L_i''. They are distinguished from the underlying languages L_i only by the greater diversity of their predicates, and the logical relations which hold between these.

The event models for L_i'' differ from those for L_i' in that the former must, for any two events $e_1 \propto e_2$ characterized by, respectively, (FQ)(a) and Q(a), have infinitely many intervening events, while the latter are required to have only one event that intervenes. By the same token a model for L_i^2 should have at least three intervening events, characterized by (BBQ)(a), (BQ)(a) and (CBQ)(a); and similarly in a model for L_i^n there ought to be at least 2^n-1 such events. Moreover, just as PDC is a natural condition on models for L_i' and its counterpart (18) a natural requirement to be imposed on models for L_i'', so we may ask of the models for L_i^n that each event characterized by a sentence of the form (BQ)(a) lies between two events characterized by (FQ)(a) and Q(a), between which there are precisely the 2^n-1 events which the preceding condition demands. Let us refer to this condition as PDC^n (thus PDC^1 is the condition PDC itself).

Note that if $\mathfrak{M}$ is an event model for L_i^n which satisfies PDC^n than there is an essentially unique[17] minimal extension $\mathfrak{M}'$ of $\mathfrak{M}$ which is a model for L_i^{n+1} and which satisfies PDC^{n+1}.

Up to this point we have considered, in connection with a given event model $\mathfrak{M}$, only instant structures which derive via Wiener's construction from completions of $\mathfrak{M}$. This carries the suggestion that the structural features of time, as determined by the model $\mathfrak{M}$, which in turn reflects a particular conceptual frame, are precisely those which all these instant structures have in common. In section 2 I already remarked, when I mentioned the issue whether time is order complete, that this fails to account for the possibility of attributing to time properties that some, or even all, of the instant structures lack. To do justice to that possibility within the general framework developed so far we must consider other ways in which a given event model $\mathfrak{M}$ can determine instant models.

The connection between the languages L_i' and L_i'' may be exploited to provide one such alternative (albeit not one which will provide us with instant structures that are order complete). For let $\mathfrak{M}$ be an event model for L_i'

---

17 By 'essentially unique' I mean that any two such models are isomorphic, i.e. there is a 1-1 map from the event set of the first to that of the second which preserves $\propto$, O and the functions $\mathcal{A}_1$ and $\mathcal{A}_2$.

which satisfies PDC. Then there is a (n essentially) unique sequence $\mathfrak{M}' = \mathfrak{M}, \mathfrak{M}^2, \ldots, \mathfrak{M}^n, \ldots$ of models $\mathfrak{M}^n$ for L_i^n, each $\mathfrak{M}^{n+1}$ determined in the obvious way by its predecessor $\mathfrak{M}^n$. The limit of this sequence (limit in a sense evident enough to obviate the need of a formal definition) is a model $\mathfrak{M}''$ for L_i'' which satisfies (18) and is complete in the sense of satisfying (17).

In this way we can associate with a given model $\mathfrak{M}$ for L_i' a set of event models $\mathfrak{M}''$ for L_i'' as the set of limits of sequences generated by completions of $\mathfrak{M}'$.[18]

As we introduced them in section 3 completions of event models for L_i' represent hypothetical situations in which all indeterminacies have been resolved. The relation between L_i' and L_i'' suggests another type of hypothetical situation in which not only the predicates of L_i' have become precise but in which moreover new predicates, those of L_i'', have been added, and thereby the means of identifying events enlarged.

The models that represent these situations must contain also the events identifiable only with the help of the extended vocabulary. In as much as the expressive capacities of L_i'' are understood as being already implicit in L_i' it would seem natural to concentrate on such models rather than on the completions with which we worked up to now. Moreover, it seems reasonable to assume that the extra vocabulary of L_i'' is inextricably connected with the existence assumptions which led us to introduce that vocabulary in the first place. Thus the models which are to replace the earlier completions may be assumed to satisfy (18), and in fact to be the unique minimal extentions, in the sense explained above, of these completions. In this way we come to the following alternative notion of an interpretation for L_i' - to avoid confusion I shall refer to such interpretations as interpretations$_2$. Each *interpretation*$_2$ is determined by an interpretation $\langle \mathfrak{M}, \mathcal{S} \rangle$ for L_i'. The interpretation$_2$ which $\langle \mathfrak{M}, \mathcal{S} \rangle$ determines is the pair $\langle \mathfrak{M}, \mathcal{S} \rangle$ where $\mathcal{S}$ is the set of all models for L_i'' which are minimal extensions, in the sense explained above, of members of $\mathcal{S}$. Clearly each class $\mathcal{K}$ of interpretations for L_i' determines a corresponding class $\mathcal{K}'$ of interpretation$_2$'s. Whoever is persuaded that the iterated construction that led to the languages L_i'' and their models is an intrinsic part of the ideas already embodied in L_i' should study metamathematical notions

---

18 Alternatively we could associate with $\mathfrak{M}$ itself, in an analogous fashion, one or more models $\mathfrak{M}''$ for L_i'' and then consider the completions of these. As far as I can see, however, the effect of this way of proceeding is the same, while it is more difficult to state.

such as validity relative to such classes $\mathcal{H}'$ rather than to the classes of interpretations from which they derive.

In general where $\langle \mathcal{M}, \mathcal{S}' \rangle$ is an interpretation$_2$ for L_i' the members of $\mathcal{S}'$ will contain many more events than $\mathcal{M}$ itself. Indeed, if $\langle \mathcal{M}, \mathcal{S}' \rangle$ is derived from the interpretation $\langle \mathcal{M}, \mathcal{S} \rangle$ and the members of $\mathcal{S}$ satisfy PDC then the instant structures derived from the members of $\mathcal{S}'$ are dense, while those derived from the members of $\mathcal{S}$ could even be finite.[19]
There are various alternatives to the semantics we have just developed. Thus we might consider interpretations $\langle \mathcal{M}, \mathcal{S} \rangle$ for L_i' where the members of $\mathcal{S}$ are reductions to L_i' of models for some language L_i^n, with $n > 1$. I doubt that such interpretations will be of more than a purely technical interest, however. On the other hand we might study interpretations $\langle \mathcal{M}, \mathcal{S}' \rangle$ for a language L_i^n where $\mathcal{M}$ is a model for L_i^n and the members of $\mathcal{S}'$ are reductions to L_i^n of models for L_i''. This kind of interpretation promises to have some conceptual significance. A natural language such as English provides a great variety of predicates, some simple and some complex, for expressing change, and some of these express changes in conditions expressed by other predicates. It is a feature typical of natural languages, and one which is often ignored in attempts to analyse aspects of natural discourse through modelling by formal systems, that devices used to form complex expressions out of simpler

19 According to the formal definition (10) of an interpretation for L_i' the members of $\mathcal{S}$ may contain more events than $\mathcal{M}$ itself; in particular, where $\langle \mathcal{M}, \mathcal{S} \rangle$ is the interpretation$_2$ derived from $\langle \mathcal{M}, \mathcal{S} \rangle$ then according to (10) $\langle \mathcal{M}, \mathcal{S}' \rangle$ is itself an interpretation for L_i'. There is a difficulty here to which I might have drawn attention earlier, but which we cannot ignore any longer now. The informal explanation that I gave of a completion might be thought to imply that if $\mathcal{M}'$ is a completion of the event model $\mathcal{M}$ then the event set E' is identical with E. This need not be so, however. For (an observation made by David Lewis) it is conceivable that because of the vagueness in some predicate Q it is undetermined whether the object a actually moved out of the extention of that predicate at one time, and back into it at a subsequent time, or whether it remained inside the extension throughout. This means that on some resolutions of the vagueness of the predicate there will be two changes, from and to Q respectively, while in others there will be none. Thus the event set of the model $\mathcal{M}$ representing such a situation cannot be identical with the event set of each of its completions.

It would seem natural to maintain that in such a case the changes are not part of the event model $\mathcal{M}$ itself, so that among its completions some will contain more events, but none fewer. Although it seems intuitively clear from this account that as a rule the models for L_i'' which extend completions of $\mathcal{M}$ should not themselves qualify as completions of $\mathcal{M}$, I do not quite know how to sharpen the formal characterization of what a completion is in such a way as to exclude them. I must leave this issue here unresolved.

ones cannot be iterated indefinitely. Certainly some, and possibly all of the devices available in English to transform a given predicate into another which expresses change with respect to the condition expressed by the first have this feature. It seems quite possible therefore that for English, as well as many other natural languages, there is an upper bound to the degree of the predicates it contains - where the *degree* of a predicate is characterized as follows:

A predicate which does not express a change with respect to a condition expressible by any other predicate of the language has degree 0; if a predicate expresses change with respect to a condition expressed by a predicate of degree n then it is of *degree* n+1. If a given language has such a finite upper bound n then that language, or at least the fragment whose constructions other than those used to express change can be represented in L_i, can be regarded as corresponding to some part of L_i^n. It is thus the semantics for that language L_i^n, rather than that for $L_i^!$, that would be relevant to the study of change as expressible in that natural language.
Whether natural languages admit of such upper bounds, and what these are, are empirical questions, to which I intend to address myself in Part II.

V. I should have liked to think that the way in which we arrived at the conclusion that instant models are densely ordered reveals the roots of the prevailing conviction that density is an intrinsic feature of the structure of time. But I am afraid that such thinking would be altogether too wishful.

There is a difficulty connected with the languages $L_i^!$ to which I must now, belatedly, draw attention: if i) there are changes intervening between times where a satisfies Q and times where it fails to satisfy Q; ii) there are means of characterizing these changes; and iii) the resulting characterization is incompatible with both the first and the second of these two conditions; then the first two conditions were not really contradictiories, but only incompatible: the very idea of a change between *contradictory* conditions which is describable as distinct from either of these appears to be incoherent.
I believe we can escape this conclusion by distinguishing between such predicates as Q and FQ on the one hand and such predicates as BQ and CQ on the other; Q and FQ, although both incompatible with the predicates BQ and CQ, would nevertheless remain contradictories, in as much as there are no predicates of the type to which Q and FQ themselves belong which are in-

compatible with both of them - that they should be incompatible with certain predicates expressing change with respect to the very conditions they express themselves is not to be taken to disprove that they are contradictory predicates. Note that this amounts to a reassessment of the concept of contradictoriness (the need for which I already announced in the footnote to p.(131)). I shall not go into a defense of this particular suggestion for revision here, but just make the following observation: The *only* way to maintain that Q and FQ are contradictories is to classify BQ and CQ as predicates of a different sort from that to which Q and FQ themselves belong, and to argue that predicates of this new sort are not relevant to the question whether predicates of the other type are contradictories or only incompatible.

Once we introduce such a subclassification of predicates, however, it becomes clear that we must distinguish between what I termed in the preceding section the 'first' and 'second' hypothesis: the hypothesis that there is a time of transition between a time where a satisfies Q and a time where a fails to satisfy Q, is distinct from the hypothesis that there is a time of transition between an event characterized as (BQ)(a) and an event characterized by the 'contradictory' condition (FBQ)(a). And precisely because the two categories of predicates are supposed to play fundamentally distinct roles we cannot assume that the second hypothesis is entailed by the first. Nor is there any entailment between the second hypothesis and the third, according to which there are times of change between e.g. events characterized as (CBQ)(a) and times where the 'contradictory' condition obtains. For to maintain that BQ and FBQ are contadictories we must classify predicates such as CBQ as belonging to yet a third category. Indeed, as I already indicated in section 4, all the hypotheses which thus arise are independent of each other; and the infinite progression which leads from $L_i^{'}$ to $L_i^{''}$ could, for all I can see, be interrupted at any point.

The need for such a progression - and therewith the conclusion that time is dense, which in the argument of the preceding section, depends on it - would be a good deal more compelling if there were grounds for assuming that

(20) each change can be characterized - or is temporally included in an event that can be characterized - with the help of predicates that are recognized from the outset as belonging to precisely the same category as those used to characterize the conditions between which that change is a change.

In case (20) holds the very same hypothesis will keep asking, again and again, for interpolating changes, which, in the limit, must lead to the event models which we have come to associate with the languages L_i''.

Note that in order that the instant structure derived from an event model be densely ordered it is sufficient that there be *some* group of predicates for which i) (20) is true; and ii) changes characterizable with the help of predicates of this group are always going on (i.e. if e_1 and e_2 are any two successive events then either one of them is a change of this sort or else they are separated by such a change.)

I believe there is such a group of predicates, vz. those used to express motion. Motion is change of position; and each individual motion of a particular body a is a change from, or to, a condition where a occupies some particular position p, to, or from a condition where a is not in that position. It is a common assumption, however, that: Whenever a is *not* in position p then it is in some position *other* than p. We assume, that is, that there is a class of predicates, which I shall represent for the sake of the present discussion as $(\lambda x)L(x,p)$, $(\lambda x)L(x,p')$,..., such that whenever an object a fails to satisfy one of them it must satisfy one of the others which is incompatible with the first. It is an equally common assumption that whenever t_1 and t_2 are successive times during which a is in, respectively, positions p_1 and p_2, there must be an intermediate event e which constitutes the transition from the first position to the second; while this event is going on a is neither in position p_1 nor in position p_2, and thus must, by the first assumption, be in a different position. Let e' be the event of a's being in some one such intermediate position p_3. e' must be temporally included in the event e. The very same hypothesis that led us to the assumption that there must be such an event as e now leads to the conclusion that there must be an event separating e' from the time t_1, and also one which separates e' from t_2; and so on.

Note, however, that this time our departure from the restricted notion of change between contradictories is irrevocable. The condition that a is in position p_1 is *not* the contradictory of the condition that a is in position p_2, for both of these are incompatible with the condition that a is in position p_3, and this last condition is of precisely the same kind as the former two. A proper analysis of the concept of motion therefore requires a more general concept of change than we have studied in this first part of the paper - the concept of an event which is a change between conditions which are incompatible but which need not be contradictories (in any sense of the word which distinguishes it from incompatibility tout court). The theory of change of this more general type will be one of the main topics of part II.

I shall conclude this part with a few short remarks, to be seen merely as prognoses of what will be discussed in the second installment of this paper.

1) Clearly the changes between conditions that are not contradictories are not in general instantaneous - in fact the short discussion of motion already bore this out, vz. where I observed that a change in a from position p_1 to position p_2 might temporally include (and of course it *must properly* include) the event of a's being in position p_3. The circumstance that such changes are often protracted causes certain metamathematical difficulties, which do not arise in connection with instantaneous changes. It has been partly out of a desire to avoid these complications that I deferred the treatment of the more general notion of change for what might appear, in the light of the discussion with which I began this section, an inexcusably long time.

2) I have failed to deal with the question of change with respect to conditions expressible with the help of complex formulae, built from atomic predicates with the help of truthfunctional connectives, quantifiers, tense operators, or possibly other devices which this paper has not even mentioned. It is in particular in this context that protracted changes are particularly troublesome and lead (I think inevitably) to a partial valued semantics - the sort of semantics, that is, the mathematical theory of which is still very poorly developed.

3) Even after we have accepted changes between conditions which are not contradictories, and thus protracted changes, as primitives, it remains an important question how they relate to the ideally instantaneous changes first discussed in section 2. At the end of section 1 I suggested that protracted changes might be analyzed as sequences of instantaneous changes. Whether, and how, that suggestion could be made good is a question which I must also defer to part II. Here I only wish to draw attention to a much more modest, but related, claim, vz. the assumption that

(21) Each change e from a given condition p to some incompatible condition temporally includes a change from p to the condition that is the contradictory of p.

In one sense (21) is incontrovertible. Surely when there is a transition from p to some other condition incompatible with p then there is also a transition from p to its contradictory. But it does not automatically follow from this that the event which constitutes the first transition must be accompanied by (or, casu quo, be identical with) an *event* which is the change from p to its contradictory. Indeed, our discussion of motion suggested that this implication might be denied in connection with motion predicates: that all changes are changes between satisfaction of one such precicate and satisfaction of another which,though incompatible with the first, is not its contradictory, while the transitions from satisfaction of such a predicate to failure to satisfy it are in some, or even all, cases 'virtual' - in the sense of there being no time at which the transition occurs. (Whether this is indeed one of the assumptions implicit in our conception of motion is a question to which I shall address myself in part II.). I should say in this connection that it may well be that different predicates require different analyses of the nature of instantaneous and protracted changes with respect to the conditions that these predicates express, and of how these two sorts of changes are connected. E.g. what we wish to say about motion may well differ substantially from what we would regard as the appropriate theory of change of colour, or pitch.

This brings me to the last major issue which must eventually be discussed but about which too I have been almost entirely silent: the means of expressing change that are available in natural languages, in particular English. This is a complex problem, and the little that I may have achieved in this first part of the paper can at best serve as a formal setting in which various aspects of that problem might be more fruitfully discussed than would be possible without it.[20]

---

20 Part of the research for this paper was done while the author was supported by DFG Projekt 'Die Beschreibung der Zeitlogik von Zeitformen und Verbalperiphrasen im Französischen, Portugiesischen und Spanischen', led by Chr. Rohrer, University of Stuttgart.

References

[1] Åqvist,L. & Guenthner,F.,1978, Verb Aspect and events in an improved tense logic. In F.Guenthner & C.Rohrer (eds.): *Studies in Formal Semantics*, North Holland.
[2] Cocchiarella, N., 1966, *Tense and Modal Logic. A study in the topology of temporal reference*. Dissertation, UCLA.
[3] Fine, K., 1975, Vagueness Truth and Logic. *Synthese*, Vol.30.
[4] Hoepelman, J., 1979, *Verb Classification and the Russian Verbal Aspect. A formal treatment*. Dissertation, Stuttgart (to appear).
[5] Kamp, J.A.W., 1975, Two Theories about Adjectives. In E.Keenan (ed.): *Formal Semantics of Natural Languages.*, Cambridge.
[6] Kamp, J.A.W., 1968, *Tense Logic and the Theory of Linear Order*, Dissertation, UCLA.
[7] Kamp, J.A.W., 1979, Instants and Temporal Reference. In: A.von Stechow (ed.), *Semantics from different points of view*, J.Springer-Verlag.
[8] Russell, B., 1956, On order in Time. In B.Russell, *Logic and Knowledge*.
[9] Wiener, N., 1914, A Contribution to the Theory of relative Position, Proceedings of the Cambridge Philosophical Society, Vol. 77.

INTERROGATIVE QUANTIFIERS

Lauri Karttunen and Stanley Peters, University of Texas at Austin

1. *Constituent questions: single and multiple*

We will begin by making a distinction between two kinds of constituent questions: single and multiple. By multiple questions we do not just mean questions like

(1) Which professor recommended which one of his students for the job.

Multiple wh-questions, such as (1), constitute only a subset of multiple questions, at least on the most plausible reading.

(2) a. Which one of his students did each professor recommend for the job?
b. What is the capital of each European country?
c. When were your oldest and youngest child born?

Multiple questions involve relations between individuals. Answers to multiple questions express propositions to the effect that such and such individuals stand in a certain relation to each other. For example, answers to questions (1) and (2)a. are of the form "Professor X recommended student Y for the job". The new information contained in the answer is the pairing of the particular individuals. Single constituent questions differ from multiple questions in that they involve properties rather than relations. They are answered by identifying the individual or individuals that have the particular property. The examples in (3) are unambiguously single questions (given that the antecedent of *his* in (3)a. is *which professor*), in spite of their similarity to the examples in (2).

(3) a. Which professor recommended each one of his students for the job?
b. What is the capital of the United States?
c. When were you and your wife married?

Answers to questions (3)a., for example, are of the form "Professor X recommended each student of his for the job."

We think it is useful to consider the distinction between single and multiple questions in this general way without focusing narrowly on questions of type (1) - - multiple wh-questions. Because of the fact that much of the work on questions has been done on English, constructions of this type have come to be regarded as paradigmatic examples of multiple questions when in fact they are not. There are many languages where there are no multiple wh-questions. Modern Irish (see McCloskey 1978), for example, is a language where the syntax of constituent questions effectively prevents two interrogative phrases from ever occurring in the same simple question. The fact that English has multiple wh-questions and that Irish doesn't is nevertheless not an indication that Irish is impoverished in its expressive capacity compared to English. Irish does in fact have multiple questions, but only of the sort illustrated under (2). We will show in this paper that the same can be said about Finnish.

Our present treatment of constituent questions builds on our earlier work on interrogatives. One important improvement is that we can now account for multiple questions where a non-interrogative quantifier (e.g. "each person") takes wide scope over a wh-phrase (e.g. "which one of his students") in the same clause. The examples in (2) are precisely of this sort. In order to show what is difficult about them we will briefly outline the treatment of constituent questions proposed in our earlier papers (Karttunen 1977, Peters 1979).

1.1 Single constituent questions

The diagram in (4) represents a simplified syntactic structure for the question "which professor recommends John". For the time being, let us think of it as a subordinate question, waiting to be embedded to some larger construction, as in "I am not sure which professor recommends John." For the sake of expository convenience we employ in this section an extended version of the "Cooper syntax" of Cooper & Parsons 1976. We have added to it rules for generating constituent questions (Q-phrases) and interrogative noun phrases (QNPs). As the diagram indicates, we assume here that the underlying syntactic structure for the question contains a place-holding dummy pro-NP that is deleted in the presence of the interrogative noun phrase.

(4)

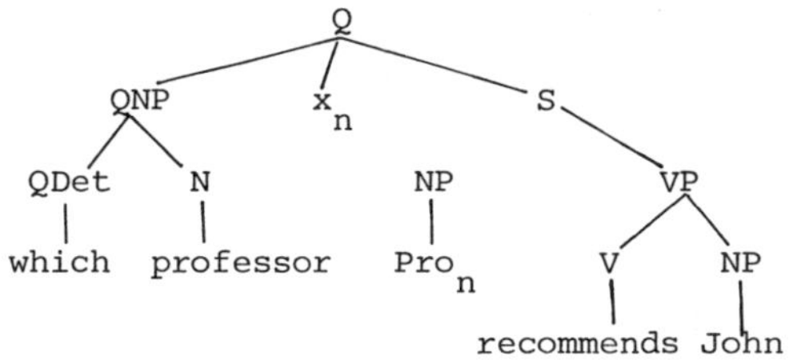

Thus the surface structure of "which professor recommends John" is obtained from (4) by deleting the NP "Pro_n" and the variable "x_n" whose role we will explain shortly.

As shown by Cooper & Parsons, a structure such as (4) can be semantically interpreted in a very straight-forward way. Each lexical item in (4) has a meaning represented by its translation to an interpreted formal language (Montague's intensional logic). The meanings of complex phrases are obtained from the menaings of their constituents by a rule appropriate for the particular construction. For example, the translation for the S-node in (4) is formed by combining the translations of the NP- and VP-nodes in accordance with rule (5).

(5) S' = NP'(VP')

The symbols S', NP', and VP' stand for the translations of the corresponding phrases. In this case, NP' is identical to the translation of Pro_n, and VP' is the result of applying rule (6) to the translations of the constituents of VP.

(6) VP' = V'(^NP')

Here V' is the same as the translation of "recommend", say, recommend', and NP' is the same as John'. Consequently, the translation for the S-node in (4) - - is the formula in (7)a. (7)a. in turn is logically equivalent in Montague's system to (7)b. (We are using Pro_n' and John' as abbreviations for $\lambda P\ P(x_n)$ and $\lambda P\ P(j)$, respectively.) Thus either formula in (7) can serve as a translation for the S-node in (4).

(7) a. Pro_n'(recommend'(^John')

b. recommend*'(x_n,j)

Up to this point the interpretation of (4) is unproblematic, the interesting problems arise in connection with the Q- and the QNP-nodes. In accordance with our earlier work on interrogatives, we assume that Q is the category for all subordinate questions. Besides constituent questions, this category also contains yes/no questions and disjunctive questions. Phrases of category Q denote sets of propositions. For example, "which professor recommends John" denotes the set of true propositions to the effect that professor X recommends John. In other words, we think of the meaning of a subordinate question as something that picks out, in each situation, the set of propositions which in that situation would constitute a true and complete answer to the question. It follows from this that the menaing of the Q-node in (4) can be represented by formula (8).

(8) $\lambda p\ \exists x\ [\text{professor}'(x)\ \&\ {}^{\vee}p\ \&\ p = {}^{\wedge}\text{recommend}^{*}{}'(x,j)]$

In Montague's intensional logic (8) designates precisely the set of true propositions expressed by sentences of the form "professor X recommends John". In a situation where professor Fudge, and no other professor, recommends John, (8) designates the unit set containing the proposition that Fudge recommends John.

In order to associate the Q-node in (4) with an appropriate translation that expresses the desired meaning, we must, first of all, decide on what the meaning to assign to the interrogative noun phrase "which professor", and then find a rule that combines the translations of the QNP- and the S-nodes in the right way to produce some expression that is equivalent to (8) above.

As for the QNP, it turns out that it is convenient to let the interrogative quantifier "which professor" have the same meaning as its existential counterpart "some professor". Thus we represent the meaning of the QNP in (4) as in (9).

(9) $\text{QNP}' = \lambda P\ \exists x\ [\text{professor}'(x)\ \&\ P(x)]$

This in turn can be obtained from the translations of the interrogative determiner and the noun "professor" in the same way as the translation of "some man" is obtained from the translations of its parts.

The semantic rule for obtaining the meaning for a Q-phrase in a structure such as (4) can now be formulated as shown in (10).

(10) $Q' = \lambda p\ \text{QNP}'(\lambda x_n\ [{}^{\vee}p\ \&\ p = {}^{\wedge}S'])$

The effect of rule (10) is to assign to the Q-phrase in (4) precisely the meaning we indicate in (8), namely a function from world-time pairs to sets of propositions of the form "X recommends John", where X designates some professor. (Note that the role of the variable x_n in (4) is just to indicate that this is the variable that gets bound when the translation of the Q-hprase is constructed by "quantifying in" the interrogative noun phrase.).

This is a somewhat simplified presentation. We would have to complicate the above rules a bit if we were to take into consideration other types of interrogative phrases -- for example, phrases like "where", "how long", and "whose student" -- but what we have said above is already enough for the purpose of introducing the problems that arise in connection with multiple questions.

1.2 Multiple wh-questions

We do not have any difficulties with multiple wh-questions, that is, with questions like "Which professor recommends which candidate". Within Cooper syntax, this could be derived from an underlying structure such as (11).

(11)

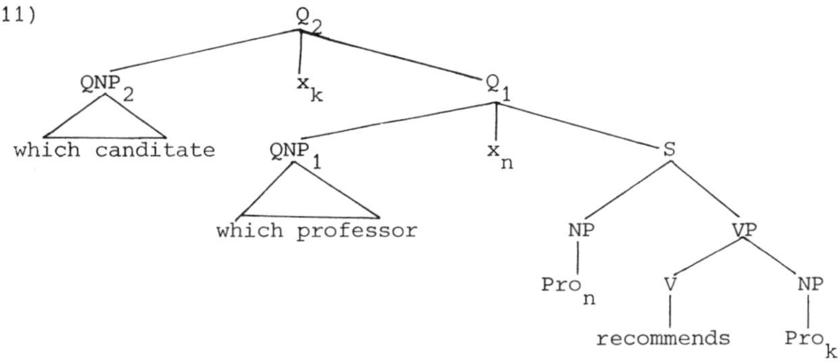

Here Q_1 is the underlying structure for the single wh-question "which professor recommends Pro_k". It differs from the Q in (4) only in that the S-node contains a second unbound pro-NP. In order to derive the surface structure of "which professor recommends which candidate" we assume here a second rule of interrogative quantification that deletes a place-holding pro-NP. That is, Pro_k in Q_1 is replaced by the QNP "which candidate". The corresponding semantic rule is given in (12).

(12) $Q_2' = \lambda p\ QNP_2'(\lambda x_k\ Q_1'(p))$

By applying (12) to the translations of the QNP "which candidate" and the Q-phrase "which professor recommends Pro_k" in (11), we obtain a translation which is equivalent to formula (13).

(13) $\lambda p\ \exists y\ \exists x\ [candidate'(y)\ \&\ professor'(x)\ \&$
$^{\vee}p\ \&\ p = {}^{\wedge}recommend^{*}{}'(x,y)]$

And formula (13) is of course just the expression that we want: it designates the set of true propositions of the type "X recommends Y", where X is a professor and Y is a candidate.

The basic idea in this treatment of constituent questions is that interrogative phrases are quantified in. The syntactic framework that we have employed above was selected in part because it makes this aspect of the analysis very perspicuous. The first incoming wh-phrase makes a single constituent question out of a declarative clause and multiple wh-questions can be formed by quantifying in more wh-phrases. There is, however, no syntactic motivation for the kind of movement rule we had to postulate for multiple wh-questions. We will show later how it can be eliminated by adopting a different translation method.

2.1 *Quantifying into questions: first attempt*

Here we come to a point where our present views differ from what was said previously by one of the authors. It was claimed in Karttunen 1977 that the approach to constituent questions which we have just outlined entails that interrogative quantifiers should always have wider scope than any ordinary quantifier in a subordinate question, unless the latter also has scope over the structure which the subordinate question is embedded in. This was supposed to follow from the observation that an ordinary quantifier could gain scope over an interrogative one only by being quantified in, and from the belief that, for semantic reasons, ordinary noun phrases could not be quantified into questions. Let us illustrate the problem by considering example (14).

(14) Which professor recommends each candidate?

Question (14) clearly has two readings which differ with respect to the scope of "each candidate". On the narrow scope reading of "each candidate", (14) is a single question; it calls for answers such as "Fudge recommends each candidate". This reading is simply the outcome of deriving the sentence in the manner shown in (15).

(15)

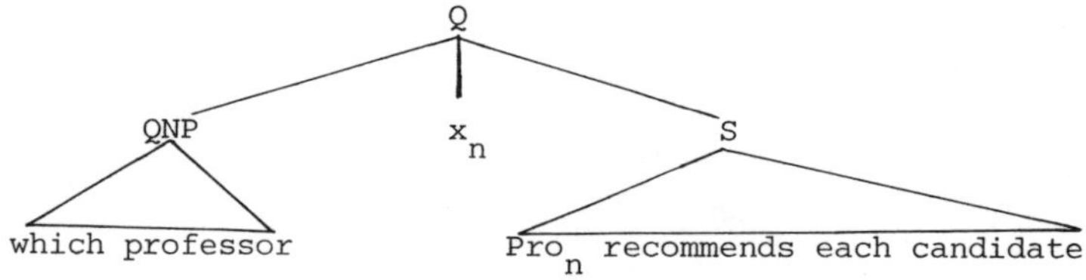

Under this analysis the QNP-phrase "which professor" is quantified into a declarative sentence. In accordance with rule (10) above it has wide scope with respect to the NP "each candidate". The problem is that (14) also appears to have a multiple reading which requires answers such as "Fudge recommends John, Dodge recommends Mary, and Hodge recommends Alvin." In some way this reading involves making "each candidate" have wide scope with respect to the wh-phrase. Karttunen 1977 briefly considered the possibility of deriving this sentence in the manner shown in (16) but came to the conclusion that there is no way to make that account to work semantically.

(16)

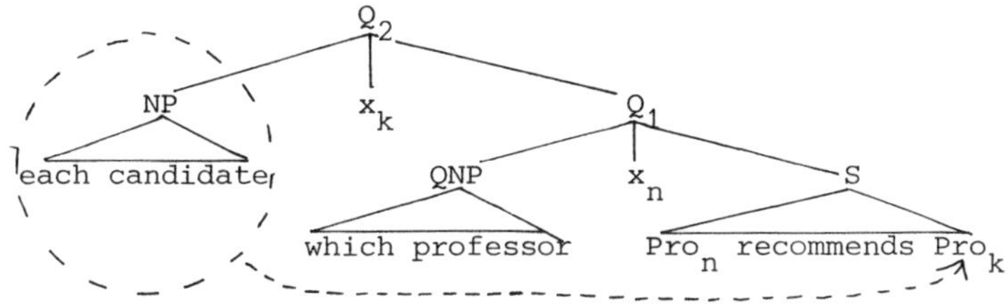

In (16) the NP "each candidate" will be quantified in to the subordinate question "which professor recommends Pro_k". Because this is a new syntactic rule, one also needs a new semantic rule that specifies how the translation of Q_2 is obtained from the translations of the nodes it dominates. The most obvious candidate for that rule is (17), but unfortunately the menaing it yields cannot be right.

(17) $Q_2' = \lambda p\ NP'(\lambda x_k\ Q_1'(p))$

The resulting translation for Q_2 in (16) is equivalent to (18).

(18) $\lambda p\ \forall y\ [candidate'(y) \rightarrow \exists x\ [professor'(x)\ \&$
$^{\vee}p\ \&\ p = {}^{\wedge}recommend^{*}{}'(x,y)]$

This cannot be the right menaing for the reason that the set of propositions it denotes must be empty in any situation where there are two candidates. -- Note that each proposition in the set designated by (18) is required to be identical to the proposition that X recommends Y for all assignments of a candidate to Y and some value of X which picks out a professor. But as the values of X and Y in "X recommends Y" change so does the proposition expressed by the sentence. No proposition is identical to the proposition that X recommends Y on more than one pair of values for X and Y. This is the reason why (18) designates the empty set whenever there exist at least two candidates. It thus fails to pick out the set of propositions which intuitively would seem to be the right ones, namely, all and only the true propositions to the effect that "X recommends Y" where X designates a professor and Y is a candidate.

2.2 *Quantifying into questions: second attempt*

For lack of any better alternative, Karttunen 1977 adopted the view that the multiple question reading of (14) could be accounted for by a version of the 'performative analysis'. One could, in particular, take the view that direct questions are to be regarded as semantically equivalent to declarative structures of the form "I want you to tell me Q", where Q is the corresponding subordinate question. That is, one would derive the wide scope reading in the manner shown in (19).

(19)

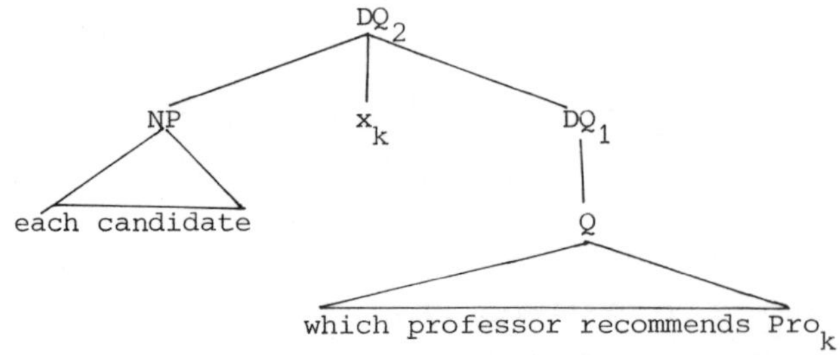

Here DQ is the category of direct questions. The members of this category would be semantically of the same type as declarative sentences; therefore, one could easily quantify in an ordinary noun phrases like "each candidate" and get an appropriate meaning. In this case the resulting translation would be equivalent to something like "For each candidate x, I want you to tell me which professor recommends x".

This solution sounds reasonable enough for direct quenstions but certain kinds of subordinate questions still remain a problem. Let us consider example (20).

(20) I wonder which professor recommends each candidate.

According to Karttunen 1977, the sentence has two available readings; a narrow scope reading for "each candidate", which implicates that all the candidates are being recommended by the same professor, and a wide scope reading, which we paraphrase here as (21).

(21) For each candidate x, I wonder which professor recommends x.

Under the view that ordinary noun phrases cannot be quantified into questions, there is no third intermediate reading, that is, no reading where "each candidate" has wide scope with respect to "which professor" but narrow scope with respect to the verb "wonder" in the higher sentence. We now think that this prediction is incorrect.

There does seem to be such a reading, although the case perhaps isn't as clear as we would like it to be. In order to argue for the existence of this reading, we will have to present a rudimentary analysis of what it is to wonder. Let us say that wondering is a compound awareness of ignorance and thirst for knowledge. Although there undoubtedly is more to wondering than just this, for our present purposes it will suffice to focus on the entailment in (22).

(22) x wonders Q ⊨ x doesn't know Q and x wants to know Q

Given (22), we can immediately derive (23) from (21).

(23) For each candidate x, I don't know which professor recommends x.

If the wide scope reading of "each candidate" in (20) is analyzed as (21) in accordance with Karttunen 1977, then the following sentence should be a contradiction.

(24) I know which professor recommended one of the candidates but I still wonder which professor recommends each candidate.

Contrary to the prediction, (24) does not strike us as being necessarily false. Knowing who recommended one of the candidates does not seem to preclude the possibility of wanting to know, for all candidates, who recommended them. If our intuition is correct, it shows that there really does exist the kind of intermediate reading for (20) where "each candidate" has wider scope than (which professor" but doesn't have scope over "wonder". Another example of the same sort is given in (25).

(25) It is unknown what each candidate has promised, although I happen to know what one of them has promised.

Again, there is no contradiction in (25) on the multiple question reading of "what each candidate has promised" as there would have to be under the assumption that the first clause of (25) has the meaning of "For each candidate x, it is unknown what x has promised". On the intended reading "each candidate" must be in the scope of "unknown" but not in the scope of "what". In the latter case, "what each candidate has promised" would be a single question presupposing that all the candidates have made the same promises.

2.3 *Quantifying into questions: third attempt*

This matter is of considerable interst to us, because we have belatedly discovered that Karttunen 1977 was simply wrong in thinking that one could not formulate a semantic rule for quantifying ordinary noun phrases into questions in the manner suggested in (16). In fact there is a very simple rule that gives the right meaning, namely, rule (26).

(26) $Q_2' = \lambda p \neg NP'(\lambda x_k \neg Q_1'(p))$

This rule assigns to the Q_2-phrase "which professor recommends each candidate", derived from (16), a translation which is equivalent to what we obtained for "which professor recommends which candidate" in (11). The two sentences both come to have the menaing expressed by formula (13) above. -- Note that our interpretation of "which candidate" is the same as that of "some candidate". Considering that "each candidate" is the dual of these, it will be evident from a comparison of rules (12) and (26) that the results are equivalent. Thus our analysis assigns the same meaning to (27)a. and (27)b. on the wide scope interpretation of "each candidate".

(27) a. Which professor recommends which candidate?
b. Which professor recommends each candidate?

We think that this is the right result because there clearly is no situation where the true answers to (27)a. would not constitute the true answers to (27)b. and vice versa. There is of course a residual difference in meaning between the two questions -- this will be discussed later on -- but we are convinced that it is a matter of conventional implicature, not part of the truthconditional aspect of meaning. (See Karttunen & Peters 1976 for an analysis of conventional implicatures of questions). Thus we maintain that (13) does adequately express the truthconditional part of the meaning for both of the questions in (27).

It is important to note that rule (26) gives sensible results only for universally quantified NPs, that is, for NPs with "each", "every", "all", or "the" as their determiner, and for conjoined NPs.
Structures similar to (16) but containing an existentially quantified NP, such as "some candidate", "a candidate" etc., or a disjoined NP, e.g. "John or Mary", are peculiar semantically. For the reasons discussed above in connection with (18), the translations associated with the examples in (28) on the putative wide scope reading of the non-interrogative NP designate nothing but the empty set whenever there are at least two candidates.

(28) a. Which professor recommends some candidate?

b. Which professor recommends many candidates?

We think that the inability of our solution to assign a sensible wide scope, i.e., multiple question, reading to these cases is a point in its favor. As far as we can tell, the kind of argument that we gave above for the existence of an intermediate reading in connection with example (24) only works for NPs like "each candidate", "Mary and Alvin", "all the candidates", etc. Examples like (29) simply do not seem to have any reading in ordinary English under which "some candidate" would have wide scope with respect to "which professor" and yet be in the scope of "wonder".

(29) I wonder which professor recommended some candidate.

Let us now summarize the discussion up to this point. We started by trying to clarify the notion of multiple question and reviewed our earlier treatment of constituent questions. One important improvement in our current description is that we can show that (27)a. and (27)b. are both multiple questions. Questions with universally quantified ordinary NPs that take scope over an initial wh-phrase are semantically in the same category as questions with multiple wh-phrases. This is an interesting result in the light of the fact that there are many languages that do not have anything corresponding to multiple wh-questions. Provided that they have questions of type (27)b. -- as Irish, for example, does -- such languages are not in any important way deprived of expressive capacity because they lack multiple wh-questions. In the next section of the paper we will examine multiple constituent questions in Finnish where the facts are quite interesting in the light of what we have found out about English.

3. *Multiple questions in Finnish*

At first glance it appears that multiple constituent questions in Finnish are very similar to multiple wh-questions in English. For example, "Tell me which boy likes which girl" translates into Finnish with two occurrences of the interrogative determiner "Mikä", as shown in (30).

```
(30) Kerro minulle Mikä poika pitää mistäkin tytöstä.
     tell  me      which boy   likes which    girl
                   (Nom.)            (Elat.)
```

Another way of translating the same English sentence into Finnish is (31).

```
(31) Kerro minulle mistä työstä mikin poika pitää.
     tell  me      which girl   which boy   likes
                   (Elat.)      (Nom.)
```

The most obvious difference between (30) and (31) is word order. In Finnish either one of the two interrogative noun phrases can be preposed, whereas in English there is no such option in this case. Another interesting feature of these examples is that in both cases the second one of the two interrogative determiners carries the suffix "-kin". -- Compare "mikä" in (30) vs. "mikin" (< "mikä"+kin") in (31); "mistä" in (31) vs. "mistäkin" in (30). -- It is a general characteristic of Finnish multiple constituent questions that all interrogative phrases except the initial one carry the "-kin" suffix.[1] Since the order of the major constituents is not as rigidly determined as in English, there are at least three ways of saying in Finnish "Tell me which speaker had what to say on which subject", as shown in (32).

```
(32) a. Kerro minulle millä puhujalla oli mitäkin sanottavaa
        tell  me      which speaker  had what    to say
                      (Adess.)           (Part.)

        mistäkin asiasta.
        on which subject.
        (Elat.)

     b. Kerro minulle mitä   milläkin puhujalla oli sanottavaa
                      what   which    speaker   had to say
                      (Part.)(Adess.)

        mistäkin asiasta.
        on which subject.
        (Elat.)

     c. Kerro minulle mistä    asiasta milläkin puhujalla oli
        tell  me      on which subject which    speaker   had
                      (Elat.)          (Adess.)

        mitäkin sanottavaa.
        what    to say
        (Part.)
```

---

1 One can also find multiple questions in Finnish where all interrogative phrases are unsuffixed, for example, "Kuka sanoi kenelle mitä?" -- "Who said what to whom?" As far as we know, interrogatives of this type only occur as independent utterances, not as embedded questions. Furthermore, they seem to be echo-questions, that is, utterances whose function in conversation typically is to ask for a repetition of a previous utterance by the addressee. We will not try to analyze this class of interrogatives in the present paper.

3.1 Overview of Finnish interrogatives

In the above examples we have consistently translated the word "mikä" in all of its various case forms as "which" or "what". We will continue to do so throughout this discussion, although it is somewhat misleading. The non-initial, suffixed occurrences of "mikä" could just as well be translated as "each". But before we proceed further with this, let us first get an overview of the Finnish interrogative system. We will use the term 'primary interrogative' for the unsuffixed interrogative determiners, noun phrases, and adverbs that occur in the initial position and 'secondary interrogative' for the suffixed forms that occur non-initially. Chart (33) contains the primary and secondary forms of some common interrogatives.

(33)

Category	Primary interrogatives	Secondary interrogatives	Gloss
NP/Det	kuka	kukin	who/which (human)
NP/Det	mikä	mikin	what/which
Loc. Adv.	missä	missäkin	where
Time Adv.	milloin	milloinkin	when
Manner & Det. Adv.	miten	mitenkin	how

Secondary interrogatives occur only in conjunction with primary interrogatives; they cannot form constituent questions by themselves, nor do they occur in declarative sentences. Thus (34)b. and (34)c. are ungrammatical in Finnish.

(34) a. Mitä hän sanoi? What did he say?

b. *Mitäkin hän sanoi?

c. *Hän sanoi mitäkin.

The secondary form of the word for "which/who" -- "kukin" -- is an apparent exception to the principle just stated. This is because interrogative twin meaning "each/each person". Unlike its determiner or as a noun phrase. Thus we can have pairs such as (35)a. -- a declarative sentence -- and (35)b. -- a constituent question.

(35) a. Kukin tarvitsee apua. Each person needs help.
each one needs help

b. Kuka tarvitsee apua? Who needs help?
who needs help

As one might expect there is an ambiguity whenever "kukin" appears in a constituent question because it could be interpreted either as a secondary interrogative "which/who" or as a non-interrogative "each/each person". An example of this is given in (36).

(36) a. Kerro minulle mistä aiheesta kukin opiskelija puhuu.
tell me about which topic each/which student talks.

b. Tell me what is the topic that each student talks about.

c. Tell me which student talks about which topic.

On the non-interrogative reading of "kukin", translated as (36)b., the sentence is a single constituent question and it implicates that all students are talking about the same topic. Under the other interpretation, (36)c., the question calls for multiple answers pairing students and topics.

The existence of these two variants of "kukin" is not a mere accicent. It is actually quite revealing of the nature of secondary interrogatives. In this respect, however, "kunkin" is an exceptional case. None of the other secondary interrogatives in (33) have non-interrogative twins; they are strictly limited to occurring in constituent questions in positions following an initial primary interrogative.[2]

---

2 All the secondary interrogatives have another use as well, which lies outside the scope of the present paper. Together with the primary question words, they form constructions which resemble free relatives in English; for example:

(i) Tuuli puhaltaa milloin mistäkin suunnasta.
wind blows when from which direction

"The wind blows (unpredictably) from different directions at different times."

Because constructions such as "milloin mistäkin suunnasta" in (i) have no parallel in English, they are difficult to translate, although their meaning is quite straight-forward. Essentially (i) says that the wind can blow from any direction. The best way to interpret such constructions may be to regard them as a universally quantified version of a type of coordination which is also found in English. Compare (i) with

(ii) Tuuli puhaltaa milloin lännestä, milloin idästä.
wind blows when from west when from east

"The wind blows sometimes from the west, sometimes from the east."

Although we have not worked out a complete analysis of this interesting class of constructions, it appears that our treatment "milloin", "mikä", etc. in interrogatives is compatible with their semantics here. (Note especially that "mistäkin suunnasta" in (i) has the meaning of a universal quantifier.).

3.2 "Mikin/kukin" vs. "mikään/kukaan"

Having now eliminated the ambiguity of "kukin" as possible source of confusion, let us now take another look at the chart in (33). The secondary interrogatives are all derived -- with minor morphological changes -- from their primary counterparts by adding the suffix "-kin". This same suffix occurs in Finnish in numerous other functions, and typically there is an alternation between "-kin" and "-kaan". This alternation is conditioned very much like the "some/any"-alternation in English with "-kaan" playing the role of the negative polarity "any".

But contrary to what one might expect, the regular "-kin/-kaan" alternation does not affect secondary interrogatives. Thus we have, for example, the form "mihinkin" -- "to which" -- in (37) in spite of the presence of negation in the same clause.

(37) Kerro minulle kuka ei vastannut mihinkin kysymykseen.
tell me who (Nom.) not answered to which (Ill.) question

Tell me who didn't answer which question.

Although the "-kin" suffix in secondary interrogatives is not subject to the "-kin/-kaan" alternation, the corresponding "-kaan" forms actually do exist. They are not interrogative, however, but ordinary determiners, noun phrases, and adverbs, similar to "any", "anyone", "anywhere", etc. in English. This is illustrated in the following examples.

(38) a. Kukaan ei tullut. — No one came.
anyone not came

b. Mikään ei toimi. — Nothing works.
anything not works

c. Juna ei pysähtynyt missään. — The train stopped nowhere.
train not stopped anywhere

d. Hän ei palannut milloinkaan. — She never returned.
she not returned ever

There are two alternative analyses for the "-kaan" forms in (38). One could regard them as expressions whose meaning involves existential quantification. In that case "kukaan", "mikään", etc. would have to be assigned narrow scope with respect to negation in (38). Alternatively, one could view them as involving universal quantification, in which case they would take wide scope over negation. The glosses and translations given in (38) were designed to bring out this uncertainty. Most of the arguments given by W. Ladusaw 1979 for treating the negative polarity "anyone" (as opposed to the free choice "anyone") as an existential quantifier apply with equal force to the Finnish "kukaan", "mikään", etc. For example, the yes/no question "Tuliko kukaan?" (Did anyone come?") has to be answered affirmatively if even one person came. If "kukaan" were analyzed as having the menaing of a universal quantifier, one would expect the question to mean "Did everybody come?" Because of this and other similar arguments, we think that the "kaan" forms in (38) involve existential quantification. This is not a satifying conclusion, because the non-interrogative "kukin", which one would expect to be the affirmative counterpart of the negative existential "kukaan", has the meaning of a universal quantifier, as we pointed out earlier in connection with (35)a. In any case, to supplement the chart in (33), we now must add a list of negative, non-interrogative quantifiers, determiners, and adverbs.

(39)

Category	Non-interrogative negative existentials	Gloss
NP/Det	kukaan	(not) anyone/any
NP/Det	mikään	(not) anything/any
Loc. Adv.	missään	(not) anywhere
Time Adv.	milloinkaan	(not) ever
Mannaer & Time Adv.	mitenkään	(not) in any way

Since there is no restriction against using the above forms in constituent questions, we have (40) as well as (37). --Note that (40) is a single question, not a multiple one like (30).

(40) a. Kerro minulle kuka ei vastannut mihinkään kysymykseen.
tell me who not answwered any question
(Nom.) (Illat.)

b. Tell me who did not answer any question.

It is puzzling that the distribution of pairs like "kukin" vs. "kukaan", "mikin" vs. "mikään" is far from being complementary, which is what one would expect given the regular "-kin/-kan" alternation in Finnish. We conclude somewhat reluctantly that "kukin" (both the interrogative and its non-interrogative twin) and "mikin" (interrogative) have the meaning of a universal quantifier, while "kukaan" (non-interrogative) and "mikään" (non-interrogative) are existential quantifiers. Besides the meaning of the non-interrogative "kukin", this conclusion rests on the observations discussed in the next section.

4. *Comparison of multiple questions in Finnish and English*

Up to this point we have worked from the assumption that multiple questions in Finnish are similar to multiple wh-questions in English. We have also assumed that the distinction between primary and secondary interrogatives is of no particular semantic import. To some extent this is certainly justified. In many cases there is no perceptible change in meaning even ef we interchange the primary and secondary interrogatives in a question. For example, the two Finnish sentences in (41) both seem to mean exactly what their English translations mean although the interrogative phrases have been switched around.

(41) a. Kuka on naimisissa kenenkin kanssa?

Kuka	on	naimisissa	kenenkin	kanssa?
who	is	married	who	with
(Nom.)			(Gen.)	

b. Kenen kanssa kukin on naimisissa?

Kenen	kanssa	kukin	on	naimisissa?
who	with	who	is	married
(Nom.)		(Gen.)		

c. Who is married to whom?

4.1 *"Which" vs. "each"*

However, this is not always the case. It is possible to construct pairs of examples of this kind which contrast in certain situations. We will illustrate this first by using English examples. Let us imagine the situation pictured in (42).

(42)

	A	E	I
customers	B	F	J
	C	G	K
	D *<clerk 1*	H *<clerk 2*	L *<clerk 3*
		check-out counters	

The diagram pictures a scene in a supermarket. The letters represent twelve customers who are lined up in front of three check-out counters waiting for their turn to be served by one of the three clerks. Suppose that we want to find out which customers are presently being served and by which clerks. In other words, we want to ask a question to which the correct answer is (43).

(43) Clerk 1 is serving customer D, clerk 2 is serving customer H, and clerk 3 is serving customer L.

Let us consider how we would elicit the desired information in English. There are of course many ways to do this, but let us compare the three questions in (44).

(44) a. Which customer is each clerk now serving?
b. Which clerk is now serving each customer?
c. Which clerk is now serving which customer?

In order to get anywhere with the first two questions, we have to rely on our addressee to be able to figure out from the context that he is supposed to interpret them as multiple questions. In other words, "each clerk" in (44)a. and "each customer" in (44)b. are intended to have wide scope. If the former, for example, were to have narrow scope, then we would be trying to get answers like "All clerks are now serving customer X". That is not a possible answer to the question we are asking here. Eliminating the unwanted single question readings in (44)a. and (44)b. leaves us with distinctions that are rather subtle and therefore difficult to verify. However, nearly all of our informants agree with us -- sometimes after prolonged reflection - that the first and the last questions in (44) are unobjectionable but that there is something wrong about using (44)b. in this particular situation. (44)b. implicates something false, namely, that all customers are being served by some clerk, when in fact only three of them are. On the other hand, it is the case that all clerks are serving some customer or other. This state of affairs is consistent with (44)c. and (44)a. on the wide scope reading of "each clerk". (44)a. and (44)c. are better questions to ask than (44)b. in this situation because there is no conflict between the facts and the conventional implicatures (presuppositions) of the question. -- Note that (44)c. is the most neutral of these three questions in the sense that it does not even implicate that all the clerks are busy. One indication of this is that the corresponding passivized question, (45), is just as good in this situation as (44)c. is.

(45) Which customer is now being served by which clerk?

In order to avoid misunderstandings, let us emphasize that we do not think that the multiple question readings in (44) express different questions -- whatever is the correct answer to one will also be the correct answer to the others. The difference in meaning lies entirely in what they conventionally implicate about the situation they pertain to.

4.2 *"Mikä" vs. "mikin"*

In the light of these English examples, it is interesting to find out that the two ways of asking the same question in Finnish, (46)a. and (47)a., contrast exactly in the manner of (44)a. and (44)b.

(46) a. Kerro munulle mitä asiakasta mikin myyjä nyt palvelee.
tell me which customer which clerk now serves
(Part.) (Nom.)

b. Tell me which customer each clerk is now serving.

(47) a. Kerro minulle mikä myyjä nyt palvelee mitäkin asiakasta.
tell me which clerk now serves which customer
(Nom.) (Part.)

b. Tell me which clerk is now serving each customer.

The secondary interrogatives "mikin" in (46)a. and "mitäkin" in (47)a. give rise to the same kind of implicatures as the word "each" in the corresponding English translations. Thus (46)a. implicates that every clerk is serving some customer and (47)a. implicates that every customer is being served. Since the latter implicature is false, only the former question would fit the situation that we have described.[3]

4.3 *"Each" vs. "mikin"*

While there is this similarity between "each" and the secondary interrogative "mikin", there is also an important difference, and for this reason, we continue glossing "mikin" as "which". "Each" forms general purpose universal quantifiers that can take scope over questions and declaratives alike. Because of this, sentences like (46)b. and (47)b. are ambiguous in English between single and multiple question readings. On the other hand, "mikin" and

3 As in English, the distinction between (46)a. and (47)a. is very subtle in Finnish. Many speakers are uncertain about what difference there is, if any, between examples of this sort. Looking at several examples of the same sort usually helps. -- The following case brings out the same contrast as the supermarket example in the text.

Assume that there are two jobs. For each job exactly one candidate has been chosen from a number of applicants. Under these circumstances, (i) is a reasonable question in Finnish, because it only implicates that both positions have been filled; (ii) is not because it implicates that every applicant has been appointed to some position.

(i) Kuka tuli valituksi mihinkin virkaan?
Who was appointed to each position?

(ii) Mihin virkaan kukin tuli valituksi?
To which position was each person appointed?

4.3 "Each" vs. "mikin"

While there is this similarity between "each" and the secondary interrogative "mikin", there is also an important difference, and for this reason, we continue glossing "mikin" as "which". "Each" forms general purpose universal quantifiers that can take scope over questions and declaratives alike. Because of this, sentences like (46)b. and (47)b. are ambiguous in English between single and multiple question readings. On the other hand, "mikin" and other secondary interrogatives of Finnish are specialized quantifiers that can only take questions in their scope. (As we noted above, "kukin" has a declarative twin.). Consequently, there is no ambiguity in (46)a. or (47)a. These can only be interpreted as multiple questions; there is no single question reading available.

Another difference is that quantifiers formed with "each" cannot be used in English to quantify into yes/no questions or disjunctive questions. We have no explanation for this, it is simply a fact that examples such as (48) are not ambiguous.

(48) Did John serve each customer?

(48) can only be interpreted as the interrogative counterpart of "John served each customer." It is not obvious why this should be so. Given the rules we introduced earlier to account for the two readings of "Which professor recommends each candidate?" -- see (16) above -- (48) should also have another interpretation. The putative second reading would be the result of quantifying "each customer" into a yes/no question "did John serve Pro_n". According to rule (26), the resulting question should have nearly the same meaning as "Which customer did John serve?". Since (48) clearly has no such interpretation, some restriction is needed in our description of English to rule out the unwanted reading.

The secondary interrogatives in Finnish behave somewhat differently. Questions corresponding to the missing interpretation of (48) are at least marginally acceptable in spoken Finnish.
Furthermore, examples such as (49)a. seem to mean exactly what they should mean on the assumption that "mitäkin" has the meaning of a universal quantifier.

```
(49) a. ?Onko    sinulla mitäkin sanottavaa?
        have+QP you     what    to say
        ("Do you have what to say?")
```

b. Mitä sanottavaa sinulla on?
   what to say you have

c. What do you have to say?

In spite of the presence of the question particle "-ko" on the preposed verb, which normally is a mark of a yes/no question, (49)a. has the meaning of a constituent question; it expresses the same question as (49)b. and its English counterpart (49)c. This is what our analysis predicts, on the assumption that secondary interrogatives in Finnish can be quantified into yes/no questions, and not just into constituent questions as "each customer", etc. can be in English. However, because the status of examples like (49)a. is unclear, this issue requires further study.[4]

5. *Quantifiers and variable binding operators*

In the last part of this paper we will go on to discuss a new method of interpreting sentences with interrogative and non-interrogative quantifiers. One of the shortcomings of the method we have used so far is that it compels us to postulate syntactic transformations whose only motivation is a semantic one. One reason for assuming the deletion and movement operations indicated in our diagrams is that the meanings that belong to the targets of these operations have to be kept apart and made to interact with the meaning of the rest of the structure in a specific way. Another reason is the requirement of one-to-one correspondence between underlying syntactic representations

---

4 Some informants report that (49)a. would not be any worse without the "-kin" suffix (?"Onko sinulla mitä sanottavaa?") and that it would mean just what (49)a. means. Another interesting problem is the connection between (49)a. and other apparent yes/no questions containing "-kin", such as

(i) Tuliko sinne paljonkin väkeä?
    came+QP there a lot of +"kin" people
    ("Did a lot of peaople come there?")

(ii) How many people came there?

Because of its form, one would expect (i) to be a yes/no question but conversationally it has the force of (ii), i.e. a cooperative addressee will respond to it as if it were the corresponding constituent question ("Kuinka paljon väkeä sinne tuli?"). It is unclear whether this is a pragmatic fact -- something having to do with the conventions of use -- or whether it is a matter of semantics proper. Does (i) actually mean what (ii) means? Whatever decision is made about cases like (i), it has some implications for (49)a.

and meanings, which is characteristic of the way formal languages are constructed. As Cooper & Parsons 1976 have argued, there is no compelling reason to uphold this principle when dealing with natural languages; it is not a precondition for interpreting a language model-theoretically.

5.1 NP-storage

In the following, we will make use of an innovation developed by Robin Cooper 1978, which is based on the recognition that in general an expression may have more than one meaning. Accordingly, the rules that assign meanings to larger constructions on the basis of the meanings of their parts are stated in terms of sets of meanings rather than single ones. Cooper's most important innovation was to provide a way for the meaning of a noun phrase to be placed off to the side so to speak, in "storage" and a variable meaning used temporally in its place; the stored NP-meaning is quantified in at some later point. In this way scope ambiguities arise in sentences such as "Everyone loves someone" without the necessity of assigning more than one syntactic structure, since the meaning of "someone" can either be stored temporarily or not stored at all. Consider the structure in (50).

(50)

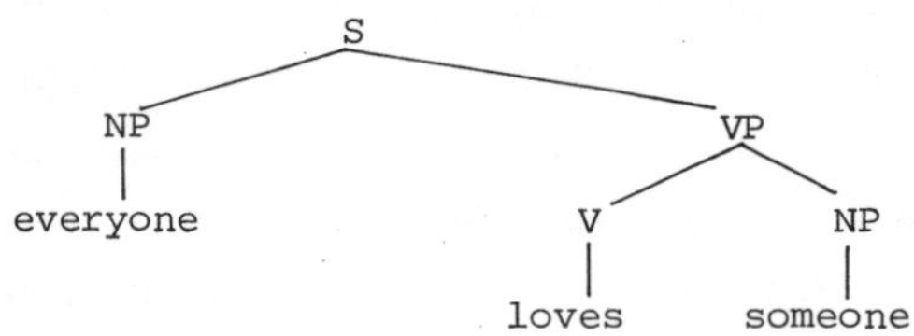

One of the meanings for this sentence can be obtained without using the storage mechanism at all. On that interpretation, the rules discussed earlier (see (5) and (6) in section 1.1) give "everyone" scope over "someone": for every person X there is some person Y whom X loves. The other reading can be derived by storing the meaning of "someone" off to the side and by using in its place a variable meaning of the appropriate type, say $\lambda P\ P(x_i)$. Along with the meaning of "someone" the store must also contain an indication of what variable the stored NP meaning is eventually going to bind. Under this interpretation, the meaning of the NP "someone" in (50) can be construed as an ordered pair consisting of a "meaning proper", the variable meaning, which interacts with the meanings of other phrases in the regular way and a store containing the yet unused noun phrase meaning with its "address". This is shown in (51).

(51) $<\lambda P\ P(x_i), <\lambda P\ \exists x\ [\text{person}'(x)\ \&\ P(x)]\ ,\ x_i>>$

The rule which interprets the VP in (50) essentially operates only on the first member of (51). The stored NP-meaning is simply carried on unchanged. At the level of the S-node the meaning of "someone" finally is brought down from the store by a convention whose effect is to take the first member of the ordered pair, abstract over the variable specified by the stored address, and apply the stored NP-meaning to the result. In this manner, the S-node in (50) comes to have a second interpretation: there is some person X such that every person Y loves X.

This storage apparatus provides a framework for the solution of many interesting problems about quantifier scope in natural language. Ladusaw 1979 has made good use of it in accounting for a wide range of polarity phenomena. We shall modify somewhat Cooper's idea about storing meanings so that it can be used in some new ways. To understand the changes we will make, it is useful to look in a bit more detail at what meaning attaches to the stored combination of NP-meaning and "address". Cooper himself does not discuss how ordered pairs such as $<\lambda P\ \exists x\ [\text{person}'(x)\ \&\ P(x)]\ ,\ x_i>$ which he places in storage can be assigned a unified meaning. Instead he regards them simply as having a pair of meanings -- the first being a quantifier (in a sense we will explain presently) and the second serving merely to identify a variable which will eventually be used to define a function by lambda abstraction. Our procedure is instead to assign a unified meaning to what is stored, treating it as a variable binding operator in its own right. This will permit the "storage retrieval" mechanism to operate in a much more uniform manner, and has other advantages also. The particular meaning that we assign to such variable binding operators (henceforth abbreviated vbo's) depends in a regular fashion on the meaning of the quantifier to which the vbo is related.

5.2 *Variable binding operators of valence <t,t>*

In order to make it clear just how such vbo's are assigned meanings, let us briefly review a few facts about meanings in general and NP-meanings in particular. A meaning for an expression $\exists$ is in general a function whose inputs are assignments of values to variables (together, in a fuller account, with assignments of values to deictic expressions such as the pronouns "I" and "you", the present and past tense, certain adverbs such as "here" and "now", etc.). This meaning function for $\exists$ yields as output the intension that $\exists$ has given the particular assignment of contextually variable factors which its input specifies; thus, for example, the meaning of "everyone loves Pro_i" maps any assignment g to the proposition that the individual $g(x_i)$ -- which g associates with Pro_i -- is universally loved. We will denote the intension of $\exists$ on the assignment g by the notation $[[\ \exists\ ]]^{g}$. Such intensions are, of course, just functions that map world-time pairs $\langle w,t\rangle$ to $\exists$'s denotation in the possible world w at the time t (the context being as specified by the assignment g). For this denotation we write $[[\ \exists\ ]]^{g,\langle w,t\rangle}$; for instance, the denotation $[[\text{everyone loves } Pro_i]]^{g,\langle w,t\rangle}$ is the truth value 0 for (for falsehood) if $g(x_i)$ is Jimmy Carter and w and t are the actual world and present time as of this writing.

We are following Montague in regarding such noun phrases as "someone" or "every sailor" as denoting restricted individual quantifiers; this is what translating them as $\lambda P\ \exists x\ [person'(x)\ \&\ P(x)]$ and $\lambda P\ \forall x\ [sailor'(x) \rightarrow P(x)]$ amounts to. (One should not confuse this sense of the term "quantifier" with another that is current in the linguistic literature, according to which quantifiers are determiners like "every" and "most", which denote relations between sets -- "every(P)(Q)" holds iff P is a subset of Q and "most(P)(Q)" holds iff the cardinality of P intersect Q exceeds that of P intersect the complement of Q.). Any quantifier, in the sense used here, must in effect denote a collection of sets of individuals since its job is to specify what sets of objects in the domain of quantification meet the condition defining that quantifier. Thus $[[\text{someone}]]^{g,\langle w,t\rangle}$ $(= [[\lambda P\ \exists x\ [person'(x)\ \&\ P(x)]\ ]]^{g,\langle w,t\rangle})$ is (technically, the characteristic function of) the set of (characteristic functions of) all sets having at least one member that is a person (as determined in context g and world w at time t). Our reasons for adopting this analysis of noun phrase semantics include those given in Cooper 1977.[5] To mention just one of them here, the object NP "a Ferrari" functions straight-forwardly as a quantifier in producing the so-called non-specific itnerpretation of

(52) John wants a Ferrari.

---

5 We are departing in a now standard way from the analysis in Montague 1973 by eliminating intensionality at two inessential places: taking the variable P to range over sets rather than properties -- Montague chose the latter purely for the sake of a minor bit of mathematical elegance -- individual concepts -- Thomason 1974 showed Montague's decision was not necessary to solve the problem that motivated it and Bennett 197 showed that it created other unfortunate problems for Montague.

On this interpretation, sentence (52) is true just in case one of John's wants would be satisfied were he to get any set of things including a Ferrari among them, i.e., to get any set that would belong to the collection denoted by the quantifier "a Ferrari". Thus one can take the verb "want" to express a relation between individuals such as John and things like the intension of the NP "a Ferrari".

References

Bennett, M., 197 , Ph.D. dissertation, Department of Philosophy, University of California, Los Angeles.

Cooper, R., 1977, "Review of Formal Philosophy", *Language.*

Cooper, R., 1978, "A Fragment of English with Questions and Relative Clauses", unpublished manuscript, University of Wisconsin, Madison.

Cooper, R., and Parsons, T., 1976, "Montague Grammar, Generative Semantics and Interpretive Semantics", in Barbara Partee (ed.), *Montague Grammar,* New York, Academic Press.

Karttunen, L., and Peters S., 1976, "What Indirect Questions Conventionally Implicate," in CLS 12: *Papers from the Twelfth Regional Meeting,* Chicago Linguistic Society, Chicago, Ill., pp.351-368.

Karttunen, L., 1977, "Syntax and Semantics of Questions", *Linguistics and Philosophy* 1:1, pp.3-44.

Ladusaw, W., 1979, Polarity Sensitivity as Inherent Scope Relations, PhD. dissertation, Department of Linguistics, University of Texas at Austin, Austin, Tex.

McCloskey, J., 1978, A Fragment of a Grammar of Modern Irish, *Texas Linguistic Forum 12,* Department of Linguistics, University of Texas at Austin, Austin.

Montague, R., 1973, "The Proper Treatment of Quantification in Ordinary English", in K.J.J. Hintikka, J.M.E. Moravcsik, and P.Suppes (eds.), *Approaches to Natural Language,* D. Reidel Publishing Co., Dordrecht, Holland.

Peters, S., 1979, "What do Questions Mean?", *Texas Linguistic Forum 13,* Department of Linguistics, University of Texas at Austin, Austin, pp.86-95.

Thomason, R., 1974, "Home is Where the Heart Is", unpublished manuscript, University of Pittsburgh, Pittsburgh, Pennsylvania.

THREE KINDS OF LINGUISTIC COMMITMENT*

Asa Kasher, Tel-Aviv University

Games people play tie their hands. When you play chess and your king is under attack, you have to defend it or move out, even if you have an otherwise attractive move elsewhere on the board. When you play football, you are committed to following the referee's instructions, even if they seem to be clearly mistaken or extremely annoying.

Institutions people operate also load them with chains. A member of a jury has to cast a vote, an adult member of a family is committed to shoulder the new-born, and participants in a rule-governed activity known under the title 'colloquium' are subjects of specific binding expectations.

Conversational games are no exception. Thus, for instance, according to an old, American "Ladies' Book of Etiquette and Manual of Politeness" — "Amongst well-bred persons, every conversation is considered in a measure confidential. A lady or gentleman tacitly confides in you when he (or she) tells you an incident which may cause trouble if repeated, and you violate a confidence as much in such a repetition, as if you were bound over to secrecy."[1]

Such a commitment to secrecy accompanies every conversation, according to that old canon of politeness in conversation. Obviously, that commitment to secrecy in polite conversation rests on maxims of politeness rather than on maxims or rules of language.

Are there any commitments which have language itself at their back? And, if there are such linguistic commitments, how are they going to be represented in a general theory of language? The present paper is a preliminary attempt to answer these questions by discussing three kinds of linguistic commitments. (Tenses and quantifiers will serve as examples.).

---

* I am grateful to E.Bach, L.J.Cohen, M.Dummett, L.Karttunen, E.Keenan, J.Raz, and C.Smith for comments on earlier versions of the paper and conversations on related topics.
Work on the present paper was partly supported by the Israel Commission of Basic research.

1 F.Hartley, *The Ladies' Book of Etiquette and Manual of Politeness*, J.S.Locke, Boston 1872, p.18.

I. A natural starting point would be the seemingly obvious observation that when a speaker promises to do something, swears or pledges under appropriate circumstances, he places himself under an obligation to behave in a certain way, by the very speech act expressing his intention to do so.

Accordingly, there is a straightforward answer to the question "Why am I under any obligation to do something if I promise to do it?" The answer is that production of voluntary obligations is the point of the game called "promising". The constitutive rules of promising establish the linguistic connection between a speech act of promising, swearing or pledging under appropriate circumstances of utterance and the produced voluntary obligations.

To see that these observations are not as innocent as they seem to be, recall another conception of promising and the completely different answer it provides for the question "Why should one keep his promises?" According to the alternative conception of promising, a promise to perform an action is an expression of a firm intention to perform it, as addressed to a person who may rely on it. A variety of social or moral considerations is then used to show that one ought to keep his promises.[2] To mention just one example; since social coordination is to everyone's benefit and since the practice of keeping one's promises enhances social coordination, a rational person should keep his promises.

There are two crucial differences between these two conceptions of promising. According to the former approach, a happy speech act of promising creates an obligation, whereas according to the latter approach such a speech act triggers a variety of reasons for not braking the promise, but not a commitment to keep it.[3] Moreover, while the obligation conception grounds the relation between a happy speech act of promising and the created commitment on linguistic foundations, the firm intention conception of promising draws a chain of reasoning for keeping one's promises in which language hardly plays a significant role.

There are three reasons for preferring the obligation approach over the firm intention conception.

---

2 Cf. P.S.Ardal, "And That's a Promise", *Philosophical Quaterly* 18 (1968), pp.225-237.

3 At this point I am indebted to J.Raz's, "Promises and Obligations", in: *Law, Morality, and Society* (eds., P.M.S.Hacker and J.Raz), Clarendon Press, Oxford 1977, pp.210-228.

First, there is the simple observation that under appropriate circumstances a person may express his firm intentions to act in a particular way and explicitly deny his sincere and literal expression of firm intention from being a promise. There seems nothing infelicitous in asserting, under standard conditions: "I have a very firm intention to visit you tomorrow, grandma, but that's not a promise. Don't hold your breath." Such an assertion poses no problem for the obligation conception, because according to the latter conception there is more to promising than firmly intending. The contrast between the expression of a firm intention and the denial of having thus made a promise manifested by the occurrence of "but" in that sentence is due to the linguistic fact that a promise may be made by sentences such as "I will visit you tomorrow" when uttered under appropriate circumstances. However, the pragmatic ambiguity of such sentences, used to assert, promise or predict, is no reason for identifying the meaning of "promise" with that of "expressing a firm intention".

On the other hand, the firm intention conception renders infelicitous the normal utterance of the sentence "I have a very firm intention to visit you tomorrow, but that's not a promise", because according to that conception this utterance amounts to an expression of a contradiction of a simple form: "I promise to visit you tomorrow, but that's not a promise."

Another reason for preferring the conception which characterizes promises in terms of rules which impose voluntary commitments rather than in terms of rules for expressing firm intentions which may be relied on, is the following, due to Joseph Raz (ibid.): Consider the case of a person *A* promising a person *B* to make a contribution, say, to a charity fund. Assume also that the promise was made in the presence of another person, *C*, who is known to be interested in that fund and who is known to be honest, naive or what have you to the extent that he may rely on *A*'s promise in making his own plans for the charity fund. Now, by promising to person *B* in presence of person *C*, the promisor *A* deliberately made person *C* aware of his firm intention to contribute to that charity fund, but even though person *C* may rely on *A*'s present firm intentions it is clear that *A* has promised *C* nothing.

Again, such a situation poses no difficulty for the obligation conception, because commitments have addresses, so to speak, and thus a promise made by person *A* to person *B* creates an obligation on *A*'s part and a right on *B*'s part only. The promissee is the addressee, that's a useful truism.
On the other hand, the firm intention conception falls short of explaining

the common intuition that person *A* has not promised person *C* anything by promising something to person *B* in *C*'s presence. A firm expression of firm intention has been made in presence of someone who may rely on it in his own plans, but the latter cannot claim to have been promised anything.
Such a combination is impossible according to the firm intention conception, but seems perfectly possible, intuitively speaking.

Our third argument for the obligation conception rests on the cluster of differences between assertions and promises. First, while the direct point of promises seems to be undertaking a restriction on future activity, the direct point of assertion of firm intentions seems to be describing a present state of mind. Secondly, whereas the rules of the assertion game allow an addressee to ask the assertor for a justification of his claim, a request for such a defence would seem inappropriate in response to a promise. Thirdly, truth seems to play different roles in assertions and promises. It makes sense to enquire whether what has been asserted in a given speech act is true, but it does not seem to make sense to enquire whether what has been promised in a given speech act is true. In case of a promise, questions of truth seem to arise just with respect to standard preconditions.[4]

It seems, then, reasonable to adopt the obligation conception of promise and other commissives, which gives us the first kind of linguistic commitments. The constitutive rules of promising are part of our language, and accordingly their standard products are linguistic. Since the rules of promising establish a commitment on the part of the speaker, when he plays his role in the language-game they define, the commitments made by promising are linguistic commitments.

Using the common distinction between the force of a speech act and its propositional part ("sense", "radical"), it would be natural to ascribe commitments of the first type to the forces of the commissive speech acts involved. The force of such a speech act fixes a scheme of commitment, a function from propositions to commitments, if you wish. Given the propositional part of a promise, the force of promising determines the *prima facie* obligation undertaken. I will call commitments of the first type "*force commitments*".

---

4 For some arguments against the opposite view (of D.Lewis, for example, in his "General Semantics", in: *Semantics of Natural Language* (eds., D.Davidson and G.Harman) Reidel, Dordrecht 1972, pp.169-218), see our "Mood Implicatures", *Theoretical Linguistics* 1 (1974), pp.6-38.

Before we turn to other types of commitment, let me sketch a way of representing such commitments in a theory of language.[5] The relation between the meaning of a promise and production of commitments can be read off lists of rules that have been put forward as characterizations of promise, but such a representation of that relation is more than a bit ad hocish when the list of rules presented as constitutive of promising does not rest on a general conception as to how should kinds of speech act be presented. It seems to me that a constitutive system of rules, which defines a kind of speech act, has what might be dubbed "institutional structure". Any institution -- be it a game, university or family -- is defined by rules which specify for it the *goals* of its activities, the *means* employed for attaining these goals, the standard *products* of actions within the institution and the *roles* played in it by different parties. Language games are also such institutions, and have such an institutional structure.

Institutional products are mainly of two kinds, in language. Products of one type are asserted propositions, orders, advices and the like. Products of another type are speaker's commitments created when speech acts are performed happily. *A description of a scheme of commitment is, therefore, part of a specification of the institutional structure of a language game.*

II. Force commitments introduce a simple speaker/hearer asymmetry. Promising to return and understanding that a return has been promised are clearly distinct at least to the extent that a commitment to return is different from a right to have someone else return.

We turn now to a presentation of some *sense commitments* and the non-trivial speaker/hearer asymmetries they accompany.

Consider the sentence "I am going to travel to a certain Greek island and write a book there," as uttered literally and sincerely by person Alpha, addressing person Beta.

We know, to use Frege's words,[6] that "if words are used in their ordinary way, what one intends to speak of is their reference." To be sure, if words are used in their ordinary way and have reference, what one speaks of is their reference. Moreover, if words are understood in their ordinary way, the

5 For a detailed description, see my "What is a Theory of Use?", *Journal of Pragmatics* 1 (1977), pp.105-120.

6 Here and in the sequel, my indebtedness to M.Dummett's views on Frege and on philosophy of language in general should be clear.

addressee understands that what the speaker intends to speak about is the reference of some of his words.

Now, words are used in their ordinary way in Alpha's mentioned utterance of the sentence "I am going to travel to a certain Greek island and write a book there." Words and phrases are assumed to have been used by Alpha in that speech act literally and seriously.[7]

Assume, furthermore, that when Alpha was using the words "a certain Greek island" he had in mind the Greek island of Hydra, and that at the same time there was nothing in the context of utterance which might have suggested to Beta, the addressee, that Hydra is the Greek island Alpha had then in mind.

Under these quite ordinary assumptions, it seems to me clear that Alpha should be taken as referring in his speech act to the Greek island of Hydra. He speaks of a certain Greek island, *namely Hydra*.

That Alpha was referring to the island of Hydra, by using the phrase "a certain Greek island", under the circumstances, might seems objectionable to some people, on different grounds.

First, one might argue that the fact that Alpha had the island of Hydra in mind when he used the expression "a certain Greek island" is irrelevant to the linguistic activity of Alpha, under the described circumstances. After all, people have all sorts of things in mind when they speak and it would be ludicrous to try to hinge on such accompanying thoughts in an attempt to specify the propositional content of an assertion. If it so happens that whenever I use the word "spire", literally and seriously, I have Oxford in mind, would that association be considered as part of my language? All the more so, if not a regular association is under consideration, but just an accidental concomitance.

To counter this argument, consider the truth-conditions of what Alpha said. If at the context of utterance, Alpha is going to travel to Hydra and write a book there, then what he said was true. If he was not going to travel to Hydra and write a book there, what he said was false. If Beta is provided with a list of all Greek islands and finds out that for each of them it is not the case that Alpha is going to travel to it and write a book on it, then Beta has thus plainly shown what Alpha said to be false. However, if Alpha had the island of Hydra in mind when he used the phrase "a certain Greek island", whereas he was going to travel to Crete and write a book there, then

---

7 If anybody is concerned about an opacity created by the phrase "I am going to", the phrase "I am travelling" can be used instead.

what he said was false, though Beta has no way to realize it under the circumstances.

To put it differently, consider the Tarski-Davidson biconditional related to Alpha's utterance: The sentence "I am going to travel to a certain Greek island and write a book there" is true, as uttered by Alpha at a context of utterance C, if and only if at context C Alpha is going to travel to X and write a book there", where "X" should be relaced by an appropriate expression of the language.

Obviously, "X" cannot stand for the expression "a certain Greek island", because that is tantamount to including an active indexical in the truth-conditions. Neither can "X" stand for the expression "the Greek island of Hydra", because that would require truth-conditions to be specified for each context of utterance separately, whereas truth-conditions are required to be finitely representable. The most natural way of phrasing that biconditional seems to be approximately as follows: The sentence "I am going to travel to a certain Greek island and write a book there" is true as uttered by Alpha at context of utterance C, if and only if, Alpha is going at C to travel to -- roughly speaking -- the Greek island he has in mind at context C, and write a book there. Which shows that what a speaker has in mind when he uses the expression "a certain Greek island" plays an essential role in the speech act he performs by using that expression or similar ones.

That was one objection to our claim that Alpha refers to Hydra under the mentioned circumstances. Another objection might take the following line. The sense of a referring expression is, to use Frege's way of putting it, a "mode of representation" of what the speaker speaks of when he uses that expression appropriately. A full understanding of such a mode of representation leads us from an expression of our language to an object or something in the world. However, it might be argued. full understanding of the phrase "a certain Greek island" would not suffice for associating in Beta's mind that expression with the island of Hydra. The latter should, thus, be deprived from serving as a possible referent of the former.

To see the mistake on which this argument rests, consider the expression "the present queen", which is undoubtedly a referring expression. Strictly speaking, the expression "the present queen" does not refer to any specific individual, and even full understanding of it does not suffice for associating it in our minds with any particular woman. When reference is made by using the expression "the present queen" to, say, Queen Elizabeth II, it is only when the expression is used in a context of utterance which fixes the values of the indices of time and state. When such an expression is used for definite reference to an object, the speaker uses it for identifying that object by means of linguistic knowledge and awareness of the context of utterance. If the island of Hydra is not a possible referent of the expression "a certain Greek island" because full understanding of this expression does not identify the island of Hydra as the object referred to, then Queen Elizabeth II is not a possible referent of the expression "the present queen", because similarly, full understanding of the expression "the present queen" does not identify Queen Elizabeth II as an object referred to. The latter dissociation is, however, absurd, which shows that the former dissociation is wrong. In both cases the speaker refers to an object by using an expression which by itself falls short of referring to the respective objects.

Notice the difference between the expressions "the present queen" and "a certain queeen". Neither of them by itself has Queen Elizabeth II as its referent, while both of them may be used for referring to her, under appropriate circumstances. However, while linguistic knowledge and awareness of the context of utterance enable one to identify the object referred to, independently of what the speaker has in mind when he uses the expression "the present Queen", linguistic knowledge and awareness of the context of utterance are not generally sufficient for enabling one to identify the object referred to by using the expression "a certain queen", independently of what the speaker has in mind when he uses this expression.

When Alpha says, under appropriate circumstances, "I am going to travel to a certain Greek island and write a book there", he is committed to being able on demand to identify the Greek island he is referring to apart from all other objects, such as other Greek islands. Thus Alpha is committed to being able to produce on demand a constructive identification of the Greek island he is speaking of -- a proper name of it or an identifying description of the island which enables any speaker of the language to know, under the circumstances, which Greek island was it that Alpha was speaking about, independently of what Alpha had in mind when he used the expression "a certain Greek island" at the time. Characteristically, Beta is generally unable to identify the Greek island Alpha was referring to apart from all other Greek islands, independently of what Alpha had in mind under those circumstances. What Beta knows is that Alpha spoke about one of the Greek islands and that Alpha is committed to produce on demand an appropriate identifying expression of that Greek island, independently of his speech act. Roughly speaking, *what Alpha says and means is what Beta understands plus what Alpha is committed to produce on demand.*

I will call commitments of that kind *"sense commitments"*, because they accompany a speaker's use of the propositional part of a speech act and not its force. It does not matter whether I assert, suggest, predict or advise that a certain bishop visits Poland, the sense commitment to an impredicative identification of the bishop is always produced by using the expression "a certain bishop" appropriately.

A few objections to this approach might arise. Usually, indefinite expressions are represented in logical forms by existential quantifiers. Why shouldn't we take the proposition expressed by Alpha and fully understood by Beta to be simply that

> there is a Greek island for which it is true that Alpha is going to travel to it and write a book there?

If this suggestion is accepted, no theoretical duty is left for a sense commitment on the part of the speaker.

However, this suggestion should be rejected for at least three reasons. First, it does not leave any room for a difference between the sentence " I am going to travel to a Greek island" and the sentence "I am going to travel to a certain Greek island", which have clearly different readings. To be sure, it would not be possible to obviate this difficulty by introducing a pair of quantifiers "there is a Greek island" and "there is a certain Greek island", since the latter quantifier could be argued to reintroduce the speaker's mentioned sense commitment.

Secondly, even if an existential quantifier is admitted to appear in the representation of the sentence "I am going to travel to a certain Greek island and write a book there", it is not clear what its scope should be. Why should the scope of the quantifier include the phrase "and write a book there"? The occurrence of the word "there" in the second conjunct, a word which refers back to "a certain Greek island" occurring in the first conjunct, does not compel us to put both conjuncts within the scope of the quantifier, as is clear from the following conversation:

> Alpha to Beta: "I am going to travel to a certain Greek island to rest a while."
> Beta to Alpha: "Will you read a lot of books there?"

It does not seem possible to put into the scope of one and the same quantifier a proposition expressed by one person and the response made by another one; all the more so, when the response is not a proposition but a question.

Obviously, it would not help our objector to introduce an existential quantifier for each conjunct separately. Neither would it be possible to restrict the scope of the quantifier to the first conjunct, excluding from it the phrase "and write a book there", not only because it leaves a dangling "there" in the second conjunct, but also because it turns out to be equivalent to the following compound, which is not well-formed: (Not every Greek island is such that Alpha is not going to travel to it) and (Alpha is going to write a book there).[8]

The inevitable conclusion from these attempts to fix a scope for the existential quantifier, which all failed, is that no quantifier is involved in the representation of Alpha's sentence.

Our third objection to the quantificational analysis is that it does not capture the referential power of the expression "a certain Greek island" which Alpha uses, as we saw, to refer to the island of Hydra.

A final argument we would like to mention which might be brought against our approach to "a certain" expressions, is the following. If the phrase "a certain book" is used for referring to a particular book, on a par with phrases such as "the book", "our book" etc., it should be possible to use the former when we use the latter ones, without any loss of meaningfulness. However, the objection goes, definite expressions may occur in sentences used for asking, ordering, advising and the like, whereas all the following sentences seem very odd:

Have you read a certain book?
Give me a certain book!
I advise you to read a certain book.

(Notice, by the way, how a quantificational analysis would misrepresent these sentences).

To meet this objection two observations should be made. First, when an utterance of any of these sentences sounds inappropriate, it is because the speaker uses the phrase "a certain book" for referring to a book which the hearer is unable, under the circumstances, to identify independently of the speaker's state of mind. If I do not know which book you are speaking of, how could I tell you whether I have read it or not? If you have in mind "The golden notebook" the answer will be in the affirmative, while if it is "The

8 I don't know to whom should the latter argument be ascribed; I have heard it on several occasions.

red booklet" that you have in mind the answer, I am afraid, will be in the negative. Thus, our approach explains the odditiy of these sentences, under appropriate circumstances.

Secondly, notice that under some other circumstances utterances of those sentences would not sound odd at all. When a speaker assumes he can rely on the addressee's awareness of the context of utterance and at the same time is interested in keeping a third party -- another hearer -- unable to identify the book under disguise, so to speak, the speaker will find it apt to use "a certain"-expression.

I would like to turn now to other examples of such sense commitments which are similarly related to speaker/hearer asymmetries.

First, indefinite articles induce ambiguities, such as the following one: the sentence "I am looking for a book on the history of Cambridge" has two readings; according to one, any book on the history of Cambridge will do -- I am just interested in there being at my disposal something which is a book on the history of Cambridge. According to the other reading I have in mind a certain book on the history of Cambridge, namely the first one ever published in England. It seems to me that the specific reading of that sentence induces a sense commitment, similar to the one related to parallel uses of "a certain" expression.

An interesting example of another kind is that of what Barbara Partee called "the deictic use of the *Past* tense morpheme".[9] Considering the sentence "I didn't turn off the stove", when uttered, say, halfway down the highway to another city, she pointed out that such a sentence clearly does not mean that there exists some time in the past of the utterance at which I did not turn off the stove. To see that, just recall the arguments used against a quantificational analysis of the conjunction "I am going to travel to a certain Greek island and write a book there", as applied to the sentence "I didn't turn off the stove, because the program I was then watching on TV was very interesting."

A speaker who utters the latter sentence, literally and seriously, refers by using the past tense inflection of the main verb phrase to some time-interval[10] during which the speaker did not turn off the stove. The identity

9 Barbara Hall Partee, "Some Structural Analogies between Tenses and Pronouns in English", *Journal of Philosophy* 70 (1973), pp.601-609.

10 Here we do not mean by "interval" what is defined by an ordered pair of real numbers on a time axis, but we cannot go into detail. See our and R.Manor's contribution in the present volume.

of that time interval is known to the speaker to an extent which sometimes is beyond what the hearer can figure out from the context.

Again, the use of such tensed expressions to refer to time intervals, commits the speaker to produce on demand more detailed information as to the identity of what has been referred to. This is a sense commitment induced by some past tense modes of representation.

There is an interesting difference between the sense commitments induced by uses of deictic past tense and the sense commitments induced by uses of the specific indefinites and "a certain" expressions. Whereas in the latter cases the speaker is committed to being able to produce an independent identification of what he was speaking of, in the former case, of past tense expressions which function, say, in describing events, the speaker is committed to being able to produce an independent description of a time interval which is smaller than the whole past of the utterance event and which includes the time interval of the described, past event. The speaker is not committed to produce an independent, uniquely identifying description of the time-interval of the event. If in response to saying "I met him once" I am asked "When was that?", there would be no infelicity involved in my answering "Well, it was last year, but I don't remember the date, or even what part of the year was it." Notice, however, that it would not do to answer that question by "I do not have the slightest idea."

These observations lead us to an interesting semantic result. *Not only hearers have to manipulate partial descriptions of what is being referred to, but speakers also do it sometimes, under ordinary circumstances of using words.* At both ends of a conversation, persons might use their language by manipulating such partial descriptions. Still, there is a possibility of the partial information at the hearer's end being just part of the partial information at the speaker's end.

Our final example of a sense commitment is related to the sentential use of the word "but".[11] Without going into details of standard analyses of the meaning of "but", we assume that they are right in pointing out the role played by a contrast between the conjuncts flanking a "but". As in previous cases, the speaker may be taken to have something important in mind when he

11 See M.Dummett, *Frege, Philosophy of Language,* Duckworth, London 1973, p.86. Ed Keenan pointed out (personal communication) an indefiniteness which is similar to that of "but" in some respect, namely that of the measure involved in any use of a comparative phrase.

says "The skies are blue but the roads are white", literally and seriously, he is committed to put forward a contrast he has in mind between the two parts of the sentence.

By no means could we take the speaker to be interested in concealing the identity of the contrast. On the contrary, it seems that he is interested in the hearer's theorizing, so to speak, about the identity of that contrast. On the other hand, for the hearer to fully understand what the speaker said, he is not required to identify the contrast independently of the speaker's state of mind. What Dummett aptly called "the indefiniteness of the contrast" creates, then, a still different type of speaker/hearer asymmetry. This time the speaker marks that he has a particular contrast in mind, he deliberately avoids identifying the contrast explicitly, but he is interested in the hearer's grasping somehow the identity of the contrast.

There are many other kinds of sense commitment, but we have pointed out cases which seem to have two related interesting ingredients: partial information and a speaker/hearer asymmetry bridged, so to speak, by the speaker's sense commitment.

III The force commitments and the sense commitments that we have considered so far are all of a purely linguistic nature, being established by the rules which govern promising, by rules which govern the use of expressions such as "but", "a certain" or by similar rules which are elements of our language. The third type of commitment I would like to briefly consider includes both linguistic and non-linguistic ingredients.

Every sincere performance of speech act commits the speaker to show that there are sufficient reasons for taking the preconditions of the speech act to obtain in the context of utterance. We will call such commitments "*precondition commitments*". Thus. for example, a speech act of ordering commits the speaker to showing the grounds for considering his position as higher than that of the addressee, in some related hierarchy.

Notice that there is a two step way from a speech act of ordering to a demonstration of an hierarchial differential. The first step is from the speech act of ordering to the precondition commitment to show the grounds for taking the hierarchial differential to hold, while the second step is from the commitment to show those grounds to a demonstration of the required difference in positions held by speaker and addressee within the relatet hierarchy.

There is a major distinction between these two steps. Whereas the first step -- from a speech act to a commitment -- is of a purely linguistic nature, the second step -- from a commitment to a demonstration -- is of a different kind. Knowledge of the language would not be sufficient for answering the question -- "What would count as a demonstration of the related hierarchial differential between speaker and addressee?" Knowledge of a clearly non-linguistic nature is required for establishing a subordination relation between two persons in a police force, a church, a university, a family or any other hierarchial social organization.

The rules of the language-game of ordering ascribe to each speech act they govern a scheme of precondition commitment with an institutional index; that is to say, a person who plays the role of a speaker in the language-game of ordering is thereby committed to showing that he has sufficient reasons for taking his position to be higher than the addressee's in the related hierarchy H. Since "H" is a free variable, what we have is not a commitment but a scheme of commitment. Now, the value of this free variable is fixed by the context of utterance; otherwise, that context would be inappropriate for ordering. Having fixed the value of the variable, we have also fixed the commitment of the speaker. This is the end of the first step, which is the linguistic part of the way. From this stage on it is a matter of establishing facts or defending beliefs about a particular hierarchial organization, knowledge of the ropes of which is not part of knowledge of one's language.

Of particular interest is a precondition commitment of assertion, namely, the commitment to back up what has been asserted.

When a speaker is challenged to support his assertion, a new institution is invoked -- an institution of backing up and criticizing. While the speaker will try to provide support for what he has said, the addressee will try to subject it to extreme criticism, so as to see whether what has been said by the speaker should be accepted, having been found to be flawless under the circumstances.

Providing support and criticizing are rule-governed activities. For example, supporting a molecular assertion is supporting a certain part of it or a certain transform of it. Thus, supporting a conjunction is supporting each of the conjuncts separately; supporting a disjunction is supporting one of its disjuncts; supporting an existential assertion is providing an appropriate example.

Such recursive clauses in the charcterizing rules backing up are of major importance, for at least two reasons. First, without such recursive clauses, finite representability cannot be guaranteed for the precondition commitment of backing up assertions. This point has been stressed by Dummett, for example in his paper on truth, where he argued for a recursive definition of use-conditions.[12] Secondly, such recursive clauses reflect the nature of what might be called "the logic of justification". Each clause is a derivation rule in such a logic. It is interesting to note that the logic of justification turns out to be intuitionistic. It seems to me that Hintikka's semantic games, under an interpretation I suggested for them elsewhere, provide a formal framework for representing that part of the institution of backing up and criticizing, but I cannot go into that here.[13]

Assuming that we know how to back up a molecular assertion in terms of backing up simpler assertions, we turn now to the problem of backing up "atomic" assertions. How does one justify assertions that Spinoza had a black beard, that Heath is a bachelork that I see the chairman, that Carter will be reelected? On reflection it should become clear that these assertions call for justifications of different types. Roughly speaking, assertions about past observations are backed up by resort to causal chains of events whereas assertions about future observations are backed up by recourse to general laws and observations about the present. Assertions about the latter are backed up in still different ways. Mastering certain systems of rules for backing up is no- required for mastering a language, but *a concept of such a system of rules of backing up* is part of linguistic knowledge. The assertor is committed to back up what he has asserted, in a way appropriate to the sentence used and to the context of utterance, but which are the rules of the appropriate justification-game is not a question to be answered by allusion to the rules of assertion solely.

Notice that the distinction between saying and meaning and backing up, at the speaker's end of the speech act, has a counterpart at the hearer's end, namely the distinction between understanding and accepting. The addressee plays a central role in the speaker's game of justification, because the addressee tries to put what the speaker has asserted to severe criticism, and in a sense it is the speaker's role to show that what he has asserted stands

---

12 M. Dummett, *Truth and Other Enigmas*, Duckworth, London 1978, pp.1-24, but also in other papers in the same volume.

13 See our "Pragmatical Representations and Language-Games: Beyond Extensions and Intensions" in: *Rudolf Carnap, Logical Empiricist* (ed. J. Hintikka) Reidel, Dordrecht 1975, pp.271-292.

to possible criticism of the type employed by the addressee. A game of justification and criticism is the theoretical bridge between understanding and acceptance.

Now, even though a speaker does not master certain games of justification as part of his learning a language, he has to master a concept of a game of justification, because he is committed to play such a game on demand when he plays the assertion game. We conclude this paper by making some remarks on such a concept of a game of justification.

A characterization of such a concept will, of course, take the form of restrictions imposed on games for them to constitute games of justification. I would like to mention just two restiririctions. First, all the recursive clauses in a definition of a justification game are constructive, i.e. given the rules and an appropriate representation of an assertion of a molecular form, a construction is provided by the rules of simpler assertions which have to be justified in order for the given assertion to be justified. Secondly, a justification game which is attached to a certain class of sentences could not render all of them unjustified. For example, the justification game for past observations could not be a game which would require conclusive evidence for an assertion to be justified, because no assertion about the past could then be justified. The philosophical defense of such a claim -- a "transcendental restriction in nature"-- is a story for another occasion.

A NEW APPROACH TO QUANTIFICATION IN NATURAL LANGUAGE*

Edward L. Keenan, UCLA and Tel Aviv University
Leonard M. Faltz, UCLA

The purpose of this paper is to present the treatment of quantification developed in *Logical Types for Natural Language* (Keenan and Faltz, 1978; henceforth *Logical Types* or LT). That treatment is distinguished from others in that it provides a general and uniform treatment of boolean operators such as *every*, *a*, *no* on the one hand, and *and*, *or*, and *not* on the other. This enables us to generate and directly interpret (rather than translate) expressions such as the following (using brackets to give a rough indication of logical constituent structure):

John[[both hugged and kissed][no student]]
[Neither John nor every student][[were working or thinking][both slowly and carefully]]
[Not[every student['s father['s employer]]]][lives in a city]
[John and[not Mary]][will sing[in[every nightclub and every bar]]]

In part 1. of this paper we give an informal summary of the logical system of LT, concentrating on the extensional fragment, which is sufficient to illustrate our treatment of quantification. In part 2. we indicate how this treatment extends in toto to the full intensional system. We refer the reader to *Logical Types* for proofs of most theorems and the treatment of many categories such as Adjectives, Adverbs, Raising, Equi, Passive, etc. which do not directly concern our treatment of quantification.

* Presented at the Stuttgart Colloquium on Quantification and Tense Logic, March 1979, and to appear in the Proceedings.

1. *Quantification in Logical Types: An Informal Statement*

1.1 *A Brief Overview of the Entire System*

Following the lead of Montague in EFL and PTQ (Montague 1970 and 1973 respectively) the intensional logic in LT is presented within a categorial framework. Categories are sets of expressions defined in a standard way by simultaneous recursion beginning with lexical categories whose members are defined, by and large, by listing. A distinction is drawn between *primitive* categories and *functional* categories. The primitive categories in this system are: *FM* (Formula), *DNP* (Determined Noun Phrase; e.g. *John, every man,* x_4, etc.), *PN* (Proper Noun), *VAR* (Variable -- of which there is only one sort), *CNP* (Common Noun Phrase; e.g. *man, not very smart or assiduous student,* etc.) and *FNP* (Function Noun Phrase; e.g. *father, mother*). *Functional* categories on the other hand have names of the form C/C' where both C and C' are category names (or in a few cases variables ranging over certain category names). Expressions in a category C/C' syntactically combine with expressions in C' to form expressions in C. And semantically the *type* for C/C', that is the set of possible denotations of expressions of C/C', is always a subset, usually very proper, of the set of functions from the type for C' into the type for C. A few of the functional categories with which we shall be concerned in this paper are: *DET* (=DNP/CNP; e.g. *every, a, the, no, one, two, three,* etc.), *VP* (=FM/DNP; e.g. *sing, sing in the park, not be believed to seem to be a crook,* etc.), *TVP* (=VP/DNP; e.g. *find, find in the park, almost persuade to leave early,* etc.), *AP* (=CNP/CNP; e.g. *tall, Albanian, skillful, fake, alleged, extremely (big and strong),* etc.), and *ADVP* (=D/D, where D is a variable ranging over {VP,TVP}[1]; e.g. *here, in the garden, slowly, must, want-to,* etc.).

---

1 The category ADVP is actually stated with much more generality than indicated here. A more accurate statement would be ADVP = D/D, where D is the union of the n-place predicate symbols defined in section 1. Semantically the type for ADVP is the set of type preserving functions from the n-place predicates into the n-place predicates, n ⩾ 1. That is, *in the garden,* etc. maps n-place predicates onto n-place predicates.

We note that in *Logical Types* itself ADVP did not include the Prepositional Phrases like *in the garden*. Rather Prepositions were treated as *Verbal Extensors* forming transitive verb phrases from verb phrases (e.g. *(sleep,in)* from *sleep*) and ditransitive verb phrases from transitive verb phrases (e.g. *(find,in)* from *find*, etc.). Faltz (in prep.) currently shows how the type for Prepositions can be treated within this framework as having the category ADVP/DNP.

The principle syntactic differences between this system and more familiar ones such as PTQ are as follows: first, there are simply many more categories in this system than in PTQ; many categories are subsets (=subcategories) of others. For example there are five (sub)categories of AP, four subcategories of FM/FM, etc. Second, the system of derivational rules forming derived members of categories is somewhat richer and somewhat more generally stated than in better known systems. Thus on the one hand there is a class of "morphological" rules which e.g. derive CNPs like *sing+er* from VPs like *sing*, APs like *allege+ed* from Sentential Transitive Verb Phrases like *allege*, and passive VPs from the corresponding (active) TVP.

Concerning the more generally stated rules we note in particular the rule of *boolean combinations* which says that for each boolean category B, if expressions e_1 and e_2 are both in B, then (e_1, and, e_2), (e_1, or, e_2), and (not, e_1) are also in B. The *boolean categories* in this system are FM, DNP, VP, TVP, AP, and ADVP.[2] Thus we directly form derived DNPs like *(not, (ever, student))*, *(John, and, (not, Mary))*, *((a, student), or, (a, teacher))*, etc. Other derivational rules are also sensitive to whether the category they apply to is boolean category B (see *Logical Types* for details). For example, if *e* is in a boolean category B and a variable *x* occurs free in *e*, then (λ, x, e) is in category B/DNP. Similarly the reflexive rule generalizes over certain of the boolean categories which meet an additional restriction. As a result we may generate *to love oneself* (=((*self, x*, (*love, x*)) and *to find* (*someone*) *in his own room* (=(*self, x*, (*find*, (*in*, (*x's room*))))) by the same rule.

Third, derived expressions in this system are formed by taking sets over simpler expressions. Thus left-right order is not defined on the expressions of the language (although constituent structure is). Consider for example the expression (1)b. below in our language. It expresses (ignoring tense and aspect, not represented in the system) the logical properties of (1)a. It's derivation in our system can be represented by (1)c., though we note that (1)c. is uniquely recoverable from (1)b. as (1)b. has only one derivation in the system.

---

2 Strictly speaking these are the boolean categories in the extensional logic in LT. In the intensional logic some of these categories are merely subcategories of others (though they are still boolean). See the discussion in section 2. of this paper.

(1) a. Not every student can either read or write.

b. ((not,(every,student)),(can,(read,or,write)))

c.

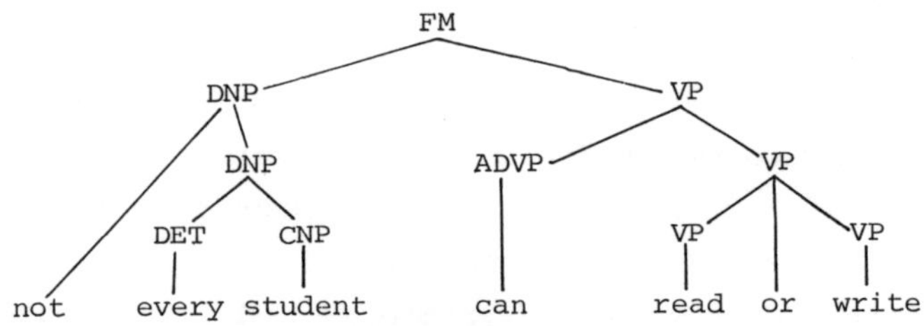

(1)b. is actually a set with two members. One member is the expression ((*not*,(*every*,*student*)) and the other member is (*can*,(*read*,*or*,*write*)). ((*not*,(*every*,*student*)) in turn is a set with two members, one member being the expression *not* the other being the two membered DNP (*every*,*student*). Similarly the major VP in (1)b. is also a two membered set, one member being the lexical adverb *can* and the other being a three membered VP (*read*,*or*,*write*). We note that all derived expressions in this system have either two or three members, and that in deference to linguistic tradition we designate sets of expressions using the left and right parentheses rather than the more usual curly brackets.

Our motivation for taking derived expressions as sets rather than strings is that we have attempted to rigidly constrain the logic so that nothing is put in the syntax of the language unless it is mentioned by the function which interprets expressions as members of their type. And in fact left-right order is not mentioned by the interpreting function. Thus it would make no difference to our semantics if English normally said *Talks John* rather than *John talks*. For the same reason we do not put verb agreement in the system.

Thus the logical structures in LT are solely designed to represent semantic properties of English sentences. As a result we can meaningfully compare the structure of surface forms in a natural language with their logical structures in this system and hope to state generalizations relating the two. In fact several such generalizations, based on the logic in LT, are proposed and empirically defended in Keenan(1979).

Finally we note that in a few cases the logical category we assign an item from English does not correspond to the category assigned to it by say PTQ. For example, VPs in our system have the category FM/DNP, that is, they combine with DNPs to form formulas, and their type will be a (very proper) subset of the set of functions from T_{DNP}, the type for DNP, into T_{FM}.

Further, prepositions in this system[3] have multiple categories. On the one hand they combine with DNPs to form ADVPs, itself a "meta"category. On the other hand, they combine directly with VPs to form TVPs and with TVPs to form expressions of the category TVP/DNP. That is, in general they may combine with n-place predicate symbols to form n + 1 place predicate symbols.

Semantically speaking LT differs more from PTQ than the use of a similar categorial notation might indicate. We should stress, however, that we have incorporated Montague's insight which enabled him to treat proper nouns like *John* and quantified noun phrases like *every man* in a logically similar way. Our treatment of quantification builds on that insight and would not have been possible without it.

The most apparent semantic differences between LT and PTQ relate to the fact that in LT the expressions of the formal English we define are directly interpreted (assigned denotations in a model structure) rather than being mapped onto expressions in some other language (the intensional logic of PTQ) and only then interpreted. Thus to each category of expression (subcategories are categories) we assign a type defined in terms of the set of truth values 0 and 1, a non-empty set U, the universe of discourse, and a non-empty set J of "possible worlds". And expressions in a category are directly interpreted as members of the type for that category.

Much of the semantics proper in LT is contained in the definition of the types. For example all of quantification is handled there. Thus the fact that (2)a. below, represented in LT by (2)b. entails but is not entailed by (2)a. follows directly from the definitions of the types for DNP and VP plus our fully general treatment of the boolean operators *and*, *or*, and *not* as well as their infinitary analogues *every* and *a*.

(2) a. A student was singing and dancing.

b. ((a,student),(sing,and,dance))

(3) a. A student was singing and a student was dancing.

b. (((a,student),(sing)),and,((a,student),(dance)))

In fact, to a significant extent it is our general treatment of the types for boolean categories which enables us to directly interpret the expression in our formal English (and it was our attempt to build a natural logic whose

---

3 See footnote 1.

expressions were directly interpreted which led us to the general treatment of the types for the boolean categories). Much of the remainder of this paper concerns our treatment of the boolean categories, and perhaps even more importantly, our treatment of categories like VP, TVP, and PossPH (DNP/DNP; e.g. *the father of, 's father*, etc.) which are interpreted as functions from one boolean type to another.

Here we shall only note without further discussion that quantification is not the only semantically interesting property directly represented in the types. Most major categories in the system (a major category is one which is not a proper subset of any other category) have a significant number of subcategories. Of course if C is a subcategory of C' then T_C, the type for C, is a subset of $T_{C'}$, almost always a proper subset. In fact, there is almost a one-one correspondence between categories and types in this system. Thus we have that for all categories C, C' (excluding the variables), $C \subseteq C'$ iff $T_C \subseteq T_{C'}$. There are thus no slash categories in this system. Further, (excluding variables), C and C' are distinct subcategories of C'' iff T_C and $T_{C'}$ are distinct. Thus logical differences between subcategories are directly represented by differences in the sort of set-theoretical objects which may interpret them. There are then no meaning postulates in this system.

For example, *cut* belongs to the category TVP_{+t} , the category of transparent TVPs, and *seek* belongs to the category TVP_{-t}, the category of non-transparent TVPs. Both of course are subcategories of TVP. The fact that (4)a. below entails (4)c. follows directly from the definitions of the types for transparent TVPs and the non-transparent TVPs. (Parentheses and commas are omitted from the expressions below where this will not affect clarity).

(4) a. (John (cut a unicorn))

b. (John (seek a unicorn))

c. ((a unicorn) exist)

In fact the logical subcategorization feature ± *transparent* is given a general (type independent) definition and used to subcategorize very many different categories, e.g. VP, AP, ADVP, FM/FM, etc. The same is true of other subcategorization features such as ± *restricting*, ± *intersecting*, ± *negative restricting*, etc. These features are in general independent and cross-classify. For example we distinguish five subcategories of AP represented lexically by *tiny, Chinese, skillful, fake,* and *alleged*. All the subcategories are interpreted by functions from T_{CNP} into T_{CNP}. Distinct subsets of this set of functions are taken as the types for the distinct subcategories of AP. For example the functions which may interpret APs from the *tiny* or *Chinese* classes must be +transparent whereas those which may interpret any of the other three categories are -transparent. This will guarantee for example that if *John is a tiny (or Chinese) doctor* is true in a possible world j, and further, in that possible world the doctors and the teachers are the same individuals, then *John is a tiny (or Chinese) teacher* will also be true in j. But given j as above, it will not follow that *John is a skillful teacher* is true in j even if *John is a skillful docter* is true in j. The reason is that the functions which interpret adjectives in the *skillful* class need not be +transparent (see section 2 of this paper for the definition of transparency). We might note here that the *skillful* class is very large, and includes all manner adjectives (*rapid, slow,* etc.) and all evaluative adjectives (*skillful, good, beautiful,* etc.).

On the other hand, the functions which can interpret adjectives in the *tiny, Chinese,* or *skillful* classes must meet the condition of being restricting, whereas those which interpret adjectives in the *fake* or *alleged* class are non-restricting. This will guarantee for example that whenever John is a tiny (or Chinese, or skillful) banker then he is also a banker, whereas no such entailments obtain if *tiny* is replaced by adjectives from the *fake* or *alleged* class. And the *fake* class is logically distinguished from the *alleged* class in that functions which interpret *fake,* etc. must meet the condition of being *negatively restricting,* whereas those which interpret *alleged* need not. So if *the painting John bought was a fake Renoir* is true in some possible world j, then also *the painting John bought was not a Renoir* will also be true in j, whereas this need not be the case if *fake* is replaced by *alleged* above. Finally, the *tiny* and the *Chinese* classes are logically distinguished in that while functions which interpret either must be both transparent and restricting, only those which interpret *Chinese* are required to be *+intersecting.* This condition says in effect that a Chinese anything is absolutely Chinese, that is, a Chinese existent. Whereas a tiny something need not be a tiny existent. Whence it follows in this system that if John is Chinese and John is a mathematician then John is a Chinese mathematician, but the analogous argument is not valid if *Chinese* is replaced by *tiny*. And of course *John is tiny* in this system does not entail *John is a tiny existent,* whereas the argument is valid if *tiny* is replaced by *Chinese* (or in fact any relative clause, e.g. *x such that S* for *S* any formula free for *x*, since all relative clauses are provably in the class of +intersecting APs). We note without argument that the *small(expensive houses)* in this system may be disjoint from the *expensive(small houses)* and both may be disjoint from the *(small and expensive) houses*.

Overall then a large number of intuitively valid arguments of English can be shown to be valid in this system, and it is in general the definition of the types which determines validity rather than say the use of meaning postulates. (Though meaning postulates could still be used to express lexical dependencies of the sort *seek = try to find;* we have not concerned ourselves with such dependencies in LT).

Finally, we should note that while the constituent structures of logical expressions in LT often corresponds one to one (ignoring tense and aspect) with English surface forms which express them, there is one systematic mis-correspondence. Namely, nothing in English seems to correspond to the use of the general variable binding operator λ. Thus for example (5)a. below arguably has two readings according as *every man* or *a woman* has wide scope over the other. Our logical expression representing the wide scope reading of *every man* is given in (5)b., and that expressing the wide scope reading of *a woman* in (5)c. The structural description of (5)c. is given in (5)d.

(5) a. Every man kissed a woman

b. ((every,man),(kiss,(a,woman)))

c. ((a,woman),(λ,x,((every,man),(kiss,x))))

d.

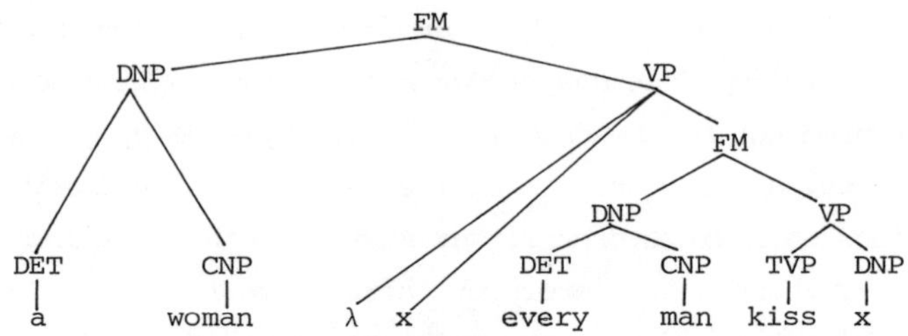

Clearly nothing in the natural constituent structure of (5)a. corresponds to the occurrence of λ in (5)d., and to that extent then (5)d. appears an "unnatural" logical structure for (5)a. We can only offer a partial defense for such structures. Namely, it happens often in this system that the most "natural" interpretation of English sentences with scope ambiguities is represented by a logical expression which does not use lambda operators, as is the case above where most people find the reading expressed by (5)b. the most natural and that expressed by (5)c. the more forced. Similarly the most natural reading of (6)a. below is given by (6)b., in which *a unicorn* is understood non-transparently (e.g. (6)b. does not entail the existence of a unicorn); whereas the transparent reading, (6)c. is less natural and its representation in LT requires the use of the lambda operator.

(6) a. John is looking for a unicorn

b. (John, (look-for,(a,unicorn)))

c. (John, ((a,unicorn),(λ,x,(look-for,x))))

In fact in this system there is a perfect correlation between wide scope of operators and function-argument structure, expressed by (7) below:

(7) Given occurrences d,d' of quantified expressions in an expression E (in LT), d has wide scope over d' iff (i) and (ii) below hold:

(i) E contains an occurrence of (D,D') in which D' is interpreted as a function taking the interpretation of D as argument and

(ii) d is part of D and d' is part of D'

This condition predicts, correctly, that neither *every teacher* or *a student* in (8) below will have the other in its scope:

(8) a. John kissed a student and every teacher

b.

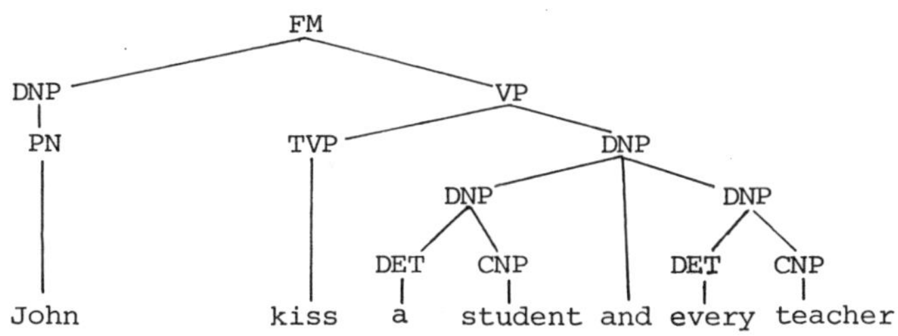

1.2 Quantification and the Boolean Categories in Logical Types

The type for each of the boolean categories (FM, DNP, VP, TVP, AP, and ADVP) in LT is not simply a set of objects of a certain sort, it is rather a set on which are defined three operations: *meet*, $\wedge$ *join*, $\vee$, and complement, '. These operations have the algebraic properties of set theoretic intersection, $\cap$, union, $\cup$, and (set-theoretic) complement, - , respectively. In fact, in the case of DNP, its type is a set whose members are all sets, and the meet, join, and complement operations are precisely intersection, union, and complement.

To see how these types represent quantification, we shall restrict our attention to the types for FM, DNP, VP, and TVP; further, the interpretation of quantification depends only on the *extensions* of expressions in these categories, so we may simplify the exposition by only considering the extensional sublogic separately and then embed it in an intensional logic. All the categories in the extensional logic are in effect categories in the intensional logic and their boolean character carries over completely. We indicate in section 2. of this paper how that is done. So in this section, when we refer to the type for FM, DNP, etc. we are referring to their type in the extensional logic of LT.

Let us consider then the (extensional) type for FM. It is the set {0,1} (0 = *false*, 1 = *true*) on which meet, join, and complement operations are defined as follows: complement, ' , is defined by the standard truth table for negation. That is, 0' is 1, and 1' is 0. Thus in defining the interpretations of complex formulas, the interpretation of (*not*,*S*) will always be the complement of the interpretation of *S*. Similarly the meet operation is essentially the truth table for conjunction. Thus the meet of 1 and 1, denoted $1 \wedge 1$, is 1; and $1 \wedge 0$, $0 \wedge 1$, and $0 \wedge 0$ are all 0. And we shall always interpret a conjunction of formulas (*S*,*and*,*T*) as the meet of the interpretation of *S* and the interpretation of *T*. Similarly the join operation is defined by the standard truth table for disjunction.

The relation indicated above concerning the meet, join, and complement operations and the interpretation of conjoined, disjoined and negated formulas is typical. The type for any boolean category B will be a boolean algebra (and thus a set on which meet, join, and complement operations are defined) and if (e_1,*and*,e_2) is an expression in B, then it will be interpreted as the meet of the interpretation of e_1 with that of e_2. Similarly, (e_1,*or*,e_2) will be interpreted as the complement of the interpretation of e_1.

We should perhaps note that the full specification of a boolean algebra requires specifying a minimal or zero element of the domain of the algebra (the domain of T_{FM} is {0,1}), the maximal or 1-element, and a binary relation $<$. Needless to say the 0-element of T_{FM} is 0, the 1-element is 1, and the binary relation is given by the following set of ordered pairs: $\{\langle 0,0\rangle, \langle 0,1\rangle, \langle 1,1\rangle\}$. We shall not in general be much concerned with the relation and the 0 and 1 elements in the algebras we treat since they are definable given a complete definition of the meet, join, and complement operations. Here we only note that the 0-element of any boolean algebra is that unique element whose meet with any element x of the algebra is 0 itself, and whose join with any element x of the algebra is x itself.

T_{FM}, the {0,1} algebra, is atypical of the algebras we shall define in only one respect: It's domain is constant and finite. That is, it does not vary with the choice of universe of discourse and no matter what model or interpretation we are considering, T_{FM} is always the {0,1} algebra. For other types however, such as T_{DNP}, T_{VP}, etc. the number of elements in the type varies with the size of the universe of discourse U. If U is infinite then these types are also infinite. Hence for these types we will want to use the infinite meet and join operations, $\bigwedge$, and $\bigvee$ respectively.

In fact, it is convenient (but not necessary) to be able to take "infinite" meets and joins in the formula algebra itself (even though we do not allow infinite conjunctions and disjunctions in the syntax). Infinite meet and join are defined on {0,1} as follows: if G is any set of truth values (i.e. any subset of {0,1}) then $\bigwedge G$ is 0 if 0 is in G and otherwise it is 1; and $\bigvee G$ is 1 if 1 is in G and otherwise it is 0. Noting that the inscriptions {0,1} and {1,0} refer to the same set the reader can verify that $\bigwedge\{0,1\} = \bigwedge\{1,0\} = 0$ whereas $\bigvee\{0,1\} = \bigvee\{1,0\} = 1$ on these definitions. Similarly, noting that the inscriptions {1,1} refers to the same set as {1} and that {0,0} refers to the same set as {0}, these definitions also yield that $\bigwedge\{1,1\} = 1$, $\bigwedge\{0,0\} = 0$; and $\bigvee\{1,1\} = 1$, and $\bigvee\{0,0\} = 0$. Thus the infinite operations coincide with their finite analogues. In general when we want to refer to the meet of two particular elements x and y in an algebra we shall write $x \wedge y$ rather than $\bigwedge\{x,y\}$. Similarly for joins.

We consider now the type for CNP. It is the set of functions from the universe of discourse U (assumed non-empty) into T_{FM}, that is into {0,1}. We shall refer to such functions as *properties* (on U) and commonly denote

T_{CNP} as P_U, the set of properties on U. Note that our use of the term *property* does not coincide with its use in PTQ (nor does it in the intensional system of LT).

Given an non-empty universe of discourse U, we shall informally refer to members of U as *entities*. Further certain properties on U will be referred to often enough to deserve special names. Thus given U, *mercy* is that property which maps each b in U onto 1; and *justice* is that property which maps each b in U onto O. (The system has a certain New Testament/Old Testament flavor to it). Further, for each b in U, we define i_b, the *identity property on b*, to be that property which maps b itself onto 1 and everything else in U onto O.

Let us consider now the type for PN (proper nouns). Following Montague's intuition, we shall interpret a proper noun like *John* as the set of properties which hold (assign 1 to) a specified element of U. More explicitly, for each b in U, define I_b, the *individual determined by b*, to be the set of properties in P_U which map b onto 1. The type for PN then will be I_U, the set of individuals, that is, the set of I_b's for each b in U. We might note that no matter what (non-empty) U we pick, for each b in U we have that i_b is a member of I_b. Further, *mercy* is a member of each I_b and is the only property which is a member of each I_b. Similarly *justice* is the only property which is a member of no I_b. Finally, note that there is a one-one onto correspondence between U and the set of individuals. Clearly any I_b, by definition, is the set of properties which holds of some b in U. And given a b in U, I_b is uniquely determined. And given b_o and b_1 distinct members of U, I_{b_o} and I_{b_1} are clearly distinct since I_{b_o} will contain i_{b_o} but will not contain i_{b_1} whereas I_{b_1} will contain i_{b_1} but not i_{b_o}. So no individual is ever a proper part of another individual. We note that there is one logical constant among the CNPs, *existent* is always interpreted as *mercy*. And since every I_b has the property *mercy* it will be the case that every individual is an existent in this system. (So we do not attempt to treat fictional or mythological individuals in this system).

Further, the type for Variables in the extensional system of LT is also I_U, the set of individuals. This is the only case in LT where disjoint categories have the same type. (In the intensional system $T_{VAR} = T_{DNP}$.). Given the one-one correspondence between U and I_U the extensional system is basically a first order system (but see the discussion of T_{AP} in *Logical Types*).

Both PN and VAR are, by rule, subcategories of DNP. What shall we take as the type for DNP? Since we want to form conjunctions, disjunctions, and negations of proper nouns, as in (*John, and, Mary*), (*John, or, Mary*), and (*John, and,* (*not, Mary*)), and we want to directly interpret them as meets, joins, and complements respectively, it will be natural to take T_{DNP} to be a set which contains each I_b, that is each individual, as a member and which is closed under the operations of intersection, union, and complement relative to P_U. We may then directly interpret (*John, and, Mary*) for example as the intersection of the I_b which interprets *John* with that which interprets *Mary*. Thus in any model, (*John, and, Mary*) will be interpreted as the set of properties which John and Mary have in common. (*John, or, Mary*) will be the set of properties which at least one of them has, and (*not, John*) will be the set which John doesn't have, that is, the set of those properties in P_U which are not in the I_b which interprets *John*).

But what about the interpretations of DNPs like *every student, a student, no student,* etc. which are not conjunctions, disjunctions, or negations of proper nouns or variables? Here we crucially need to use infinite intersection and union. We may think for example of an individual student as some I_b which contains the property which the CNP *student* is interpreted as. And we shall interpret *every student* as the set of properties which the individual students have in common. Intuitively then *every student* will denote $I_{b_1} \cap I_{b_2} \cap \ldots$ where each I_{b_i} has the student property. But of course there is no way to say in general how many individual students there will be since U may be as large as we like, and *student* is interpreted as an arbitrary property. We might under some interpretation for example take student to be *mercy* (in which case every individual in that interpretation would be a student).

Somewhat more formally then, *every student* will be interpreted as $\bigcap\{I_b: s \text{ is in } I_b\}$, where s is the property that *student* is interpreted as. Recalling from the definition of the I_b's that a property, such as s is in an I_b iff $s(b) = 1$, we shall in general refer to the interpretation of phrases like (*every, student*) by $\bigcap_{s(b)} I_b$. And we might note that *every* itself is interpreted as that function A from P_U ($=T_{CNP}$) into T_{DNP} such that for every property p, $A(p) = \bigcap_{p(b)} I_b$. Similarly the determiner *a* will be interpreted as that function E from P_U into T_{DNP} which is such that for all properties p, $E(p) = \bigcup_{p(b)} I_b$. Thus (*a, student*) will be the set of properties p such that at least one individual with the student property has p. And *no* will be that function N such that for each property p, $N(p) = \bigcap_{p(b)} (I'_b.)$.

Thus (*no*, *student*) will be the set of properties q such that each individual with the student property does not have q.

In very many respects the definition of T_{DNP} is the cornerstone of our semantics, so we shall give the explicit definition here:

Def: T_{DNP} is the least subset of P^{*}_{U}, the set of all sets of properties on U, such that (i) through (iii) hold:

(i) each I_b is in T_{DNP}

(ii) if Q is in T_{DNP} then Q' (= P_U - Q) is in T_{DNP}

(iii) if G is any subset of T_{DNP} then $\bigcap G$ and $\bigcup G$ are both in T_{DNP}.

In fact we can easily prove that T_{DNP} as defined above is P^{*}_{U} the set of all sets of properties on U. (For typographical reasins we denote by X the power set of X). We may then regard P^{*}_{U} (= T_{DNP}) as the boolean algebra of P_U. Meet, join, and complement are $\bigcap, \bigcup$, and ' (relative to P_U) respectively. The 0-element is of course $\emptyset$, the null set, the 1-element is P_U, and the $\leqslant$ relation is just the subset relation.

The most important fact about T_{DNP} in what follows is that any set of properties Q is identical to some boolean function of individuals, that is some combination of intersections, unions, and complemtns of I_b's.

To see this more formally, it is sufficient to show that for each property p, {p} is expressible as a boolean function of individuals. Then since any set of properties Q is identical to $\bigcup\{\{p\} : p \text{ in } Q\}$ it will follow that any set of properties Q is a boolean function of I_b's. We show then:

Thm 1: for all properties p,

$$\{p\} = \bigcap\{I_b : p \text{ is in } I_b\} \cap \bigcap\{I_b' : p \text{ is not in } I_b\}$$

Proof: Clearly p is in each I_b which it is in, so it is in their intersection. And clearly p is in the complement of any I_b which doesn't contain p (since the complement of such an I_b is the set of properties not in that I_b) and hence it is in the intersection of those complements. Thus p is a member of the set denoted by the right hand side of the equation above since it is in each of the two sets over which the intersection is taken. Hence {p} is a subset of the set denoted by the right hand side of the equation.

Going the other way, let q be any property in the set denoted by the right hand side of the equation. Then q is in each I_b which contains p. That is, for all b in U, if p is in I_b then q is in I_b. But p is in I_b iff p(b) = 1. Ditto for q. Thus we have that for all b in U, if p(b) = 1, then q(b) = 1. Similarly, for all b in U, if p is in

I_b' then so is q (since q must be in each of the sets over which the intersection is taken on the right above). But p is in I_b' iff p is not in I_b iff p(b) = 0. Ditto for q. So we have that for all b in U, if p(b) = 0 then q(b) = 0. Combining these two results we have that for all b in U, p(b) = q(b), which is to say that p is q. But q was arbitrary. Thus any property in the set on the right above is p, so the set on the right above is a subset of {p} and the proof is completed.

The significance of this theorem will be apparent when we consider the type for VP, which we shall now do.

We shall discuss the type for VP in some detail, as the method used in defining this type used in extends almost without change to types for all categories which take DNPs as arguments, e.g. TVPs, n-place predicates in general, Prepositions, and Possessive Phrases (e.g. *the father of*, *'s father*, etc.).

VP is the category FM/DNP, so we expect that a member of VP will be interpreted as some function from T_{DNP} (= P_U^{**}) into T_{FM} (= {0,1}). But is any such function an appropriate interpretation for a VP? Clearly not. Consider for example the formulas (9)a. and (9)b. below.

(9) a. ((John, and, Mary),sleep)

b. ((John,sleep), and, (Mary,sleep))

Clearly any function which is a possible interpretation for *sleep* must be such that if it assigns 1 to the individual which interprets *John* and it assigns 1 to the individual which interprets *Mary* then it must assign 1 to their intersection. Otherwise we would have interpretations in which *John is sleeping* is true and *Mary is sleeping* is true but *both John and Mary are sleeping* is false.

So we want to restrict the possible interpretations for *sleep* to functions which map a meet (intersection) of sets of properties onto what we get when we apply the function separately to each of the sets over which the meet is taken and then take the meet of the results in the formula algebra. Loosely then, functions which are possible interpretations for VPs must *preserve meets*.

By similar reasoning we also want such functions to preserve joins. That is, we want to guarantee that the truth value in any interpretation of *Either John is sleeping or Mary is sleeping* is the same as that of *Either John or Mary is sleeping*. Note that it is infinite meets and joins which must be preserved. Thus if *every student is sleeping* is true in some interpretation then it must be the case that each individual with the student property is sleeping. And if *a student is sleeping* is true it must be the case that for at least one individual with the student property that individual is sleeping.

Finally, we also want to require that possible VP functions preserve complements. That is, the truth value that *sleep* for example assigns to the interpretation of (*not*,(*every*,*student*)) must be the opposite of the one it assigns to the interpretation of (*every*,*student*). So a possible VP function f must map Q', the complement of a set of properties Q, onto the truth value obtained by applying f directly to Q and taking the complement of that truth value. That is, f(Q') must always be identical to (f(Q))'.

Functions which map one boolean algebra into another in such a way as to preserve the boolean operations are, by definition, the (boolean) *homomorphisms* from the one algebra into the second. We shall then take the type for VP to be the set of homomorphisms from T_{DNP} into T_{FM}, which we shall denote by $H_{T_{FM}/T_{DNP}}$. Note that each of these homomorphisms is a *function* from T_{DNP} into T_{FM}. That is, it assigns a truth value to each set of properties Q. But it doesn't assign truth values to Q's arbitrarily. Thus whenever Q is for example the intersection of two other sets of properties, Q_1 and Q_2, the truth value that a homomorphism h assigns to Q must be the same as the meet of the truth value it assigns to Q_1 and the truth value it assigns to Q_2. Similarly for unions and complements. This is a very strong condition for a function from T_{DNP} into T_{FM} to meet. Most functions in that set (which we denote by $F_{T_{FM}/T_{DNP}}$) will not meet these conditions, no matter what non-empty U we pick. Consider for example the function which maps each set of properties onto 0. It must map the interpretation of (*every*,*student*) onto 0 and also it will map the interpretation of (*not*,(*every*,*student*)) onto 0. This function is not a homomorphism since it does not preserve complements.

It is perhaps enlightening to consider just how many of the functions in $F_{T_{FM}/T_{DNP}}$ are homomorphisms. Consider for example the near trivial case where the universe of discourse U contains exactly two members. There will be then $2^2 = 4$ properties on U, $2^4 = 16$ sets of properties, and $2^{16} = 65{,}536$ distinct functions from T_{DNP}, the set of all sets of properties, into {0,1}. But in fact only 4 of the 65,536 functions are homomorphisms! Note that in this example the number of VP homomorphisms is identical to the number of properties on U. This is not an accident of the example. No matter what U we pick, there is a one-to-one correspondence from the set of properties on U onto the set of VP homomorphisms. This point will become evident below.

First, however, let us turn our attention to a somewhat more practical problem. We shall on several occasions want to define particular VP homomorphisms. For example, the logical VP *exist* cannot be interpreted by just any homomorphism. The homomorphism which interprets *exist* must for example always assign value 1 to each I_b. But since a VP homomorphism assigns truth values to each set of properties, to define such a homomorphism we must say what value it assigns to each set of properties, not, it would appear, just the I_b's. Note further that the set of I_b's is a very small subset of the set of all sets of properties. For as we have seen the I_b's are in a one-to-one correspondence with the members of U. So the number of I_b's is always the same as the cardinality of U. But the number of properties on U is always 2^u (where u denotes the cardinality of U), so the number of sets of properties is 2^{2^u}. Thus the cardinality of T_{DNP} is always two powers higher than that of I_U (the set of I_b's). For example if U has 3 members then there are only 3 I_b's but $2^{2^3} = 2^8 = 256$ distinct sets of properties. If U has 4 members then there are 4 I_b's but $2^{2^4} = 2^{16} = 65{,}536$ sets of properties. To define an arbitrary function from T_{DNP} into $\{0,1\}$ for such a U we would have to give the values of the function for over 65,000 arguments. But the homomorphisms are not arbitrary functions; they are ones which meet very special conditions, and it turns out that to define a VP homomorphism it is sufficient (and necessary) to state its values on the I_b's. Once done, its values on all the other sets of properties is uniquely determined. Note first:

Thm 2: Let B be *any* boolean algebra, and h_o and h_1 homomorphisms[4] from T_{DNP} into B. Then $h_o = h_1$ iff for each I_b, $h_o(I_b) = h_1(I_b)$.

The proof of Theorem 2, as well as that of other theorems we cite, is given in LT. But lets note why the theorem is plausible. To say that $h_o = h_1$ is to say that they have the same value when applied to any set of properties. But given that they are homomorphisms, if they assign the same value to any I_b, then they must assign the same value to the complement of any I_b because the value of a homomorphism applied to a complement of a set of properties Q is

---

4 Here and elsewhere when we refer to homomorphisms we only consider the *complete* homomorphisms, that is, those which preserve infinite meets and joins. All the algebras between which homomorphisms are defined in this system are complete, that is, infinite meet and join exist for any subset of the algebra. We shall not continue to repeat this restriction of homomorphism to complete homomorphism.

uniquely determined by its value applied to Q. Similarly if h_o and h_1 have the same values on each I_b then they will have the same value on the intersection of any set of I_b's since the value of a homomorphism applied to the intersection of any collection of sets is uniquely determined by its value on each of the sets in the collection. Analogous claims hold for unions. And by Theorem 1 we know that any set of properties Q is identical to a certain combination of unions, intersections, and complements of I_b's. So if h_o and h_1 have the same values on those I_b's they will have the same value on Q.

Theorem 1 tells us then that in defining a VP homomorphism, once we have specified its values on the I_b's we have specified its values everywhere. But can we specify the values on the I_b's arbitrarily? That is, given any wierdo function g from I_U ($= \{ I_b$: b in U$\}$) into $\{0,1\}$ is there a homomorphism which has just those values on the I_b's? The Justification Theorem for VPs below says the answer is yes.

Thm 3: *(Justification Theorem for VPs)*

Let g be any function from I_U into $\{0,1\}$. There exists a unique homomorphism h_g from T_{DNP} into $\{0,1\}$ such that for each I_b , $h_g(I_b) = g(I_b)$.

Thus we may define a VP homomorphism solely by specifying its values on the I_b's, and we may do that in any way we like. For example, we may define the interpretation of *exist* as that member of T_{VP} which maps each I_b onto 1.

Notice that given any property p, we may define a VP homomorphism h_p by saying that h_p assigns to each I_b the value which p assigns to b itself. Thus we could equally well have defined the interpretation of *exist* to be h_{mercy}, that is, that unique homomorphism which has the same value on any I_b as *mercy* does on b itself. (*mercy* recall is that property which assigns each b in U value 1). Notice further that not only does each property p determine a VP homomorphism h_p, but each VP homomorphism h is determined by some property p. That is, given a homomorphism h, define p_h, *the property determined by* h, to be that property which assigns to each b in U the same value that h assigns to I_b. Clearly $h = h_{p_h}$, since for any I_b (which by Theorem 2 is all we need consider), $h_{p_h}(I_b) = p_h(b)$, and p_h was so defined that for each b in U, $p_h(b) = h(I_b)$. Thus $h_{p_h} = h$.

We may infer then that the VP homomorphisms are in a one-to-one correspondence onto P_U, the set of properties on U. And this is completely intuitive.

In a standard first order logic we define a one place predicate (i.e. a VP) simply by stating what truth values it assigns to the members of the universe of discourse. And that is all we do in this system to define a VP homomorphism. Given a property p, h_p is uniquely determined and there aren't any other VP homomorphisms besides those which are determined by properties.

But treating the VPs as homomorphisms gives us a bonus: Once we have specified a property, and thus a homomorphism, we have not only specified the truth values of the homomorphism on the individuals (the I_b's) but we have determined its values for intersections, unions, and complements of I_b's, that is for all possible denotations of negative and quantified noun phrases.

Finally, consider a last case in which we want to define VP homomorphisms. VP is itself a boolean category, so we want to directly interpret e.g. (*sing, and, dance*) as the meet of the homomorphism which interprets *sing* with that of the one which interprets *dance*. Analogously for (*sing,or,dance*). And we want to interpret (*not,sing*) as the complement of the homomorphism which interprets *sing*, etc. That is, given h_o and h_1 in the set of homomorphisms from T_{DNP} into T_{FM}, we want to define $(h_o \wedge h_1)$ to be itself a homomorphism in that set. By Theorem 2 and the Justification Theorem we may define such a homomorphism merely by specifying its values on the I_b's, and we may do that in any way we like. For example, we may define $(h_o \wedge h_1)$ to be that VP homomorphism which assigns to any I_b the truth value obtained by applying h_o to I_b and h_1 to I_b and taking the meet of the results in the formula algebra. And our linguistic intuitions support the correctness of this definition. Clearly we want a sentence like *John is both sleeping and snoring* to have the same truth value in any interpretation as the conjunction of *John is sleeping* and *John is snoring*. Extending the definition to the infinite case, we may define for each set H of VP homomorphisms, $(\bigwedge H)$ is that VP homomorphism which assigns to any I_b the meet in the formula algebra of the set of $h(I_b)$ for each h in H.

Note that we define the meet of H solely by referring to the values each h in H has on the I_b's. We shall informally refer to such definitions as being done *pointwise on the* I_b's. Similarly we define joins of VP homomorphisms, $(\bigvee H)$, by $(\bigvee H)(I_b) = \bigvee\{h(I_b)\text{: in } H\}$. This will guarantee for example that *John is sleeping or snoring* will always have the same truth value as *John is sleeping or John is snoring*. Similarly we define complement by $(h')(I_b) = (h(I_b))'$, so (*John*, (*not,sleep*)) will always have the opposite truth value from (*John, sleep*), that is, the same truth value as (*not*, (*John*, *sleep*)), which is intuitively correct.

We shall then take the type for VP to be the set of homomorphisms from T_{DNP} into T_{FM} regarded as a boolean algebra under the operations $\wedge$, $\vee$, and ' as defined above.

Further, it is important to realize that the definitions of the boolean operations on T_{VP} are done pointwise on the I_b's, not pointwise on arbitrary sets of properties. We shall illustrate this by showing that *John is singing or dancing* is logically equivalent in this system to *John is singing or John is dancing*, but that the corresponding equivalence fails if *John* (which is necessarily interpreted as an I_b) is replaced by say *every student* (which will denote an intersection of I_b's and so will not in general be an I_b. E.g. in a model in which there are two (or more) I_b's each of which have the student property, the intersection of all the I_b's with the student property will be a proper subset of each I_b with the student property, and cannot be itself an I_b since, as we have seen, no I_b is a proper subset of another. The subset must be proper since it will not contain any i_b's, and each I_b contains its i_b.).

We may represent the interpretation of a formula in this system by a *semantic tree* whose root node gives an algebraic expression of the truth conditions of the formula. For example, the tree for *John is singing and dancing* would be:

(10)

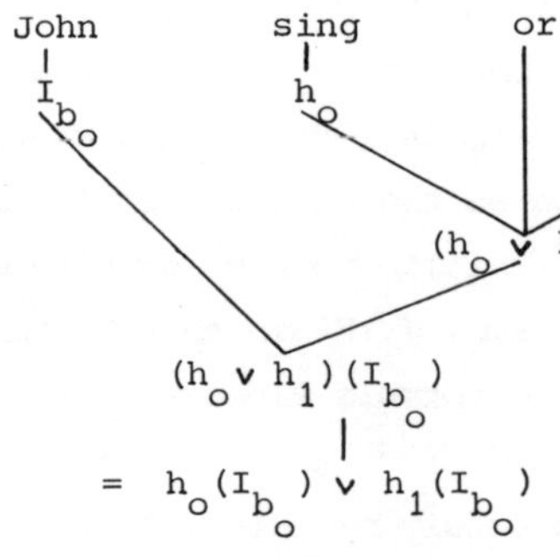

by the definition of join in the VP algebra.

In this tree, I_{b_o} represents an arbitrarily chosen individual, and h_o and h_1 are arbitrarily chosen members of T_{VP}. Notice that if the argument of the next to the last line in the tree were not an I_b we could not appeal to the definition of join in the VP algebra to derive the last line since the definition of join in the VP algebra only says that a join of VP homomorphisms applied to an I_b is the same truth value as join in the formula algebra of each of these homomorphisms applied separately to that I_b.

The fact that (10) is logically equivalent to (11) in LT is seen to be trivial, since the root nodes of the two trees are identical.

(11)

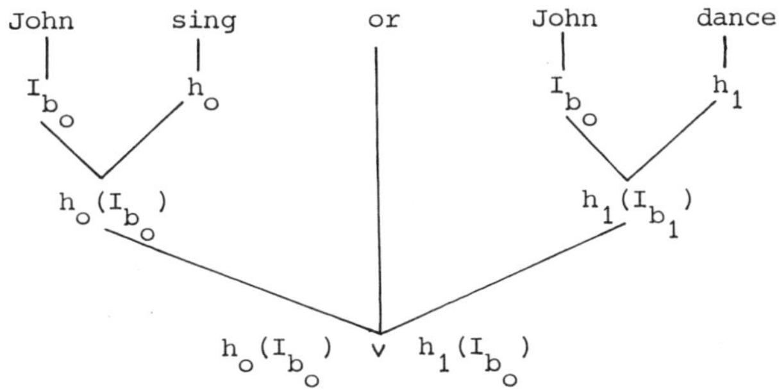

But consider now the semantic trees for *every student is singing or dancing* and *every student is singing or every student is dancing.*

(12)

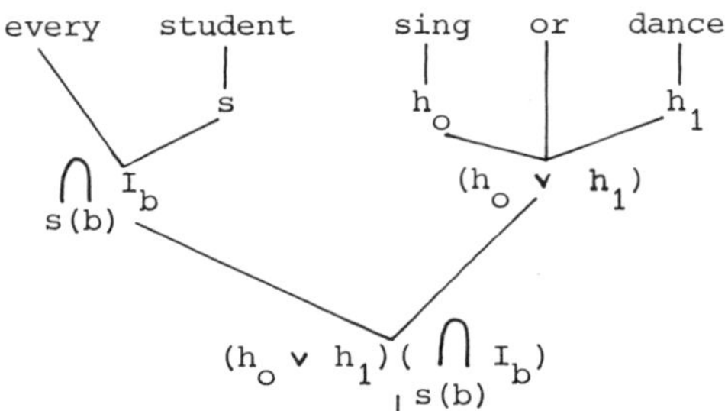

$$= \bigwedge_{s(b)} [(h_o \vee h_1)(I_b)]$$ since $(h_o \vee h_1)$ is a VP homomorphism

$$= \bigwedge_{s(b)} (h_o(I_b) \vee h_1(I_b))$$ by def. of the VP algebra.

The last line in (12) is a meet of truth values. It has value 1 just in case each value in the set over which the meet is taken is in fact 1. But each value in the set is expressed by $(h_o(I_b) \vee h_1(I_b))$ where b has been chosen arbitrarily from those which the student property s assigns 1 to. And the join will have value 1 for a given choice of b just in case at least one of h_o or h_1 assigns the I_b value 1. In other words, (12) is true in a given interpretation iff for each I_b which has the student property (i.e. such that s is true of b) it is the case that that I_b is either singing or dancing.

But consider now the tree for *every student is singing or every student is dancing:*

(13)

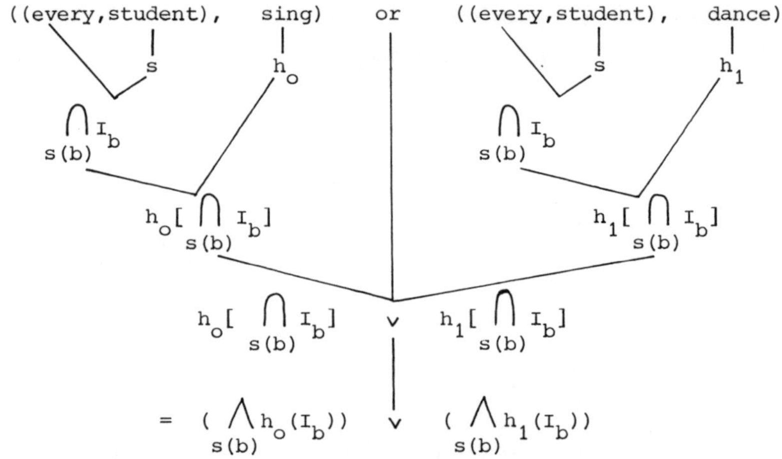

Note that the left hand element in the next to the last line may be replaced by the left hand element in the last line since h_o is a homomorphism and thus preserves meets. Similarly for h_1.

Now the last line in (13) is a join of two truth values, and so has value 1 iff at least one of the values over which the join is taken has value 1. But the left hand expression will have value 1 iff $h_o(I_b)$ has value 1 for each I_b with the student property, i.e. for each b such that $s(b) = 1$. Similarly the right hand side has value 1 iff $h_1(I_b)$ is 1 for each I_b with the student property.

It is now immediate that the last line of (13) does not necessarily denote the same truth value as the last line of (12). For the last line of (13) says in effect that either h_o assigns each I_b with the student property value 1 or h_1 assigns each I_b with the student property value 1. Whereas the last line of (12) says in effect, pick any I_b with the student property. Either h_o assigns that I_b value 1 or h_1 does. To formally construct a model in which (12) is true and (13) false it is sufficient to take $U = \{b_o, b_1\}$, interpret *student* as *mercy*, *sing* as that VP homomorphism which assigns I_{b_o} value 1 and I_{b_1} value 0, and *dance* as the complement of *sing*, i.e. *dance* assigns I_{b_o} value 0 and I_{b_1} value 1. Then clearly for each I_b with the student property, that is, for each I_b, either *sing* assigns that I_b value 1 or *dance* does, so (12) is value 1. But neither *sing* nor *dance* assign each I_b value 1, so (13) has value 0.

By a similar use of semantic trees the reader may verify that (2) and (3) given earlier in this paper are not logically equivalent.

Let us step back for a minute and consider the general method employed in constructing the type for VP. Overall it is terribly simple. We take the type for VP to be the homomorphisms from T_{DNP} into T_{FM} and define a boolean structure on that set pointwise on the I_b's.

We may construct the type of TVP, and indeed for n-place predicates in general, in direct analogy to that for VP. Thus TVP is the category VP/DNP and so its type will be some set of functions from T_{DNP} into T_{VP}. Noticing that each of these types is a boolean algebra we might reasonably guess that the type for TVP will be the set of homomorphisms from T_{CNP} into T_{VP}. And we might expect to define a boolean structure that set of homomorphisms pointwise on the I_b's just as was sone for T_{VP}. Both steps seem reasonable.

Thus we want e.g. (*kiss*, (*John*, *and*, *Mary*)) to be the same VP function as ((*kiss*, *John*), *and*, (*kiss*,*Mary*)). That is, if an idividual kissed John and that individual also kissed Mary then he kissed both John and Mary; and if he kissed both of them then he kissed each one. Which is to say that a TVP like *kiss* should map a meet of I_b's onto the corresponding meet in the VP algebra. Analogous claims hold for joins and complements. Briefly, if John kissed either Susan or Mary then John either kissed Susan or kissed Mary, which by the definition for join now in the VP algebra says that John kissed Susan or John kissed Mary.

We shall then take the type for TVP to be the set of homomorphisms from T_{DNP} into T_{VP}. And we shall define meet, join, and complement operations on TVP pointwise on the I_b's. Thus (*hug*, *and*, *kiss*)(*Mary*) will be interpreted as the same VP function as ((*hug*,*Mary*), and, (*kiss*,*Mary*)). Analogous definitions will obtain for *or* and *not* applied to TVPs. Note that the definitions will guarantee that while (*hug or kiss*) *Mary* is the same VP as (*hug Mary*) *or* (*kiss Mary*), they will also guarantee that (*hug or kiss*)(*every student*) will not in general be the same VP as ((*hug*(*every student*)) *or* (*kiss*(*every student*)), which is correct.

Clearly this approach generalizes immediately to a general definition of n-place predicates for $n \geqslant 0$. We may define them inductively as follows: P_0, the 0-place predicates, is simply T_{FM} as already defined. And for all $n \geqslant 0$, if P_n is defined then P_{n+1} is defined to be the set of homomorphism from T_{DNP} into P_n, regarded as a boolean algebra under the operations of $\wedge$, $\vee$, and ' defined pointwise on the I_b's.

The reader might notice, however, that to define each P_n in this way we need a theorem analogous to the Justification Theorem for VPs ($=P_1$). For example, to define P_2, namely the type for TVPs, we must prove that every function from the I_b's into P_1 ($= T_{VP}$) extends uniquely to a homomorphism. This is in fact the case. The two theorems we need to show this are given below:

Thm 5: The justification Theorem[5]

Let g be any function from I_b into B, where B is any boolean algebra isomorphic to a power set algebra. Then there is a unique (complete) homomorphism h_g from T_{DNP} into B such that for each I_b, $h_g(I_b) = g(I_b)$.

Thm 6: For each $n \geqslant 0$, P_n (the n-place predicates) is isomorphic to the power set algebra U^n.

(We assume in Theorem 6 that $U^0 = \{\emptyset\}$).

This generalized form of the Justification Theorem is (believe it or not) linguistically suggestive. We have only used the theorem to justify our treatment of n-place predicates. And this in itself is not insignificant. We have covered all the classical uses of quantification (divorced from variable binding) in standard first order logic, and obtained the same results. But our logical structures conform much better than those of standard logic to natural language surface forms, and further they make it clear that there is no inherent link between quantification and variable binding. We have not used variables at all in our definitions of the n-place predicates.

But notice that the Justification Theorem does not mention specifically the n-place predicates. Rather it only refers to homomorphisms from T_{DNP} into a boolean algebra B, required only to be isomorphic to a power set algebra, i.e. to an algebra which is isomorphic to the algebra of the set of all subsets of some set. This suggests that any category in our system of the form C/DNP, may be treated as set of homomorphisms from T_{DNP} into T_C, provided T_C is a boolean algebra isomorphic to a power set algebra. Of course a suggestion is only a suggestion, not a logical necessity. We did after all justify on linguistic grounds that we wanted to interpret e.g. VPs like *sleep* as meet, join, and complement-preserving functions.

---

5 We are indebted to Jonathan Stavi and John Barwise for pointing out an error in the original statement of the Justification Theorem in LT. The restriction of the theorem so that the range algebra is isomorphic to a power set algebra rectifies the error.

There are nonetheless other categories in LT of the form C/DNP, C boolean, for which the homomorphism analysis does seem appropriate. In fact, for all the categories of that form so far investigated the homomorphism analysis seems linguistically appropriate.

Consider for example the category DNP/DNP. This category includes expressions like (*'s, father*), (*the-of, father*), etc. In LT all such expressions are derived, (though we could equally have treated expressions like *father* as being in DNP/DNP and treated at least the *'s* as a surface case marking phenomenon). Thus expressions like (*'s, father*) will combine with DNPs like *John*, (*every, student*), etc. to form further DNPs like (*John*, (*'s, father*)), ((*every, student*), (*'s, father*) etc. It is natural then to test the hypothesis that PossPHs (= Possessive Phrases, i.e. the category DNP/DNP) are homomorphisms from T_{DNP} into T_{DNP} itself. Since T_{DNP} is defined as a power set algebra, the subsets of P_U, we may define homomorphisms from that set into itself simply by specifying their values on the I_b's.

To be sure, Possessive Phrases in English cover a multitude of logical sins, so we have only attempted in LT to capture the most elementary type, those formed from primitive function nouns like *father* and *mother*. Thus we interpret such function nouns as partial functions on U. (f is said to be a partial function on U iff there is a subset U_f of U such that f is a function from U_f into U). Then *'s* (and *the-of*) is taken as a syncategorematic symbol interpreted as a function mapping partial functions on U onto homomorphisms from T_{DNP} into T_{DNP} as follows: for each I_b, $('s(f))(I_b) = I_{f(b)}$ if b is in the domain of f. Otherwise $('s(f))(I_b) = \emptyset$ (the null set). (Recall that the null set is the O-element in the DNP algebra). To take an example: Suppose in some interpretation that *father* is interpreted as a partial function f on U such that say $f(b_o) = b_1$, and that further *John* is interpreted as I_{b_o}. Then (*John*, (*'s, father*)) will be interpreted as I_{b_1}. On the other hand, if b_o were not in the domain of f, and so naively, the individual determined by b_o does not have a father, then (*John*, (*'s, father*)) would be interpreted as the null set. This latter case in fact receives a rather interesting, if contentious, treatment in LT.

Note that a homomorphism h from a boolean algebra B_1 into a boolean algebra B_2 must preserve O-elements; that $h(O_B) = O_{B_2}$. So if (*John*, (*'s, father*)), (or to take a more famous example, *the king of France*, is interpreted as the null set, that is, the O-element of the DNP algebra, then each VP homomorphism must map it onto the O-element of the FM algebra, that is, O. In such a case then ((*John 's father*), *sing*) would be false and so would ((*John 's father*),(*not,sing*)). The result is not too unintuitive. After all if John was the product of a virgin birth then the claim that his father is definitely singing is surely false, and the claim that his father is definitely not singing hardly seems more true. Notice, however, that in this case VP negation, e.g. ((*John,* ('*s,father*)),(*not,sing*)) yields a different truth value from sentential negation (*not,* ((*John,* ('*s,father*)),*sing*)). The former is, as we have just shown, false in the case where John has no father. But since ((*John,* ('*s,father*)),*sing*) is also false, it follows that its complement, expressed by (*not,* ((*John,*('*s,father*)),*sing*))) is true. So *John's father isn't singing* will be false, but *it is not the case that John's father is singing* will be true. Thus we have a clear distinction in this system between "internal" negation and "external" negation, which seems about the best we can do in a strictly two valued system.

Consider now the advantage of taking *'s father* as a DNP homomorhism. The interpretation of ((*a,student*),('*s,father*)) will be represented as in the tree below:

(14)

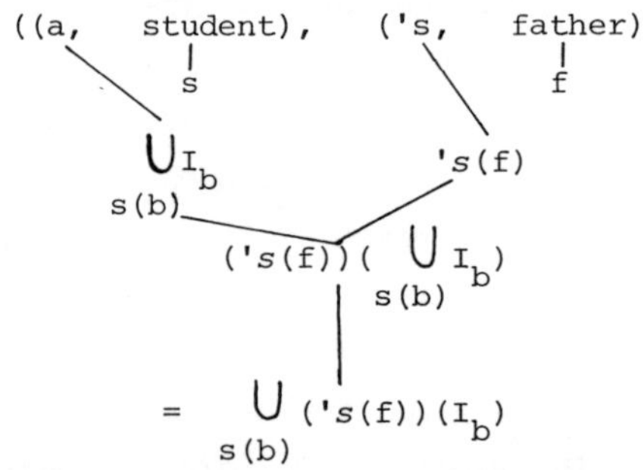

(We have used *'s* itself to refer to the function it is interpreted as.). The step to the last line above is justified since *'s*(f) is a DNP homomorphism and thus maps unions onto unions. Now, just what set of properties does the last line refer to? What it says is, take each individual I_b which has the student property and form that (mixing levels) I_b's father. That will either be another I_b or the null set. Now take the set of I_b's fathers and form their union. If a VP homomorphism assigns value 1 to that union it must assign 1 to at least one of the I_b's fathers over which the union was taken. That is, *A student's father called me up yesterday* will be true iff for some individual with the student property, that individual's father called me up yesterday. And this is intuitively correct. Similarly, *Every country's president attended the meeting* will be true iff for each country, it's president attended the meeting. And this seems basically correct.

Further, as per the Scope Condition in (7), *Every country's president rented a car* will, on its natural reading, say that *every country* has wide scope over *a car*. And that also seems correct. Further, it is not forced that every country have the same president, which is clearly correct. But this latter claim is less convincing if *every country* is replaced by a simple conjunction of proper nouns. Does *France and Germany's president rented a car* have a reading on which France and Germany are understood to have the same president? Our intuitions are not clear on this point. The case for sameness seems stronger in *John and Mary's father rented a car* (in all cases of course the conjoined NPs are assumed to be under the scope of *'s*, i.e. we are not considering the reading, naturally represented in LT, of the above sentence on which it means *Mary's father and John rented a car*). On the other hand, our judgment of sameness of father seems weaker if the conjoined possessive is not a subject. Thus *I talked to both John and Mary's father yesterday* seems not too unnatural.

Overall then the homomorphism analysis of possessives seems reasonably well justified on first pass, but clearly we have merely begun a full analysis of possessives.

Finally there is a last, rather more complex category in our system which maps DNP meanings into a boolean algebra; Namely, the Prepositions. For simplicity let us restrict our attention to PPs like (*in*, (*the*, *garden*)) etc. to their use as VP modifiers. That is, Preps will combine with a DNP to give us something of the category VP/VP. We are motivated to define a boolean structure on this latter category, since we want to directly interpret conjoined, disjoined, and negated members of that category. That is, we will form derived VPs like (*work*, (*slowly*, *or*, *carefully*)), (*work*, (*here*, *and*, (*in*, *Chicago*))), even (*work*, (*not*, *carefully*)). The latter will receive the same interpretation as (*work*, *and* (*not*, (*work*, *carefully*))), just as in the AP algebra, the property ((*not*, *smart*), *man*) will be true of a b in U iff *man* is true of that b and (*smart*, *man*) is false of that b. In fact the algebra on the restricting adverbs is isomorphic to that defined on the restricting APs, which is perhaps not surprising. More surprising perhaps is that these algebras are in fact isomorphic to a power set algebra. Thus by the Justification Theorem we may take homomorphisms into them solely by defining their values on the I_b's. And we are motivated to do so.

Thus we want (*in*, (*every*, *city*)) to mean the same as, roughly, *in* $city_1$ *and in* $city_2$ *and* .... That is, *in* will map an intersection of sets of properties onto the corresponding meet in the Adverb algebra. Similarly we want (*in*, (*Chicago*, *or*, *New York*)) to mean the same as ((*in*, *Chicago*), *or*, (*in*, *New York*)). Ultimately this will mean that *John is living in Chicago or New York* will have the same truth value as *John is living in Chicago or in New York*, which in turn will be the same as *John is living in Chicago or is living in New York*, and this in turn, by the VP algebra already defined, will have the same value as *John is living in Chicago or John is living in New York*.

Finally, we should point out, as mentioned at the beginning of this paper, that the semantics of the stative locative PPs like *in the garden* provides a nice example of a linguistic generalization in this system based on the high degree of logical similarity of VPs and TVPs (both are homomorphisms from the DNP algebra, they differ only with regard to the range algebra). Skipping formal details, and noting that although tense and aspect are not represented in this system we are forcing simple sentences to have the meaning they would have in English if understood at a single point in time, we may characterize the semantics of the locative PPs as follows: They define in effect a VP which is understood to hold of the argument of the function they apply to, regardless of whether that function is itself a VP or a TVP one. Thus *sing in the garden* will assign to any I_b the same value as *sing* does if that I_b is in fact in the garden. But if that I_b is not in the garden then *sing in the garden* will assign to that I_b the same value as does the 0-element of the algebra in which the argument (*sing*) of the PP takes its denotations. Since *sing* is interpreted as a member of the VP algebra, and the 0-element of that algebra is the VP homomorphism which assigns each I_b value 0 we have that *John is singing in the garden* will be false if John is not in the garden, whereas if John is in the garden then the value of *John is singing in the garden* will be the same as that of *John is singing*.

The analysis of a derived TVP like *see in the garden* is the same. That is *see in the garden* will assign an I_b the same VP meaning as *see*(I_b) if in fact I_b is in the garden. If not, then *see in the garden*(I_b) has the same interpretation as $0_{see}(I_b)$ where 0_{see} is the 0-element in the algebra in which *see* takes its denotations. (That element is the one which maps each I_b onto the 0-element of the VP algebra). This will mean for example that if John wasn't in the garden then *No individual saw John in the garden* will be true. Which is intuitively correct.

It follows from this analysis (see the more extensive discussion in LT) that *John sang in the garden* entails *John was in the garden*, whereas *John saw Mary in the garden* entails *Mary was in the garden*. In this respect then "absolutives" (Subjects of intransitive verbs and objects of transitive verbs) are logically alike. In fact, holding the point of time constant, *John sang in the garden* is logically equivalent to *John sang and John was in the garden*, whereas *John saw Mary in the garden* is equivalent to *John saw Mary and Mary was in the garden*.

To conclude this section let us observe that many of the categories in the logic of LT are ones which take quantified NPs among their arguments and which semantically treat them homomorphically. Thus, within in effect a traditional first order semantics we may state the value of any function taking quantified noun phrase interpretations as arguments solely by treating these functions as homomorphisms and stating their values on the individuals, as must be done in a classical first order system. Further the boolean treatment of *and*, *or*, and *not* allows us to interpret conjoined, disjuncted, etc. members of homomorphic categories directly as homomorphisms of the right sort without having to map them into conjoined, disjuncted, etc. formulas, as is done traditionally in generative grammar (e.g. so called Conjunction Reduction rules) as well as in PTQ in some cases. That is, in both these traditions *John is singing and dancing* would be interpreted by in effect mapping *sing and dance* onto a conjunction of formulas. In the generative grammar case the entire sentence *John is singing and dancing* would itself be interpreted as *John is singing and John is dancing*. But to our knowledge no satisfactory approach along these lines has ever been able to account for the cases in which the subject was a quantified NP. That is, we do not want to interpret *A student* (*or no student*) *is both singing and dancing* by *A student* (*no student*) *is singing and a student* (*no student*) *is dancing*. The translation of *sing and dance* in PTQ is less massive, but it is still the case that to interpret *sing and dance* we must form the conjoined formula *sing*(x) *and dance*(x) and then reconvert it to a VP with the use of a lambda operator, (λ,x,(*sing*(x) *and dance*(x))).

Our treatment of quantification does, however, disassociate quantification from variable binding, the latter being treated, as in Montague Grammar by a lambda operator in the general case (though we do use special operators for reflexives and relative clauses for reasons of linguistic naturaleness -- formally they are not necessary as we could make lambda do all the work). Notice, however, that the lambda operator itself can be stated with considerable generality using the boolean types. Thus if e is an expression in any boolean category B and e has a free occurrence of x, then (λ,x,e) is an expression in the category B/DNP and is interpreted as a homomorphism from

T_{DNP} into T_B.

Before proceeding to indicate how the extensional system of LT extends to an intensional system, preserving the boolean properties of the original, let us briefly countenance one objection the reader may have to the homomorphism approach to quantification we have developed. Namely, it might be objected to our analysis of say VPs that not all ordinary VPs in English in fact behave homomorphically as we have indicated. And the point is correct, though not we feel an objection to our approach, but rather an advantage of it.

Consider first some classes of VPs in English which cannot be interpreted by (first order) homomorphisms in the way that e.g. *sleep*, etc. can. All reciprocal VPs are in this category. Thus to say that *John and Mary love each other* is not to claim that *John loves each other* and *Mary loves each other*. Indeed the latter two "sentences" don't make any sense at all. Furthermore, quite a variety of logically reciprocal predicates can be formed in English. Thus many adjectives are inherently reciprocal, as in *parallel lines, antogonistic forces, adjacent houses, similar proposals*, etc. and cannot, obviously, be interpreted by homomorphisms. The last sentence above does not mean that *this line is parallel and that line is parallel*, sentences which are at best highly elliptical. Other reciprocal predicates can be productively formed by the use of rather logical sounding "adjectives" like *same, different, similar*, etc. E.g. *John and Mary live in the same town, go to different schools*, etc.

A rather different category of non-homomorphic VPs are illustrated by predicate copular constructions in which the following DNP is numerically plural. E.g. *John and Mary are the two students I like best* cannot, obviously, mean *John is the two students I like best and Mary is the two students I like best*. Similarly for *John and Mary were the only survivors*, etc.

Another rather different type of case is cited in Gil (1977) and concerns certain derived nominals, as in *John and Mary are a bore*, which does not entail *John is a bore and Mary is a bore*, although in this case the attempted "homomorphic paraphrase" at least makes sense. Somewhat similar claims hold for VPs formed from adverbs like *together, between them*, etc.

It seems to us an advantage of our system that it provides a clear test of the logical differences between say reciprocal predicates and, let's call them distributive predicates, like *sleep*. If we wanted to add reciprocal predicates to LT we could just add a new subcategory of VP, much as in the intensional logic we have already two subcategories of VP, the +transparent ones and the -transparent ones. And in fact, the -transparent ones do not behave homomorphically. Thus non-transparent VPs like *must be a citizen*, *be required to be over 35*, etc. are clearly not interpretable homomorphically. Thus *either the President or the Vice-President must be over 35* will not entail, much less be equivalent to, *Either the President must be over 35 or the Vice-President must be over 35*. It does appear incidentally that most non-transparent functions on DNP meanings are nontheless additive, in the sense that they do preserve meets (but not joins or complements). Thus in the intensional logic of LT *The President and the Vice-President are required to sign the bill* will be equivalent to *The President is required to sign the bill and The Vice-President is required to sign the bill*. By the same restriction on non-transparent TVPs we have e.g. that *John is looking for a white unicorn* (non-transparently) will entail (non-transparently) *John is looking for a unicorn*. Still, preserving meets does not make a homomorphism.

There is then no problem adding non-homomorphic subcategories to the system. Indeed the system is designed for subcategories. What is of interest here is that reciprocal predicates, etc. must, at the very least have a different logical subcategory from the distributive VPs. For if we enlarge the class of functions which can interpret *sleep*, etc. we will admit non-homomorphisms into the class and hence we can find cases where e.g. *Not every student is sleeping* will have the same truth value as *Every student is sleeping*. We count it as an advantage of our approach that it makes clear in what respects the distributive VPs behave differently than say reciprocal ones.

Further, it seems to us doubtful that the extensional reciprocal VPs will simply be another class of first order VPs and take their denotations among the non-homomorphisms in the set of functions from T_{DNP} into T_{FM}. Intuitively a VP like *love each other* should at least be able to take *sets* of I_b's as arguments. So *John and Mary love each other* predicates something of the set whose two members are John and Mary. Similarly *every student and every teacher were insulting each other* seems to predicate something of the set whose two members are denoted by *every student* and *every teacher* and seems (roughly) to say that each member of either of those sets was insulting each member of the other set (but not necessarily each other member of its own set). If further investigation reveals that the arguments of reciprocal VPs should be sets of subsets of P_U^{**} then they will have to be drawn from a much larger set of functions than the one from which the distributive predicates are drawn, since there are 2^n subsets of P_U^* where n is the cardinality of P_U^*

We are currently working on an extension of the system in LT which is properly higher order and takes individual collectivities as the starting point. (E.g. given B a subset of U, I_B, *the individual collectivity determined by* B, will be the set of functions from U^* into {0,1} which assign B value 1). The results so far are suggestive, but many problems remain.

2. *The Intensional Logic of LT*

Space will not permit a thorough statement of the intensional system of LT here, and fortunately that is not necessary for understanding how the homomorphism approach to quantification of the extensional logic carries over to the intensional logic. We shall here then simply present informally the types for certain of the "core"primitive categories and functional categories indicating how the algebraic structure of the extensional logic is "lifted" into the intensional logic and how the properly intensional, that is, non-transparent, categories are treated.

Syntactically speaking the intensional logic is presented in the same categorial framework as the extensional logic. The only major difference between the two is that there are many more categories and subcategories in the intensional logic. Every category in the extensional logic, however, corresponds to an identifiable category in the intensional logic, though it may now be merely a subcategory of a larger category. In addition there are a few new syncategorematic symbols and thus a few new derivational rules, such as the ones which derive *ly* adverbs from adjectives (e.g. *slow+ly* from *slow*), *-ed* adjectives from sentential transitive verbs (e.g. *allege+ed* from *allege*), "Raising Predicates" from sentential transitive or intransitive verbs (e.g. *seem+to* from *seem*, *believe+to* from *believe*), and "Equi Predicates" from sentential transitive and ditransitive verbs (e.g. *hope+to* from *hope*, *persuade+to* from *persuade*), etc.

We shall preceed directly to the discussion of the types for the core categories.

In the extensional logic the types were defined relative to a non-empty universe of discourse U (and of course the set of two truth values 0 and 1). In the intensional logic types will be defined relative to a non-empty U and a non-empty set J, called the *possible world index set*. Members of J will be referred to as *possible worlds*, though it is more useful to think of the j's in J as indices for interpretations of the language in the extensional sense.

So for C a category in the intensional logic, T_C will, as before, denote its type. But the definition of T_C will now depend on J as well as, usually, on U. In the case where C is also a category in the extensional logic and we want to refer to its type in the extensional logic we shall denote it by T^e_C.

The types for the primitive categories (FM, PN, VAR, DNP, CNP, FNP) in the intensional logic are almost trivially constructed from J and their types in the extensional logic as follows: if C is a primitive category in IL (the intensional logic) then, with a few restrictions to be noted below, T_C is simply the set of functions from J into T^e_C, denoted in general by $F_{T^e_C/J}$. So for example T_{FM} is now the set of functions from J into the {0,1} algebra as already defined. We will refer to these functions as *propositions*. Similarly T_{CNP} is now $F_{P_U/J}$, the set of functions from J into P_U, the set of properties on U. T_{DNP} is now $F_{P^*_U/J}$, and moreover, T_{VAR} is now the same as T_{DNP}, in contrast to its treatment in the extensional logic where it was identical to T_{PN}. T_{PN} is now simply the set of functions from J into T^e_{PN} (= $\{I_b$: b in U$\}$), although we might consider restricting T_{PN} to be the set of constant functions from J into T^e_{PN}.[6]

Let us consider more explicitly T_{FM} in the intensional logic (IL). As in the extensional logic (EL) we shall want to interpret conjunctions of formulas as meets of the interpretations of the conjuncts, etc. So we still want T_{FM} to be a boolean algebra. And it is obvious which algebra it should be. Clearly we want say *Nero fiddled and Rome burned* to be true in j (that is,

---

6 The interest of this restriction, brought to our attention by Barbara Partee, is at least in part, that we could perhaps provide a more adequate semantics for counterfactuals in that in such (and other) cases the denotations of definite descriptions seem to be able to vary in a way in which those of proper nouns do not. Assuming the restriction on the type for PN, the semantics for (i) and (ii) below would be intuitively slightly different.

(i) If John were a linguist he wouldn't say such silly things
(ii) If the President were a linguist he wouldn't say such silly things

(i) would be true in a possible world j if in some possible would j' the individual we call John in j had the property which *be a linguist* determines in j' and the consequent holds in j'. But (ii) would be true if the interpretation of *the President* in j' has the property denoted by *be a linguist in j'*.

We were not specifically concerned for such sentences in LT and did not impose this restiriction on the type for PN. It is, however, completely unproblematic to add it.

to map j onto 1) iff both *Nero fiddled* is true in j and *Rome burned* is also true in j. More formally, given two propositions p,q we shall define their meet, $(p \wedge q)$, to be that proposition whose value at a possible world j is the meet (in the {0,1} algebra now) of the value of p in j and that of q in j. Similarly we want the negation of a formula *S* to have the opposite truth value in any j from the one *S* itself takes in j. Generalizing now to the infinite case, and ignoring $\bigvee$ whose definitions are always copies of those for meets replacing $\bigwedge$ by $\bigvee$, we may define T_{FM} to be the set of functions from J into T^e_{FM} , (the {0,1} algebra) regarded as a boolean algebra under the operations of $\bigwedge$and ' as defined below:

(15) a. for P any set of functions from J into {0,1},$(\bigwedge P)$ is that member of that set of functions which is such that for

all j in J, $(\bigwedge P)(j) = \bigwedge_{p \in P} p(j)$

b. for p any function from J into {0,1},(p') is that function from J into {0,1} such that for

all j in J, $(p')(j) = (p(j))'$.

We must of course prove that T_{FM} so defined is in fact a boolean algebra, but the proof is merely an elementary exercise.

Notice that to define T_{FM} in IL as a boolean algebra we have simply "lifted" the algebra of T^e_{FM} onto the set of functions from J into T^e_{FM} . The completely analogous definitions lift the DNP algebra in EL onto the set of functions from J into T^e_{DNP} , that is, T_{DNP} in the intensional logic. To actually give the definition simply replace {0,1} in (15)a. and (15)b. above by P^*_U . This will say for example that (*John, and, Mary*) is interpreted by that DNP intension, that is, that function from J into the set of all sets of properties on U, which assigns to any j in J the intersection of what the interpretation of *John* assigns to j and what the interpretation of *Mary* assigns to j. For the primitive categories then there is no problem assigning them a boolean structure in a natural way. The more interesting problems arise when we consider the functional categories, in particular ones which are not, naively, transparent (extensional). To treat these categories let us consider first somewhat more closely the nature of the types for the primitive categories. We take T_{CNP} as an example.

T_{CNP} is the entire set of functions from J into P_U (= T^e_{CNP}). We may refer to members of T_{CNP} as *intensional* properties, in distinction to members of P_U which may now be called *extensional* properties. And in general for C any category in IL, members of T_C may be referred to as *intensions of type C*. Given p in T_{CNP} , p maps each j in J onto some extensional property, p(j). Intuitively we shall want to consider p(j) to be the *extension of* p *in* j. And since the type for any primitive category is some set of functions with domain J, any member d of the type for any primitive category will have an extension in each possible world j.

Somewhat more formally, given d in the type for a primitive category, and given j in J, we define d_j, *the extension of* d *in* j, to be simply d(j). In referring to members of primitive types then we will use the notation d_j and d(j) interchangeably. The reason for making the distinction in notation is that we shall also want to refer to extensions in j for members of types of some (but not all) functional categories. And the types for these categories are not functions with domain J, so the notation d(j) would not make sense there.

Furthermore, we may say that a *type* itself *is extensional* iff each of its members has an extension in each possible world, that is, iff d_j is defined for each d in the type and each j in J. And if T is an extensional type we will denote by T_j the set of its extensions in j, that is the set of d_j for all d in T.

So let's pick a j in J and consider $(T_{CNP})_j$. It is by definition $\{p_j: p \text{ in } T_{CNP}\}$. Recalling that $p_j = p(j)$, we have that each p_j is a member of $P_U = T^e_{CNP}$, and so $(T_{CNP})_j$ is itself a subset of P_U. Moreover, for any q in P_U there is a p in T_{CNP} such that p(j) = q. To see this, recall that T_{CNP} is the entire set of functions from J into P_U. So pick that function, call it f_q, which maps each j_o in J onto q. Then in particular $f_q(j) = q$. So each q in P_U is a member of $(T_{DNP})_j$, whence P_U is a subset of $(T_{CNP})_j$. Combining these two results and noting that j in J was arbitrarily selected, we have that $(T_{CNP})_j = P_U = T^e_{CNP}$ no matter what j in J is chosen. And the results here are typical of all the categories in the intensional logic. That is if C is a category whose type is extensional, then the extension of T_C in any possible world j is simply the type for C in the extensional logic. Of course not all types in IL are extensional, so those types correspond to nothing in the extensional logic.

Finally, returning to T_{CNP}, let us note that in the same way that it was convenient to distinguish certain extensional properties in EL so in IL it is convenient to distinguish certain intensional properties. For example, we may define *Mercy* in T_{CNP} as that function from J into P_U which maps each j in J onto *mercy* (that member of P_U recall which maps each b in U onto 1).

We now consider some representative functional categories, specifically VP (= FM/DNP). As in the extensional logic (but not as in PTQ for example) the type for any category in IL of the form C/C' will be some subset of the functions from $T_{C'}$ into T_C. So in particular T_{VP} will be some set of functions from T_{DNP} into T_{FM}. In IL, however, we distinguish two subcategories of VPs, VP_{+t} and VP_{-t}, called the *transparent* VPs and the *non-transparent* VPs respectively. The types for each of these two categories will be a set of functions from T_{DNP} into T_{FM}, but the sets will be different. T_{VP} itself will simply be the union of these two sets.

Consider first the type for the transparent VPs. Syntactically the transparent VPs include all the VPs treated in the extensional logic. And for the reasons given earlier we still want them to preserve the boolean operations. E.g. we still want ((*John,and,Mary*),*sleep*) to have the same truth value in each possible world as the conjunction of (*John,sleep*) with (*Mary,sleep*) has in that world. This result we can guarantee as before. Namely, since both T_{DNP} and T_{FM} are boolean algebras in IL we shall require members of the type for VP_{+t} to be homomorphisms from T_{DNP} into T_{FM}. We shall of course not place this requirement on the functions which may interpret non-transparent VPs since, as we argued earlier, they do not preserve joins ("or") and so cannot be homomorphisms.

But there is another difference between the two types. Namely we want the extension (truth value) in a possible world j of a sentence formed from a transparent VP to depend only on the extension of the subject in j. So if *John* and *the President* are the same individual in j, (Loosely, *John*(j) = (*the President*)(j)) then (*John,sleep*) and ((*the President*),*sleep*) will have the same truth value in j. On the other hand, if the VP is selected from the non-transparent class, e.g. *must be a citizen*, etc. then this will not be so. So we are going to restrict further the class of functions which may interpret transparent VPs to those which meet the Transparency Condition below. Furthermore, the feature ±transparent will be used to logically subcategorize many other categories besides VPs, so we shall give the type independent definition below and illustrate it with the VPs and the APs.

(16) *The Transparency Condition*

Let X and Y both be extensional types (and thus for each d in either X or Y, d_j, the extension of d in j, is defined for each possible world j in J). Then a function f from X into Y is transparent iff for each d,g in X and each j in J, if $d_j = g_j$ then $(f(d))_j = (f(g))_j$.

Noting that both T_{DNP} and T_{FM} are extensional types, we may define the type for transparent VPs as follows:

Def: $T_{VP_{+t}} = \{f \varepsilon H_{T_{FM}/T_{DNP}} : f \text{ is transparent}\}$

(Recall that for X and Y boolean algebras, $H_{Y/X}$ denotes the set of homomorphisms from X into Y).

To illustrate the definition, take *sleep* as a member of the transparent VPs. Consider a particular interpretation of the language in which *John* is interpreted as p and (*the,president*) as q, and *sleep* as one of the transparent homomorphisms in the type for the transparent VPs. Note that p and q may be very different functions, that is, there may be many j in J such that $p_j \neq q_j$. But suppose there is one j in J, call it j_o , such that $p_{j_o} = q_{j_o}$. Then since *sleep* is interpreted by a transparent function, call it Sleep, then it follows that $(\text{Sleep}(p))_{j_o} = (\text{Sleep}(q))_{j_o}$. Loosely, if the extension of *John* in j_o is the same as that of *the president* in j_o then the extension (truth value) of *John is sleeping* in j_o will be the same as that of *The president is sleeping* , in j_o.

On the other hand non-transparent VPs like *must be over 35, be required to be a citizen,* etc. will be interpreted by functions which are not required to meet the transparency condition. Hence *John is required to be over 35* and *The President is required to be over 35* may have different truth values in j_o even if the extension of *John* and *the president* is the same in j_o.

Given that the functions in the type for the non-transparent VPs are not required to be transparent or to be homomorphisms, we might wonder if they have any substantive properties at all (aside from being different from transparent VPs in the ways indicated). And it does appear (though more investigation is clearly called for here) that the non-transparent VPs should be required to preserve meets. Thus if both the President and the Vice-President are required by law to address the Congress then it does seem to follow that The President is required by law to address the Congress and so is the Vice-President. We may define a function f from one boolean algebra into another to be *additive* iff it preserves meets, i.e. $f(\bigwedge P) = \bigwedge_{p \in P} f(p)$ and we shall require that the type for the non-transparent VPs consist of functions from T_{DNP} into T_{FM} which are additive.[7] It is worth noting here, though we shall not justify it in detail, that additivity covers the case of intensional inclusion. E.g. taking *is being sought by the police* as a non-transparent VP, it follows from the additivity condition that if *A man with a scar on his face is being sought by the police* is true in some j then *A man is being sought by the police* is also true in j. This seems reasonable.

Although there is more to be said about the VPs, let us consider for a moment how the Transparency Condition applies to APs. They will be functions from T_{CNP} into T_{CNP}, and since T_{CNP} is an extensional type we may distinguish functions in this class according as they are transparent or not. *tall* and *skillful* belong to different subcategories of AP, and the functions which interpret the former are required to be transparent whereas those which interpret the latter are not. Hence in a model in which say *docter* and *teacher* have the same extension in say j_o it will follow that *tall docter*

---

7 More consideration should be given to the actual definition of the type for the non-transparent VPs. In addition to being additive, we required in LT that the functions which interpret non-transparent VPs be not transparent. That is, they must actually fail the Transparency Condition. A weaker claim, and probably the more justifiable is that the functions which interpret non-transparent VPs simply not be required to meet the Transparency Condition. This would allow that in a given interpretation a non-transparent VP might in fact behave like a transparent one, although we would be guaranteed that in other interpretations this would not be so. The stronger condition requires that in any interpretation there be at least one j in J such that the interpretation of a non-transparent VP fails the Transparency Condition in that interpretation.

The stronger condition seems the more intuitive if J is selected sufficiently large. That is, to our mind the most sensible interpretation (informally speaking) for J is that it be an index set for the entire class of extensional interpretations. We have in general in LT treated J as though it were this class, although not very much depends on that assumption, except perhaps the interpretation of *possible* and the issue of requiring non-transparent operators to be non-transparent in each interpretation or merely not requiring them to be transparent in each interpretation. As these issues go beyond that of quantification proper so we shall not consider them in detail here. It is perhaps worthwhile to note, however, that in LT *necessary* is interpreted as universal quantification over possible worlds, and *possible* as existential quantification. Thus:

(i) for all j in J, $(\textit{necessary}(\varphi))(j) = \bigwedge_{j \in J} (\varphi)(j)$ and

(ii) for all j in J, $(\textit{possible}(\varphi))(j) = \bigvee_{j \in J} (\varphi)(j)$

and *tall teacher* also have the same extension in j_o, though the inference is not valid if *tall* is replaced by *skillful*.

Notice, however, that functions which interpret either *tall* or *skillful* are required to be restricting, which we might define for APs as follows:

Def: f in $F_{T_{CNP}/T_{CNP}}$ is *restricting* iff for all p in T_{CNP}, all j in J and all b in U, if $(f(p))_j(b) = 1$ then $p_j(b) = 1$.

(As with transparency, we define a more general notion of restricting in LT, but the more general definition need not concern us at the moment). It is clear that being restricting is independent of being transparent, for even if $p_j = q_j$ and f is merely restricting, we may only infer that the set of b in U which $(f(p))_j$ holds of is a subset of those which p_j itself holds of, and similarly for $(f(q))_j$; but we may not infer that the two sets are the same as they would have to be if f were transparent (for then $(f(p))_j$ would have to be identical to $(f(q))_j$).

Finally, let us complete our analysis of the transparent VPs. As in the extensional logic we want to take boolean combinations of transparent VPs, e.g. we want to form (*sing, and, dance*), (*not,sing*), etc. So we expect to define the corresponding boolean operations on the type for the transparent VPs. Furthermore, we shall want to define specific members of the type for the transparent VPs such as the interpretation of *exist*. In distinction to the extensional logic, however, there is an additional requirement we want the transparent VPs to meet. Namely, we want to make a transparency distinction in the TVPs (*hit* vs. *seek*,etc.). We expect that the type for the transparent TVPs will be transparent functions into the type for the transparent VPs. But to use the definition of transparency, we will need the type for the transparent VPs to itself be extensional. Then we can say of a transparent TVP that the VP extension obtained by applying it to a member p of T_{DNP} depends only on the extension of p.

So let us first define extensions for the transparent VPs. It will turn out that once this is done, we may define the boolean operations on $T_{VP_{+t}}$ in terms of their extensions, in a way similar to that in which the algebras on T^e_{FM} and T^e_{DNP} were lifted onto the corresponding types in the intensional logic. Note further that basically we know what the extension of f in $T_{VP_{+t}}$ should be at any $j \varepsilon J$: it should be simply a member of the type for VP in the extensional logic, that is, a homomorphisms from P^*_U into {0,1}.

Consider what we know about a function f in the type for the transparent VPs. Given any two members p,q of T_{DNP} (that is, functions from J into P^*_U) if, for some j in J, $p_j = q_j$ then $(fp)_j = (fq)_j$. Both p_j (= p(j)) and q_j (= q(j)) are members of P^*_U. Similarly, both $(fp)_j$ (= (f(p))(j)) and $(fq)_j$ are members of {0,1}. So if we pick a j in J and hold it constant, the function f determines a correspondence between members of P^*_U and {0,1}. Namely $p_j \rightarrow (fp)_j$, $q_j \rightarrow (fq)_j$, etc. We may in fact (though it looks a little sneaky) define f_j, the extension of f in j, by this correspondence. Let us state the definition and then prove that it is a well definition.

Def: Given f in $T_{VP_{+t}}$ and j in J, $f_j =_{df}$ that function from P^*_U into {0,1} such that for all p in T_{DNP}, $f_j(p_j) = (f(p))_j$.

P^*_U of course is T^e_{DNP} and {0,1} is T^e_{FM}. We must show that f_j above is in fact a function whose domain is P^*_U and whose range is included in {0,1}. Clearly for each p in T_{DNP}, that is, each function from J into P^*_U, $p_j = p(j)$ is a member of P^*_U. Furthermore, each Q in P^*_U is in fact a p_j for some p in T_{DNP}. Since T_{DNP} is the set of all functions from J into P^*_U we may simply select p to be f_Q, that member of T_{DNP} which maps each j_o in J onto Q. So in particular $f_Q(j) = (f_Q)_j = Q$. So the domain of f_j above is clearly P^*_U. Further, since for each p in T_{DNP} and each j in J, $(fp)_j = (f(p))(j)$ is in {0,1} the range of f_j is clearly included in {0,1}. It remains to show that f_j is in fact a function. That is, if P and Q are any members of P^*_U we must show that if P = Q then $f_j(P) = f_j(Q)$. Let g_P be that member of T_{DNP} which maps each j_o in J onto P. Similarly for g_Q (i.e. it maps each j_o in J onto Q). Specifically then for the j mentioned in the definition, we have that $(g_P)_j = (g_Q)_j$ since P = Q by assumption and $(g_P)_j = (g_P)(j) = P$ and $(g_Q)_j = (g_Q)(j) = Q$. But since $(g_P)_j = (g_Q)_j$ then, since f itself is transparent, we have that $(f(g_P))_j = (f(g_Q))_j$. But by the way f_j was defined, we know that $(f(g_P))_j = f_j((g_P)_j) = f_j(P)$ (since $(g_P)_j = P$). And similarly $(f(g_Q))_j = f_j((g_Q)_j) = f_j(Q)$. So, reading the equalities backwards, we have that $f_j(Q) = (f(g_Q))_j = (f(g_P))_j = f_j(P)$, which is what we wanted to prove.

We have shown then that for each f in the type for the transparent VPs and for each j in J, f_j, *the extension of* f *in* j, is a function from P^*_U into {0,1}. But is each such f_j a homomorphisms from P^*_U into {0,1} ?
Not surprisingly the answer is yes. The proof is largely a matter of

keeping one's j's straight. We will illustrate part of the proof by showing that each f_j preserves intersections. We illustrate the proof with finite intersections for simplicity of presentation, but generalizing to infinite intersections is only a matter of notation.

Given any f_j and any P,Q in P^{*}_{U} we are to show that $f_j(P \cap Q) = f_j(P) \wedge f_j(Q)$. We assume the definitions of g_P and g_Q as given above. We have than:

$f_j(P \cap Q)$	$= f_j((g_P)_j \cap (g_Q)_j)$	by the def. of g_P and g_Q
	$= f_j((g_P \wedge g_Q)(j))$	by def. of the algebra on T_{DNP} i.e. $(g_P \wedge g_Q)(j) = (g_P(j) \wedge g_Q(j)) = ((g_P)_j \wedge (g_Q)_j)$
	$= (f(g_P \wedge g_Q))_j$	by def. of f_j
	$= (f(g_P) \wedge f(g_Q))_j$	because f is a homomorphism from T_{DNP} into T_{FM}
	$= (f(g_P))(j) \wedge (f(g_Q))(j)$	by def. of algebra on T_{FM}
	$= f_j((g_P)_j) \wedge f_j((g_Q)_j)$	by def. of f_j
	$= f_j(P) \wedge f_j(Q)$	by def. of g_P and g_Q

qed

(The proof that $f_j(Q') = (f_j(Q))'$ is largely showing that $g_{Q'} = (g_Q)'$). So each f_j then is a member of the extensional type for VP. Thus the extension of T_{VP} in any j in J is a subset of T^{e}_{VP}. Is it in fact identical to it? Again, not surprisingly, the answer is yes. To show this we must establish that for each j in J, and for each h in T^{e}_{VP} there is an f in T_{VP}^{+t} such that $f_j = h$. The proof is essentially as follows:

Given j and h as above, define a function f from T_{DNP} into T_{FM} such that for all p in T_{DNP} f(p) is that proposition such that for *all* j in J, $(f(p))(j) = h(p)(j))$. We note without proof that this does define a *function* from T_{DNP} into T_{FM} and that further for each p in T_{DNP}, f(p) is indeed a proposition, i.e. a function from J into {0,1}. We must show three things: First, that f is a homomorphism from T_{DNP} into T_{FM}; second that f is transparent; and third, that f_j does in fact = h.

We illustrate the proof that f is a homomorphism by showing that it preserves complements. We must show that for all p in T_{DNP}, $f(p') = (f(p))'$, which is to say, for all j in J, $(f(p'))(j) = ((f(p))')(j)$. Let j be arbitrary in J. Then,

$$
\begin{aligned}
(f(p'))(j) &= h(p'(j)) && \text{by def. of } f \\
&= h((p(j))') && \text{by def. of complement in the } T_{DNP} \text{ algebra} \\
&= (h(p(j)))' && h \text{ is a homomorphism in } T^{e}_{VP} \\
&= ((f(p))(j))' && \text{by def. of } f \\
&= (f(p))'(j) && \text{by def. of the } T_{FM} \text{ algebra}
\end{aligned}
$$

qed

f is easily shown to be transparent. Let p,q in T_{DNP} and j in J such that $p_j = q_j$. We must show that $(f(p))_j = (f(q))_j$. We have, from the definition of f that $((f(p))_j = h\,(p_j) = h\,(q_j) = (f(q))_j$.
And the proof that $f_j = h$ is straightforward. We must show that for any Q in P^{*}_{U}, $f_j(Q)$. We have,

$$
\begin{aligned}
f_j(Q) &= f_j((g_Q)_j) && \text{by the def. of } g_Q \text{ as above} \\
&= (f(g_Q))_j && \text{by the def. of } f_j \\
&= h(g_Q(j)) && \text{by the def. of } f \\
&= h(Q) && \text{by the def. of } g_Q
\end{aligned}
$$

qed

We have shown then that for each j in J and each f in the type for the transparent VPs, f_j is defined. So $T_{\substack{VP\\+t}}$ is an extensional type, and moreover, for each j in J, the extension of that type in j is T^{e}_{VP}, the type for VPs in the extensional logic. How may we use these facts to define specific members of $T_{\substack{VP\\+t}}$, in particular to define meets, joins, and complements of members of that type. To do that we must note that any member of the type for the transparent VPs is uniquely determined by specifying its extension in each j, and moreover any way of specifying these extensions will do.
Specifically we show:

Thm: *The Extensionality Theorem for Transparent VPs*
Let h be any function from J into T^{e}_{VP}. There is a unique f in $T_{\substack{VP\\+t}}$ such that for each j in J, $f_j = h(j)$.

Proof: *Uniqueness*
Uniqueness: Let f,k in $T_{\substack{VP\\+t}}$ such that $f \neq k$. Then there is a p in T_{DNP} such that $f(p) \neq k(p)$. Therefore there is a j in J such that $(f(p))(j) \neq (k(p))(j)$. But $(f(p))(j) = f_j(p_j)$ by the definition of f_j.

Similarly for k. Thus $f_j(p_j) \neq k_j(p_j)$, whence $f_j \neq k_j$. So if $f_j = k_j$ for all j in J then f = k.

Existence:

Given h as in the theorem, define f in the set of functions from T_{DNP} into T_{FM} such that for all p in T_{DNP} and all j in J, $(f(p))(j) = (h(j))(p(j))$. Clearly f is in that set of functions and for each p in T_{DNP}, f(p) is in T_{FM}.

We must show that f so defined is a homomorphism, that it is transparent, and that for each j in J, $f_j = h(j)$.

f is a homomorphism

We illustrate the proof by showing that f preserves meets. Let P be any subset of T_{DNP}. We must show that for each j in J,

$(f(\bigwedge P))(j) = (\bigwedge_{p\in P} f(p))(j)$. Let j in J arbitrary. We have:

$(f(\bigwedge P))(j)$	$= h(j)((\bigwedge P)(j))$	by the def. of f
	$= h(j)(\bigwedge_{p\in P} p(j))$	by def. of T_{DNP}
	$= \bigwedge_{p\in P} (h(j)(p(j)))$	since h(j) is a homomorphism in T^e_{VP}
	$= \bigwedge_{p\in P} ((f(p))(j)$	by the def. of f
	$= (\bigwedge_{p\in P} f(p))(j)$	by the T_{FM} algebra

qed

Again the proof that f is transparent is almost immediate:
Given j in J and p,q in T_{DNP} such that $p_j = q_j$ we have that:

$(f(p))_j$	$= h(j)(p(j))$	by the def. of f
	$= h(j)(q(j))$	by assumption
	$= (f(q))_j$	by the def. of f

qed

And the proof that for each j, $f_j = h(j)$ is again almost immediate. For Q arbitrary in P^{*}_{U}, we have that $f_j(Q) = f_j(g_Q(j)) = (f(g_Q))_j = h(j)(g_Q(j)) = h(j)(Q)$. qed.

The Extensionality Theorem for transparent VPs says then that we may define a member of that type merely by specifying in whatever way we like its extension in each j in J, that is, saying for each j in J what member of

T^{e}_{VP} it denotes. Thus we may define the interpretation of *exist* as that member *Exist* of the type for transparent VPs which is such that for each j in J, $Exist_j$ is the 1 element of T^{e}_{VP}, that is, that homomorphism which maps each I_b onto 1. Similarly we may define a boolean structure on $T_{VP_{+t}}$ as follows:

Def: a. for each h in $T_{VP_{+t}}$, h' is that member of $T_{VP_{+t}}$ such that for each j in J, $(h')_j = (h_j)'$ and

b. for H any subset of $T_{VP_{+t}}$, $(\bigwedge H)$ is that member of $T_{VP_{+t}}$ such that for each j in J, $(\bigwedge H)_j = \bigwedge_{h \in H} h_j$

Clause (a) says for example that the complement of a transparent VP h is that transparent VP whose extension in any j is the complement of the extension of h itself in j.

As an illustration of this definition consider the logical representation for *Either Peter or Martha are both singing and dancing*, represented by (19)a. below. (19)b. give its semantic tree.

(19) a. ((Peter,or,Martha),(sing,and,dance))

b.

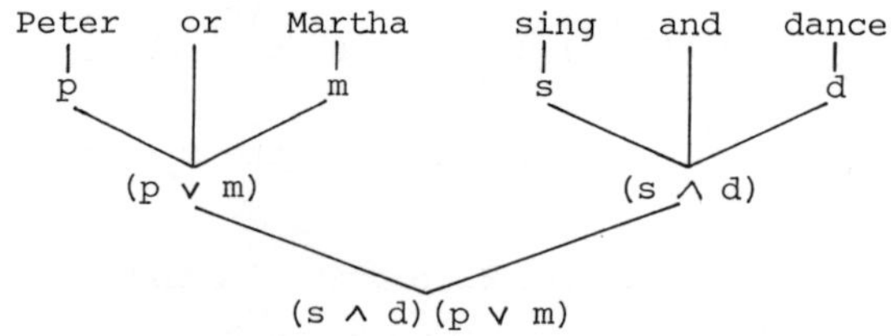

Consider the extension (truth value) of $(s \wedge d)(p \vee m)$ in any j. It is given by:

$((s \wedge d)(p \vee m))(j)$

$=(s \wedge d)_j(p \vee m)_j$ — by the def. of the f_j in $T_{VP_{+t}}$

$=(s \wedge d)_j(p_j \vee m_j)$ — by def. of the T_{DNP} algebra

$=(s \wedge d)_j(p_j) \vee (s \wedge d)_j(m_j)$ — since $(s \wedge d)_j$ is a homomorphism in T^{e}_{VP}

$=(s_j \wedge d_j(p_j) \vee (s_j \wedge d_j)(m_j)$ — by def. of algebra on T_{VP}

$=(s_jp_j \wedge d_jp_j) \vee (s_jm_j \wedge d_jm_j)$ — by def. of algebra on T^{e}_{VP}

Now this last expression is 1 iff one of the values over which the join is taken is 1. Which is to say, either *Peter is singing and Peter is dancing* has value 1 in j, or *Martha is singing and Martha is dancing* has value 1 in j.

Note that we defined only the type for transparent VPs to be a boolean algebra here. And in fact in LT only the transparent VPs (among the VPs in general) were boolean. That is, we did not form conjunctions, disjunctions, and negations of VPs other than transparent ones. There is however no particular problem in doing so, but more investigation into the semantic properties of boolean combinations of arbitrary VPs would be needed before a proper semantics could be proposed. Assuming it were done, then the extension of the current system would proceed along the following lines. A conjoined VP (d,and,g) would be interpreted as the meet of the interpretations of d and g if both d and g were in VP_{+t}. If not, then, in the worst of cases (d,and,g) would be interpreted as an arbitrary member of the unions of the types to which the interpretation of d and that of g belong. Such a weak proposal seems unlikely to be correct, however. For many choices of DNPs and boolean combinations of VPs which include non-transparent ones there is a natural interpretation. For example, *The President is not required to be over 40* should surely have the same truth value as *it is not the case that the President is required to be over 40*. Perhaps a better guess at interpreting (d,and,g) when not both d and g are transparent VPs would be to say, in effect, that for any DNP intension p, (d,and,g)(p) = (d(p), and, g(p)) (confusing levels for simplicity of expression here). This seems in many cases not unnatural when both d and g are in the non-transparent VPs, but it ignores an interaction effect when one is transparent and the other non-transparent. In some such cases at least, the transparent member seems to force the non-transparent one to be understood transparently, something we can represent in LT as it stands, by using the lambda operator, which forces the function it operates on to be interpreted transparently. In any event, we leave for later a serious investigation of these problems.

In summary, let us note that lifting the boolean structure of the extensional logic into the intensional one proceeds in a somewhat more general way than our discussion of the transparent VPs would suggest. Basically we prove one fat generalized Extensionality Theorem which says that whenever X and Y are extensional types then the set of transparent functions from X into Y is an extensional type; furthermore, any transparent function f in that set is uniquely determined by the set of its extensions f_j for each j in

J. Moreover, if X and Y are boolean algebras and f a transparent homomorphism from X into Y then for each j in J, f_j is a homomorphism from X_j into Y_j and f itself may be defined simply by specifying for each j a homomorphism from X_j into Y_j. (Note that X_{j_o} and X_{j_1} are identical for any j_o, j_1 in J).

References

Gil, D., 1977, *The Semantic Interpretation of Noun Phrases,* M.A. Dissertation, Dept. of Linguistics, Tel Aviv University.

Faltz, L. (in prep), "*On the Treatment of Adverbs and Prepositional Phrases in Logical Types*", Dept. of Linguistics, UCLA.

Keenan, E. L., 1979, "On Surface Form and Logical Form". To appear in *Studies in Linguistic Sciences,* special issue on the Forum Lectures, the Linguistic Institute 1978.

Keenan, E. L. and Faltz, L., 1978, *Logical Types for Natural Language*, UCLA Occasional Papers in Linguistics, No. 3., Dept. of Linguistics, UCLA.

Keenan, E. L. and Faltz, L., 1979, "Passive: A Case Study in Markedness", presented at the GLOW conference, April 18-22, 1979, Pisa, Italy.

Montague, R., 1970, "English as a Formal Language" in B.Visenti et al *Linguaggi nella Societa e nella Tecnica*. Milan, Edizioni di Cummunitã, 1970.

Montague, R., 1973, "The Proper Treatment of Quantification in Ordinary English" in J. Hintikka, J. Moravcsik, and P. Suppes (eds.), *Approaches to Natural Language,* Dordrecht, Reidel.

ON THE CONTEXT-DEPENDENCE OF THE PROGRESSIVE IN ENGLISH

Ekkehard König (Hannover)

0. *Introduction*

The meaning and use of the progressive or expanded form *(He is eating)* as opposed to the simple form *(He eats)* of the verb in English has been treated in so many investigations that it seems hardly possible to contribute anything strikingly new to the existing literature.[1] However, hardly any of these analyses has given sufficient thought to the problem of formulating criteria of adequacy for the language of description and the description in general. Many analyses attribute semantic properties to the aspectual forms in English which are totally unrelated and sometimes even incompatible. The most serious shortcoming of the older analyses is that they have not been made precise to such a degree that they are falsifiable. It is precisely at this point that some recent analyses of the meaning of the expanded form (EF) carried out in the framework of formal semantics open up new possibilities. The analyses I am referring to are Scott (1970), Bennet and Partee (Ms), Dowty (1977) and Taylor (1977). The aims and the scope of these formal analyses are certainly more restricted than those of most traditional descriptions. Their aim is to account for only some of the principles that underlie the use of aspectual forms in English, but in a manner that is coherent and meets certain criteria of adequacy. Within formal semantics, the description of the meaning of a word, an expression or form of a word amounts to a description of the contribution the word or expression makes

---

1 Good bibliographies can be found in SCHOPF (1974) and SCHEFFER (1975).

to the truth conditions of the sentences in which it occurs. This then enables us to account for the logical relations (entailment etc.) that hold between sentences of the language investigated.

For reasons which will be elaborated upon below, Dowty's analysis seems to be the most interesting and most adequate of the four analyses mentioned above. In the following paragraph I will briefly summarize Dowty's analysis and then examine whether it can account for certain well-known phenomena. The main objective of the present paper is to show that Dowty's analysis is incomplete in an important respect: it does not cover a major use of the progressive. In other words, an important feature, particularly apparent in the use just mentioned, is not taken into account in his analysis. The use in question will be described in detail. Finally, a few suggestions will be made as to how Dowty's analysis could be modified and supplemented in order to make it a more general and more adequate theory.

1. Dowty's Semantic Analysis of the EF

Dowty's starting point is the so-called 'imperfective paradox': (1) logically implies (2) but the inference from the parallel (3) to (4) fails. In (3) and (4) an accomplishment verb phrase replaces the activity verb phrase of (1) and (2) (cf. Vendler, 1967):

(1) John was pushing a cart.
(2) John pushed a cart.
(3) John was drawing a circle.
(4) John drew a circle.

Dowty's aim is to account for the fact that with accomplishment verb phrases the entailment from the 'progressive tenses' to the simple tenses fails and he tries to achieve that by giving a semantic analysis both of accomplishment predicates and the expanded form. The characters *(Aktionsarten)* of predicates are explicated in terms of a componential analysis. Thus, Dowty postulates that all accomplishment verb phrases, indeed all predicates which denote a change of state, contain the one-place operator BECOME. Truth conditions for an expression of the form

[BECOME [ϕ]] where ϕ stands for a tenseless sentence, are defined in the following way (cf. Dowty, 1977:53):

(5) [BECOME[ϕ]] is true at an interval I iff (a) there is an initial boundary interval J for I such that $\neg$ is true at J, (b) there is a final boundary interval K for I such that ϕ is true at K, and (c) there is no non-empty interval I' such that I'$\subset$I and such that (a) and (b) hold for I' as well as for I.[2]

Dowty follows Scott (1970) and Bennet & Partee (Ms.) in analyzing the EF syntactically as a one place sentential operator.[3] However, as far as its semantics is concerned, the EF is not analyzed as a temporal operator as in Scott (1970) or Bennet & Partee (Ms.) but as a "kind of mixed modal-temporal operator".[4] Thus, truth conditions for expressions with the EF - Dowty uses the abbreviation PROG - are defined as follows:

(6) [PROG ϕ] is true at I and (a possible world) w iff there is an interval I' such that I$\subset$I' and there is a world w' for which ϕ is true at I' and w', and w is exactly like w' at all time preceding and including I.

In that particular case in which ϕ has the form [BECOME[ψ]] the situation in which the relevant expression [PROG [BECOME[ψ]]] is true can be illustrated by the following diagram (cf. Dowty, 1977:57):

---

2 Dowty is undecided as to whether condition (c) in (5) should be regarded as a truth condition or as a felicity condition based on conversational rules.

3 Cf. Cresswell (1977:14) for some arguments concerning the syntactic status of the progressive. Additional support for the assumption that it functions as a sentential operator comes from sentences like the following:

(i) He doesn't answer the phone.
(ii) He is not answering the phone.

According to Hatcher (1974:215 n.57) the second sentence implies a decision not to answer, as is more clearly shown by the following example from I. Murdoch, *Under the Net*. (Penguin, p.85):

(iii) In any case the first thing to do was to go to Holborn to see if he was hiding there and just not answering the phone.

The negation in (ii) and (iii) cannot be constituent negation, since the sentences do not imply that the person referred to was doing something else. It rather suggests that he was engaged in not answering the phone. So it is reasonable to postulate the following structure for (ii):

(iv) [PROG [Not [He answers the phone]]]

4 Scott (1970) and Bennett and Partee (Ms.) cannot account for the imperfective paradox.

(7)

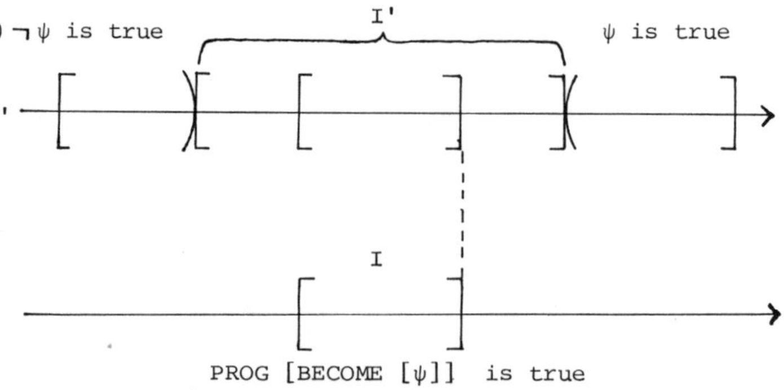

In contrast to the analysis in Scott (1970) which stipulates that [PROG[ϕ]] is true at I iff there exists an open interval I' such that IcI' and ϕ is true at I', Dowty postulates that ϕ must be true (in w') before and during I. During the subinterval of I' , however, which follows I ϕ may be true - if the previous course of events continues - but need not be true. In the absence of any evidence to the contrary the expresion [PROG [BECOME [ψ]]] allows the inference that [BECOME [ψ]] is true after I and that the state ψ is reached. However, this is only an implicature or invited inference due to conversational rules (Dowty, 1976:59). The sentence *John was dying when I last met him*, for example, certainly permits the inference that John is dead now if no further information is given. However, the sentence could be continued as follows: *But recovered again*, which would exclude this inference. Dowty's analysis thus accounts for the 'imperfective paradox' and gives us the right results with regard to admissible and non-admissible inferences in the case discussed.[5]

Clearly, Dowty's analysis also incorporates and formalizes many traditional insights. There is, first of all, the claim very frequently made that the EF expresses (limited) duration and that it focuses on the midpoint of an event ('introspective') and disregards the beginning and the end. It is shown in Scheffer's summary of the relevant literature (1975:21 ff.) how frequent this analysis is. Dowty's condition IcI' captures this traditional insight. The equally frequent insight that the EF denotes 'possible incompleteness' in

---

5 A summary of Dowty's analysis must not omit his reasons for including (c) in (5). This sentence excludes a sentence like *John is dying* from being true in a situation where the person referred to is perfectly well. Condition (5c) requires that a BECOME-sentence is true only of the smallest interval over which the appropriate change of state takes place. And the truth conditions for the EF require that somebody has already entered this smallest interval (cf. Dowty, p.58).

connection with certain predicates is the central objective of Dowty's formalization, as pointed out above. Dowty can moreover point out that his modal analysis is also required for sentences that do not contain an accomplishment or achievement predicate. The claim that the EF expresses '(limited) duration' is normally illustrated with examples such as the following:

(8) a. John was watching television when Bill entered the room.
b. Bill was working at eight o'clock.

Without a co-text which might provide evidence to the contrary one may indeed conclude on the basis of conversational principles that the actions mentioned continue after the point in time specified in the sentence. For cases like these, Scott's analysis therefore seems quite adequate. However, this inference is only vaguely plausible in cases like (9) and must be ruled entirely for (10), because of the co-text:

(9) Mary was preparing dinner when John left.

(10) a. John was watching television when he fell asleep.

b. The second pair (of pursuers) were just catching up when he broke away.

Dowty's basic idea is already foreshadowed in two earlier analyses, which, however, do not attain his standard of explicitness and formalization. The analyses I have in mind are Joos (1964) and Hirtle (1967). According to Joos the EF expresses the probability of validity at a moment principally referred to. At this moment there is a hundred percent probability of validity whereas in the environment of this moment it diminishes towards zero (1964:108 f.). Even more reminiscent of Dowty's analysis is Hirtle's (1967:26 f.):

> "The opposition between simple and progressive is therefore basically one between whole and part. An event whose material significate strikes the mind as being complete, as permitting of no further additions, will be expressed by the simple form. One which gives the impression of lacking something, of leaving room for something to come, will be expressed by the progressive. In grammatical terms this opposition can be stated thus: the simple form is *perfective*, the progressive *imperfective*."

Thus, Dowty's analysis can be regarded as an explication and formalization of many insights which appear in previous descriptions as well as an explanation of many problems so far unsolved.[6] Dowty concedes that his analysis still has some shortcomings. His attempt to formalize the notion of identity of possible worlds up to a given time leads to serious difficulties, regardless of whether time is seen as linear or as branching. Another problem is that next to nothing is said about the alternative world w' in (6). The only cases where any specification can be given for this world are the ones where the context excludes w' from being the actual one (cf. (10)). In these cases the negation of the relevant sentence must be true in w'. In spite of these shortcomings Dowty's analysis certainly accounts for central aspects of the use of the EF in English. In what follows I want to show, however, that Dowty's analysis cannot be applied to another important use of the progressive and that a property most obvious in this use turns out to be a general property of the English progressive.

2. *Multiple description of actions through the choice of aspect*

The first to have drawn attention to the use we want to analyse here was Jespersen (*MEG*. 12.8), who made the following comment with regard to examples like (11):

"... Here the expanded form implies *identity of the two acts* ..."

(11) If I should go to one of the tea-parties in a dressing gown and slippers, I should be insulting society, and eating peas with my knife.

The most detailled and comprehensive discussion of this use of the EF can be found in Buyssens (1968:135 ff.). Buyssens notes that the use under discussion can typically be found in constructions where a simple form is followed by an expanded form. As far as the meaning is concerned "the same fact is mentioned

---

6 In some cases the translation of English sentences into their 'logical form' (or into a formal language which can then be interpreted) is not as straightforward as Dowty seems to suggest. Consider the following sentence, which could be uttered in October if the relevant figures for that year are just about as high as the figures for the previous years:

(i) More people are dying of cancer this year than in previous years.

In contrast to examples like *John was dying* the verb phrase in (i) does not correspond to German *im Sterben liegen*. The following paraphrase seems to be a reasonable approximation to the 'logical form' of the sentence:

(ii) The number of people who die of cancer this year is getting to be greater than in previous years.

twice... but the second time it is interpreted, it is presented in a new light" (1968:139). Therefore he employs the term 'interpretation' to describe this use of the EF. In order to illustrate the range of this use of the EF and in order to characterize the constructions in which it occurs we have ordered the following examples into various groups on the basis of mainly syntactic criteria:

(i) *in* + participle + main clause

(12) In recommending a separation of context and content I am not suggesting that there is no interaction between them.

(13) But students of Saussure such as Martinet ... actively repudiated this notion and urged that linguistic explanation be confined to the interrelations of internal, structural factors. In so doing they were certainly following the spirit of Saussure's doctrine.

(*Labov,* 1970:32 n.2)

(ii) *by* + participle + main clause

(14) By doing this - or trying to do this - you are creating in the most marvellous way, because you are using life itself as the raw material.

(*Buyssens*:150)

(iii) Conditionals

(15) If we choose to place the dividing line between the two at the county boundary, then we are basing our decision on social rather than linguistic facts.

(*Trudgill,* 1974:15)

(16) Should you insist on calling me Frederick or Fred, then you are not only ignoring the custom of this and other English gentlemen's houses, but you are also intruding into my private life, sir.

(*Buyssens*:142)

(17) A pupil, a proof-reader, and an oculist's patient might all be told to read carefully a certain paragraph: the pupil will be disobeying his instructions, if he notices the misprints but not the arguments.

(*Ryle,* 1949:137)

(iv) temporal clause + main clause

(18) As soon as you start asking what education is for, what the use of it is, you are abandoning the basic assumptions of any true culture, that education is worth while for its own sake.

(*Buyssens*:150)

(v) coordination, parataxis

(19) Back in 1963, Cleanth Brooks published *William Faulkner: The Yoknapatawpha Country*. A major Southern critic was writing about a major Southern novelist. (*New Statesman* 7/21/78, p.90)

(20) 'But yours is well?' I spoke to both of them, but once more I was asking Margaret.

(C.P. Snow, *Homecomings*, Penguin, p.209)

(21) The speaker was the reporter's girl friend. She had spoken to him as one would expect, in casual key. In class she automatically ... shifted to consultative. She was adjusting her speech to the situation ...

(*Gleason*: 360)

(22) At about ten to eight last Friday morning a grey-haired gentleman ... could be seen walking along Holland Park Avenue, carrying a large bundle of newspapers. He was clearly engrossed in what he was reading ... The eyes of the grey-haired gentleman searched hurriedly through each of the newspapers. They were plainly looking for something. (*New Statesman*, 6/2/77, p.766)

(vi) main clause + relative clause

(23) ... for the amendment means that those on the roll who do not vote will in effect be voting 'No'. (*New Statesman*, 6/2/77, p.766)

(24) Thus it is plain that any politician who claims that the physical well-being of British citizens ... can be attributed to our wonderful NHS ... is talking rubbish.
(*New Statesman*, 10/28/77, p.576)

(25) He must have known therefore, that Stancliffe, had no right of seizure and that anyone who bought the cats from him was buying a property that belonged to someone else ...

(*Buyssens*: 139)

(vii) parataxis - direct speech

(26) A. 'Do you smoke?'
B. 'No, thanks.'
A. 'Oh, I wasn't offering. I just wanted to know if you did.'

(Kemelman, *Tuesday the Rabbi Saw Red*, Fawcett Crest Books, p.180)

(27) A. 'I'm glad you have the courage to admit it.'
B. 'I'm not admitting anything. I'm telling you that I did it.'
(*Hatcher*: 205 n. 45)[7]

---

7 In her discussion of (24) Hatcher had already noted that in this and similar examples we have a 'reference to the immediate past'.

This list of the different types of constructions may not be exhaustive, but it certainly enumerates the most typical syntactic frames of the aspectual meaning under discussion. Schopf (1969:210 ff.) also speaks of an 'interpretive' use of the progressive and draws specific attention to examples like (26) and (27) in connection with his discussion of the use of aspectual forms in performative utterances.

Can Dowty's analysis also be applied to examples like (12) - (27)? This question, so it seems, must be answered in the negative.
According to Dowty, an expression of the form [PROG [ϕ]] is true at I and w iff ϕ is true at I' and w' where, *inter alia*, the following condition must hold: IcI'.
This condition, however, does not seem to be met by any of our examples. This is probably clearest in those cases where the same verb is employed twice, as in (11), (20), or (22). The additional conditions of (6) which concern the different possible worlds and the likeness between them are also irrelevant here.[8] In trying to describe the meaning of the EF in the examples listed above we therefore have to start with Jespersen, Buyssens and Schopf and try to explicate their intuitions.
The examples listed above are reminiscent of a problem that has played an important role in the philosophy of action in the last twenty years: the problem concerning the identity of actions or individuation of actions. The problem was brought into the philosophical discussion by Austin in his William James Lectures, by Ryle (1949), and by G.E.M. Anscombe (1957,1976[2]),

---

8 Hirtle, whose analysis is somewhat similar to Dowty's though not as explicit, and who also briefly discussed examples like the ones listed above seems to disagree with us here. Hirtle discussed almost exclusively cases where the EF occurs in conditional constructions and comments that 'in such cases the link between the event in the protasis and that in the apodosis is felt to be so close that the realization of the first necessarily brings about the realization of the second' (p.106). Jespersen's example (11) is described in the following way: 'The link between the event seen as a condition and that seen as a necessary consequence may be so close as to suggest something in common' (p.106). Then he goes on and tries to subsume this use under the description quoted above: 'The suggestion of identity arises from the impression that the realization of *go* necessarily realizes the first step in insulting; *go* is in fact identified with the accomplished portion of *insulting*. But that an insult involves more than merely going to a tea-party is implied by the fact that insulting leaves a part of the event yet to be accomplished.' I find this far from convincing.

whose treatment is the most detailled. The problem can be illustrated by the following simple case: if I sign my name and (in doing so) issue a proclamation, have I performed one act or two? This and similar questions are at the centre of a philosophical controversy which is still going on today. Anscombe (1976^2:45 f.) discusses a more complex case where a man (intentionally) moves his arm, operates the pump, replenishes the water supply of a house and poisons the inhabitants. In her view, the person concerned has only performed one single act.

> "For moving his arm up and down with his fingers round the pump handle is, in these circumstances, operating the pump; and, in these circumstances, it *is* replenishing the house water-supply ... So there is one action with four descriptions, each dependent on wider circumstances ..."
>
> (*Anscombe,* 1976:46)

Many philosophers (e.g. Ryle, 1949:130 ff., Davidson, 1969) have expressed similar views.[9] Goldman (1971), on the other hand, has taken issue with this claim. In his view, the fact that a natural order obtains with regard to the four descriptions of Anscombe's example is incompatible with the view that only one action is performed or that the four acts described are identical. This natural order is revealed as soon as we change the order of the arguments of the predicate *is* in the quotation above. The following would not be true: replenishing the water-supply is, in these circumstances, operating the pump. Goldman argues that several actions are described in cases such as these. However, he captures the unity which is felt to exist between the acts in the pumping case and in similar cases in terms of a by-relation. This relation is irreflexive, asymmetrical and transitive and thus a much weaker relation than logical relations like entailment etc.[10] It should also be mentioned that H.J. Heringer (1974) has developed a semi-formal theory of action which may be

---

9 Davidson (1969:222 f.), however, has pointed out that if Jones apologized by saying 'I apologize' it is misleading to say that 'Jones apologized' and 'Jones said 'I apologize'' describe the same action. We should rather say that the same event may make the two sentences true.

10 Goldman also introduces a slightly broader relation, under which the ordinary by-relation is subsumed and which is called 'level generation'. A complex of acts related in this way can be represented by a single act tree, where each of the nodes on the tree represents either a basic act or an act that bears the by-relation (or Goldman's more general relation) to another node.

regarded as a synthesis of the two opposing views just sketched. Heringer's equivalent to Goldman's by-relation, the *indem*-relation, holds between act-patterns *(Handlungsmuster)*. Heringer also introduces another relation, symbolized as 'α' and read as 'of (pattern)' *(nach)*, which holds between act-tokens and act-patterns. Thus, in the two cases described above one single act would be of several act-patterns.

Our examples (12) - (27) fall squarely into the range of examples which have given rise to the discussion concerning the identity or individuation of actions. In our examples (12) - (15) we find the preposition *by*, which provided the name for Goldman's asymmetric relation, as well as the preposition *in*, which is very similar in meaning. The other examples could be paraphrased by a construction involving the latter preposition. The resultant sentences are at least very similar in meaning to the original sentences.

(18') In asking what education is for ... you are abandoning the basic assumptions of any culture.

In the examples given above the EF belongs to those expressive devices which indicate that a by-relation (or Goldman's more general 'level generation') holds between the two acts or, alternatively, that the relevant sentences or clauses can be made true by the same event (in Davidson's view). In examples such as (20) - (21) the fact is signalled exclusively by the EF. How illuminating is Buyssens's term 'interpretation' as a description of the use of the EF in the examples given above? It is certainly true that sentences such as these could be used to point out to the hearer or audience what an action means or amounts to in certain circumstances. However, the term is not very satisfactory for cases like (19) where the second clause is negated and where 'rejection of an interpretation' seems to be a more appropriate description. Moreover, any description of an action can be considered an interpretation in which certain features are taken into account whereas others are neglected. Thus Buyssens's description certainly needs further explication.

In the philosophy of action a distinction is often drawn between more or less 'fundamental' or 'immediate' actions (cf. Anscombe, 1976[2]:86) as well as between more or less 'basic' actions (cf. Goldman, 1971). As pointed out in White (1973:10), an answer to the question 'In doing what, was he X-ing?' could be said to give a more 'fundamental' description of what was done than X gives, and 'By doing what, was he X-ing?' could be said to give a more basic description. Since the preposition *in* is much more characteristic of our examples than *by*, it is the former distinction, if any, which is relevant here. Applying this distinction to our examples we could say that it is the clause with the simple form of the verb - usually the first in the sequence - which gives the more fundamental description, i.e. the description which is nearer to the merely physical and is less dependent on wider circumstances.

Let us take a closer look at the prepositions *in* and *by*. Austin (1962:122 ff.) was the first to point out a marked difference in the use of the progressive in examples such as (12) as opposed to (14). He considered the following sentence frames as helpful tests in his attempt to distinguish illocutionary from perlocutionary verbs:

(28) a. 'In saying x I was doing y' or 'I did y'.
b. 'By saying x I did y' or 'I was doing y'.

The data given in Buyssens as well as my own data confirm Austin's view that *in* is more commonly used with the EF whereas *by* normally combines with the simple form of the verb. There is only one example in Buyssens where the EF combines with a participial construction introduced with *by*. Since the open sentences (37)a. and (37)b. do not provide a clear test for the distinction between illocutionary and perlocutionary verbs, Austin suggested a further test (1962:125):

> "...where we can put the y-verb into a non-continuous tense ... instead of the continuous tense, or equally where we can change the 'in' into 'by' while keeping the continuous tense, the y-verb is not the name for an illocution. Thus, for 'In saying that he was making a mistake', we could put, without a change of sense, either 'In saying that he made a mistake' or 'By saying that he was making a mistake': but we do not say 'In saying that I protested' or 'By saying that I was protesting'.

How can we explain these collocational restrictions between the EF and illocutionary verbs on the one hand, and the EF and *in* on the other? Partly, the reason for this restriction seems to be that only in the case of illocutionary verbs and the preposition *in* does there seem to be no doubt with regard to the simultaneity of the actions in question. Perlocutionary and other effects may be felt later than the action producing them. This explanation, however, would not hold true of all examples with *by*.

Before we go on and try to relate the 'interpretive' use of the progressive to the one analysed in Dowty, there remains an important question to be answered: Is there a clear opposition between EF and SF in the examples discussed above or is there just a tendency to prefer the EF in a context where both are equally possible? In order to answer this question tests were carried out in which my informants, five students from Britain, had to evaluate alternative texts which only differed in the aspectual form of the verbs.

Whenever both the EF and the SF were possible, the informants were asked to comment on any difference in meaning. There was a great deal of agreement in the responses of my informants even though their judgement varied depending on the type of example. In all the examples given above and in many similar ones, the version actually attested, i.e. the version with the progressive, was found to be the better version, even though the SF was regarded as acceptable in some cases. In examples like (26) and (27) the SF was ruled out entirely. In the examples under (v) my informants saw a clear difference of meaning between the version attested and a version where the SF replaces the progressive. In (19) and (20) one would expect a continuation if the EF is replaced by the SF. The simple form would introduce a new event. An analogous effect could be observed in the two examples following. The simple forms *adjusted* and *looked* would not be related to the temporal contexts introduced by *shifted* and *searched* respectively. They would describe new events rather than 'explain' or 'interpret' the events described in a preceding sentence.

These reactions of my informants show that the SF and EF are not in free variation in examples like (12) - (27). The EF is the preferred form in all examples even though the contrast between the two forms seems to be a slight one in some cases. There is, however, a very clear semantic contrast between the two forms in all cases where a paratactic relationship holds between the relevant clauses or sentences. Here the EF is the only formal means of marking Goldman's by-relation. On the basis of Buyssens's findings one can assume that the 'interpretive' use of the EF is a more recent phenomenon in the history of English. It seems to have been rare in previous centuries and is only now becoming more deeply entrenched in the structure of English.

3. *The Context-dependence of the Progressive*

How is the use of the EF as it is discussed in the present paper related to the use analysed in Dowty (1977)? Is it possible to extend his analysis in such a way that it also covers our data? In order to answer these questions we have to first of all extend the scope of our analysis from single sentences to combinations of sentences, i.e. to texts. The necessity for this is most apparent in cases such as (20) - (22) and (26) - (27). Let us first look at an example that could be subsumed under Dowty's analysis but that also requires such an extension of the analysis to texts. Many analyses have commented on the distinction between the two following texts:

(29) John looked at Mary. She smiled.

(30) John looked at Mary. She was smiling.

It has often been pointed out that the combination SF - SF expresses a sequential relationship, which is interpreted as a causal relationship *(post hoc ergo propter hoc)* whenever this is possible.[11] The combination SF - EF, on the other hand, does not advance the plot (cf. Schopf, 1974:256 f.). (30) implies that Mary had begun to smile before John looked at her. How would Dowty's analysis apply to the second clause of (30)? Well, the sentence would be true at I and w iff there is a I' (IcI') and a w' (exactly like w up to and including I) and [*She smiles*] is true at I' and w'. The interesting question in this case, however, is where does the I come from? Quite obviously it is exactly the interval at which the preceding sentence is evaluated. In other words, the interval at which the sentence with the progressive is interpreted is introduced by a preceding simple form.

Let us now turn to a similar example from our own list. In order to avoid all the problems concerning the interpretation of prepositions, conjunctions, etc., we select an example with a paratactic relationship between the two relevant components.

---

11 Cf. Hymes (1974:99) for a discussion of this maxim. In Geis and Zwicky (1971) this tendency to interpret a temporal sequence as a causual relationship is analysed as one instance of a more general phenomenon, which they call 'invited inferences'. In contrast to Gricean conversational implicatures these 'invited inferences' are not based on violations of conversational maxims.

(31) Back in 1963, Cleanth Brooks published *William Faulkner: The Yoknapatawpha Country*. A major Southern critic was writing about a major Southern novelist.

Like in the previous case the combination SF - EF does not advance the plot. To interpret the second clause of (31) we have again to go back to the interval introduced by the first sentence. Following Davidson in taking events as primitive entities, we could simply say that the two sentences can be made true by the same event.
On the basis of our observations with regard to (30) and (31), we can now formulate the following general hypothesis:

(31) The interpretation of the progressive (EF) depends on the context. In contrast to the simple form of the verb the EF does not introduce a temporal context but depends for its interpretation on a temporal context independently specified. Such a specification may be given inter alia by a simple form in the co-text of the sentence with the EF.

Let us now test this general hypothesis, established on the basis of (30) and (31) and clearly supported by all the data discussed so far, against some of the most frequently made and more or less generally accepted observations on the use of the EF. Such an examination will show that this hypothesis is supported by many more facts and that it enables us to explain a large number of phenomena which, up to now, have been thought of as unrelated.

I have already mentioned the observation frequently made in the literature that a combination of two SFs in the past tense, in contrast to the combination SF - EF, frequently expresses a sequential relationship. This cannot only be observed in cases like (29) and (30) but also in cases where we find a hypotactic relationship between two clauses:

(33) a. When I came home Mary made some tea.
b. When I came home Mary was making some tea.

According to our hypothesis this is due to the fact that a simple form introduces a temporal context. A temporal sequence can only be expressed if the two clauses specify their own temporal context. In (33)b. one has to go back to the temporal context of the first clause in order to interpret the second. A combination of SF - EF may express a temporal sequence as in the following cases, but this can only happen under special conditions:

(34) a. They all had stopped and she turned round and was looking at Winterbourne. (H. James, *Selected Short Stories,* Penguin 1967:156)

b. Late afternoon of the third day the telephone rang and I was talking to a man who said his name was Howard Spencer. (R.Chandler, *The Long Good-Bye,* Penguin, p.70)

c. Walter mentioned the Fast - a success - and then made a clumsy sort of transition which Jack missed and was talking of Lourdes. (D.Lessing, *The Story of a Non-Marrying Man,* Penguin 1976:266)

The conditions for this use include the conjunction *and*, which prevents the EF from being interpreted at the temporal context of a preceding SF. Note also that the events expressed by the progressive in (34) are always natural consequences of the preceding events so that the impression of immediate succession blends into that of simultaneity. Even examples like (34) may not be exceptions to our claim concerning the context-dependence of the EF.

In connection with the plot-advancing quality of the simple form it is noted in several descriptions of tense and aspect in English that the EF often serves as a background for the plot-advancing events (cf. Joos, 1964:127; Weinrich, 1977^3:124 ff.; Hirtle, 1967:66). According to Joos this function of the progressive can only be observed in the past tense, whereas Weinrich takes this to be the basic function in all tenses. Again, this particular property of the progressive can be explained and related to other facts on the basis of our hypothesis (32). The effect that events and states described in the progressive can serve as a setting or background to a foreground in the SF can be observed whenever an event described in the progressive is assigned to the temporal context(s) introduced by one or several SFs.[12]

---

12 In examples like (35) the claim that the EF does not introduce a temporal context does not seem to square with intuition. However, it is an undeniable fact that the progressive is evaluated at the same temporal contexts as the SFs that follow. Furthermore, texts like (35) typically occur at the beginning of a chapter so that this particular narrative situation may also create a temporal context. Exactly the same problem arises in conditionals whenever a progressive form precedes a simple form:

(i) If Marx were reviewing Halbwachs, he might tell you that 'the time has come to exchange the sociology of suicide for the suicide of sociology' (*New Statesman,* 6.2.78, p. 745)

(35) Through the underground corridor of the hospital ... Margaret and I were finding our way to Geoffrey's office. Along the passage ... went mothers with children. At a kind of junction or open space sat a group of women ... Nurses, their faces high-coloured and opaque moved past them ... (C.P. Snow, *Homecoming,* Penguin, p. 314)

Another fact we want to consider in the light of our hypothesis (32) is the incompatibility of the progressive with explicitly performative formulae. Whenever the EF occurs with the first person of a performative verb, as in (27) or the following examples, the utterance in question is not used to perform the speech act denoted by the performative verb:

(36) a. I'm not insisting, I'm only suggesting. (Hatcher:205)
b. Life is ... one long procession of assorted shocks, major or minor as the case may be. Not the slightest use protesting about it. I am merely adding this complaint to the various other ones I wish to make ... (*N.S.* 11.3.1978, p. 581)

Here we often find a reference to the 'immediate past' (Hatcher) and the speech act performed is one of explanation. Utterances like (27) or (36) are used to explain the function of an utterance previously made. This again reveals the context-dependence of the EF. (27) was in fact taken as an instance of the interpretive use analysed above. Looking at the facts from the opposite angle we can say that it is only natural that an explicitly performative formula should introduce its own temporal context.

Our hypothesis also provides an explanation for the fact that the progressive invariably occurs after verbs denoting 'mental acts' *(discover, notice, realize, find, reflect, resent, know,* etc.*)* as well as after epistemic modals, or verbs like *seem, appear,* and adverbs like *obviously,* which express conclusions arrived at through these 'mental acts'.

(37) a. Suddenly she noticed that one of the boys was not working.
b. To her horror she discovered that she was shaking with sobs.
c. Tony realized that she was trying to assert herself.
d. He certainly knew what he was doing.
e. I must be seeing things.
f. Then on Christmas Day she moved about a lot and seemed to be moving her right hand and arm quite freely. (Buyssens:139)

In examples like (37) it is quite obvious that the progressive is interpreted at the temporal context introduced by a preceding simple form.

Thus, our hypothesis enables us to account for a large number of facts which have never been shown to be related in any way. These facts include the 'interpretive' use of the EF, the use of two or more SFs to express a temporal sequence, the incompatibility between the EF and explicitly performative formulae, the explanatory quality of utterances where a performative verb in the first person takes the EF, the use of the EF to characterize the setting or background, and the use of the EF in examples like (37).

4. *An Extension of Dowty's Analysis*

How can we modify and extend Dowty's analysis so that it also applies to examples like (12) - (27), i.e. to the interpretive use of the progressive? It is certainly no problem to add the central claim of this paper to the truth conditions Dowty gives for the progressive: All we have to do is to add to (6) the proviso that I is provided by the context (i.e. the moment of speech, an adverbial of time, a SF of the co-text, etc.). As our discussion in the last section has shown, this condition seems to be required not only by the interpretive use but also by a large number of other uses of the EF. The problem, however, is whether the conditions of (6), plus the above addition, can account for the interpretive use? This does not seem to be the case. As pointed out above, (6) as it stands, does not cover cases like (12) - (27). The condition $I \subset I'$ is not met by these examples. This is clearly the opinion of Buyssens, who introduces his discussion of the interpretive use with the following statement:

> "Ce chapitre est consacré aux choix entre les deux aspectifs lorsque l'aspect est total, c'est-à-dire lorsque l'on considère un fait ... du début à la fin." (p.135)

One way of remedying the situation would be to relax the requirement that I be a proper subinterval of I' and to simply require that $I \subseteq I'$. This condition, however, is crucial for uses other than the interpretive one. So the conclusion seems inescapable that whereas the requirement of context-dependent interpretation is necessary for most if not all, uses of the EF, the truth conditions given by Dowty are not applicable to the interpretive use. As a consequence one might want to regard the interpretive use of the EF as a special use with truth conditions different from the standard use. Let us investigate in more detail what these truth conditions would look like. The first problem is one of specifying the structures in which the relevant use may occur. As is shown by our list of examples, truth conditions must be given for two clauses or sentences which may exhibit various syntactic relations: coordination or subordination; and the latter with either a temporal, conditional or 'instrumental relationship' between main and subordinate clause. All of our examples show that the interpretive use can only occur when the relevant clauses or sentences have co-referential subjects.[13] This is a necessary but by no means sufficient condition. The following examples are certainly not instances of the use we are investigating but can be subsumed under Dowty's analysis:

(38) (Suddenly M. spoke to me.) 'He's absolutely right', she said. She was smiling, she was trying to speak easily. (C.P. Snow, *Homecomings*, Penguin:212)

(39) He depressed the receiver arm, released it, was calling urgently: 'House exchange ...' (Hirtle, 1976:67)

The difference between these two examples is that in (32) the interval at which *he was calling* must be evaluated is not the one that is introduced by the preceding predication but is an adjacent interval. Examples such as these show that we must add a further condition: the two clauses or sentences must be verifiable by the same event (Davidson), there must be a specific form of Goldman's by-relation between the two actions or act-patterns in question (Goldman, Heringer). Given these conditions it is easy to state the remaining one in more precise terms:

(40) Given a text of the form $\phi[\text{PROG}[\psi]]$, where ϕ and ψ are sentences, and ϕ is a sentence with the simple form of the verb
$[\text{PROG}[\psi]]$ is true at I and in w
iff: ψ is true at I and in w
and ϕ is true at I and in w

In the examples under analysis the EF does not have a modal character but simply specifies that the sentence in which it occurs must be evaluated at the same interval as the 'preceding' sentence.[14]
It is only this latter point which the type of use under analysis and the use discussed in Dowty have in common. The progressive form of the verb does not introduce a point in time or interval. The interval needed in order to evaluate a sentence in the progressive may be specified by an adverbial expression (cf. (41), (42)), by the point of speech (43) or by another sentence with the simple form of the verb.

---

13 In examples like (26) the relevant identity must hold between the subject (and speaker) of the sentence in the progressive and the speaker of the sentence in the simple form. This shows that the relevant condition cannot be stated in purely syntactic terms.

14 The reason for putting *preceding* into inverted commas is that the relevant clause need not precede the sentence with the progressive or be in its immediate environment (cf. (19)).

(41) He was sleeping at eight o'clock.

(42) Marian appeared in the half darkness and in a moment he was holding her tightly in his arms.
(Murdoch, *The Unicorn,* Penguin:178)

(43) Can't you see I'm working.

There is thus an interesting parallelism in the interpretation of first and second person pronouns as opposed to the third person and the use of the progressive in the present tense as opposed to its use in the past tense. In the former case the situation of speech provides referents for the pronouns and a time interval for the progressive, whereas in the latter it is the co-text that provides a referent and a time interval. Note finally that condition (40) does not follow from the requirement that the two sentences or clauses be verifiable by the same action. So, at least, some philosophers would argue. In Davidson (1969:299) it is claimed that actions may be identical even if they do not consume identical stretches of time. Some of the examples given in Hirtle (1967:106) are very interesting in this connection. In the following example, for instance, the second event cannot be simultaneous with the first, if the mother of the person referred to is not present:

(44) If you arrest him now, you will be breaking his mother's heart.

This, however, opens up a vague possibility of subsuming the interpretive use of the progressive under Dowty's analysis. It was pointed out above that in examples like (12) - (27) the more fundamental description is given with the SF, whereas the EF gives the description that is dependent on wider circumstances. If these wider circumstances also include larger stretches of time, the cases in point could be regarded as falling under Dowty's analysis. Note that the examples in the present tense are unproblematic. In these cases it is hardly ever possible to draw a clear distinction between the interpretive use and the standard use Dowty is trying to account for. Consider the following dialogue and the contrast between the last utterance and the preceding ones:

(45) I. - You called British politicians 'pigmies'. Were you also referring to Mrs. Thatcher when you said that?
P. - I was making a very general comment.
I. - Were you including her?
P. - I am excluding no one.
(*New Statesman,* 10/13/78, p. 461)

The last utterance clearly refers to an earlier one and can thus be regarded as a case of the interpretive use of the progressive. On the other hand, this sentence also expresses the speaker's current, and possibly future, opinion and falls therefore under Dowty's analysis. There are thus no problems in the present tense. Any attempt, however, to subsume the interpretive use in the past tense under Dowty's analysis does not look very convincing. It seems that conditions as formulated by Dowty are irrelevant in precisely those cases where two clauses or expressions can be made true by the same event. Knowledge of this kind can certainly be assumed to be part of our ability to speak a language. It is crucially involved in our ability to translate non-finite constructions like gerunds and participles of languages such as Russian, Italian or English into languages such as German where the use of non-finite constructions is severely restricted. As is well-known, the gerund in Italian or Russian may express a great variety of relationships to the proposition of the containing clause which have to be expressed by sub-ordinating conjunctions in German. Whether or not a gerund in these languages can be translated by *indem* ... in German, depends on whether or not the two components of the relevant sentence, i.e. gerund and main clause, can be made true by the same event (cf. Pusch (1979)).

5. Conclusion

My aim in this paper has been to draw attention to a use of the progressive which is almost completely ignored in the literature. I have examined this use in some detail and tried to show that a condition for the use of the progressive which is most conspicuous in the interpretive use also applies in nearly all other uses of the *EF*: in contrast to the simple form of the verb, the progressive does not introduce a temporal context but depends for its interpretation on a temporal context independently established. Several possibilities to extend Dowty's analysis in such a way that it also covers the interpretive use were examined in this article but the question of whether Dowty's truth conditions are relevant for the interpretive use has not received a definitive answer.

Finally, a word of explanation should be offered for the fact that nothing has been said about the use of the progressive in the present perfect. As is well-known, the present perfect has defied most attempts to characterize the contributions of the two components to the overall meaning of the present perfect-progressive construction. It seems that in combination with the perfect the context-dependence of the progressive shows up in the fact that this construction always denotes a stretch of time up to and including the moment of utterance during which the state of affairs described by the verb holds.

Bibliography

Anscombe, G.E.M., 1963[2], *Intention*, Blackwell, Oxford.

Austin, J.L., 1962, *How to Do Things with Words,* Clarendon, Oxford.

Bennett, M. and Partee, B., ms. 'Toward the Logic of Tense and Aspect in English'.

Buyssens, E., 1968, *Les deux aspectifs de la conjugaison anglaise au XXe siècle,* Presses Universitaire de Bruxelles, Brussels.

Cresswell, M.J., 1977, "Interval Semantics and Logical Words", in C. Rohrer (ed.) *On the Logical Analysis of Tense and Aspect,* Gunter Narr, Tübingen.

Davidson, D., 1969, "The Individuation of Events", in N. Rescher (ed.) *Essays in Honor of Carl G. Hempel,* Reidel, Dordrecht.

Dowty, D.R., 1977, "Toward a Semantic Analysis of Verb Aspect and the English 'Imperfective' Progressive", *Linguistics and Philosophy* 1, 45-77.

Geis, M.L. and Zwicky, A.M., 1971, "On Invited Inferences", *Linguistic Inquiry* 2, 561-566.

Gleason, H.A., 1965, *Linguistics and English Grammar,* Holt, Rinehart and Winston, New York.

Goldman, A.I., 1971, "The Individuation of Events", *Journal of Philosophy* 68, 761-774.

Hatcher, A.G., 1951, "The Use of the Progressive Form in English, a New Approach", in Schopf (ed.): 1974, pp. 177-216.

Heringer, H.J., 1974, *Praktische Semantik,* Klett, Stuttgart.

Hirtle, W.H., 1967, *The Simple and Progressive Forms. An Analytical Approach,* Les Presses de l'Université Laval, Québec.

Hymes, D., 1974, *Foundations in Sociolinguistics,* University of Pennsylvania Press, Philadelphia.

Jespersen, O., 1954[2], *A Modern English Grammar on Historical Principles,* Allen & Unwin, London.

Joos, M., 1964, *The English Verb,* The University of Wisconsin Press, Madison and Milwaukee.

Labov, W., 1970, "The Study of Language in its Social Context", *Studium Generale* 23, 30-87.

Pusch, L.F., 1979, *Kontrastive Untersuchungen zum italienischen 'gerundio': Instrumental- und Modalsätze und das Problem der Individuierung von Ereignissen*. Niemeyer, Tübingen.

Ryle, G., 1949, *The Concept of Mind,* Penguin, Harmondsworth, Middlesex.

Scheffer, J., 1975, *The Progressive in English,* North Holland Publishing Company, Amsterdam.

Schopf, A., 1969, *Untersuchungen zur Wechselbeziehung zwischen Grammatik und Lexik im Englischen,* De Gruyter, Berlin.

Schopf, A., 1974, *Der Englische Aspekt,* Wissenschaftliche Buchgemeinschaft, Darmstadt.

Scott, D., 1970, "Advice on Modal Logic", in K. Lambert (ed.), *Philosophical Papers in Logic,* Reidel, Dordrecht.
Taylor, B., 1977, "Tense and Continuity", *Linguistics and Philosophy* 1, 99-220.
Trudgill, P., 1974, *Sociolinguistics,* Penguin Books, Harmondsworth.
White, A.R., 1973, "Introduction" in White: 1973, *The Philosophy of Action,* O.U.P., London, S. 1-18.

ORDERING ADVERBS LIKE *ERST, ZUVOR, DARAUF, DANACH* ETC. AS A CASE OF A LEXICAL FIELD*

Peter Rolf Lutzeier (Freie Universität Berlin)

Introduction

Before I start I must tell you right away that some of the problems which are relevant to this study will not be mentioned at all or only be treated superficially. The area in question is simply too vast.

Still I intend to discuss a set of German ordering adverbs in their temporal reading; i.e. German adverbs which can be used to order events, states or actions in a temporal sequence. From the point of view of *sentence semantics* I am aiming at insights which are relevant to a formulation of restrictions for our interpretations of given fragments. These interpretations are part of *formal semantics*. However, the source for my paper, which is part of the process of getting relevant insights for these interpretations, is not formal semantics , but what I would call intuitive or *natural semantics*. Our formal semantics can at best be an approximation to our natural semantics. I must ask you to bear this in mind for the whole of my lecture.

The very fact that we have a label for the set I want to deal with already suggests that its members have something in common. This should emerge more clearly if we treat them together rather than individually. In addition, such a treatment seems to be the only way in which we can capture the relations which intuitively exist within the set. Please do not get me wrong, I do not want to advocate such a *holistic view*, as I want to call it from now on, to the exclusion of individual investigations of lexical items, which would be according to an *atomistic view*. On the contrary I think both approaches have their merits and both have a definite place in word semantics. Nevertheless, I think that for the sort of question I am interested in a holistic view is more helpful than an atomistic one.

---

* I am especially grateful to my wife Elizabeth C. Lutzeier for checking my English.

Now I want to show that it is relatively easy to arrive at such insights in the sense of a holistic view by treating the ordering adverbs as a *lexical field*. Of course this in itself should not be too surprising because the idea of lexical fields is very much in the tradition of holistic views. It is unfortunate that this tradition goes hand in hand with the idea of lexical fields as something notoriously vague. Obviously, I cannot be satisfied with this situation because I definitely need a clear understanding of the notion 'lexical field' before my attempt can have any chance of success.

First criteria for lexical fields

So I must invite you to make a little detour. I shall be looking into the idea of lexical fields before I come back to the ordering adverbs.

Intuitively we know that lexical fields are special paradigms deriving from verbal contexts. The elements of such paradigms represent choices expressive of intentions which are possible in the given verbal context. In other words, such a paradigm is the repertoire of choices in the language with respect to a given verbal context and lexical fields have to be special types of such paradigms.

In order to be precise about the notion 'lexical field' I ought to start with definitions for 'verbal context' and 'paradigm', but I do not want to bore you with them. I shall assume that we all roughly know what is meant by these two notions. In case of the verbal context you have to make sure to rule out contexts which lead to phonological paradigms or paradigms which consist of inflections. Paradigms are then based on a substitution process according to wellformedness criteria.[1]

The problem we are facing now is that of which paradigms should we consider as candidates for lexical fields? A few conditions are fairly obvious: First we do not want a paradigm consisting just of *wordforms* but we do want one consisting of *words*. By the way, this transition from wordforms to words can be explained in precise terms too but I shall not go into that detail either.[2]

---

1 You can find slightly more information about these two notions, not my final word about them though, in my article Lutzeier (1978), pp.241-246.

2 Details about this transition are in the following articles: Lieb (1977), Matthews (1965a) and Matthews (1965b).

Secondly, all members of the paradigm should belong to the same *syntactic category*. Surprisingly enough, this is already quite a step forward towards our intuitive idea about lexical fields, but it does not suffice yet. The restriction to one syntactic category will not keep us from results we would prefer not to have. I suppose for a lexical field we are not just interested in choices expressive of intentions which are possible in the given verbal context, but also in choices expressive of intentions which are possible in the given verbal context and which are relative to a certain *semantic aspect*. Of course, this introduction of a semantic aspect is nothing else than what has traditionally been called 'field value'. The only difference, and it is by no means a trivial one, is that my semantic aspect is given in advance and does not establish itself in a mysterious way out of the field.

Role of dimensions

We have now reached a point whre we have to make a decision about what sort of *structure* we want to assign to lexical fields, otherwise there would not be any further room for more conditions on paradigms.

What one inevitably has is the assignment of some *dimensions* to the set. Each dimension causes a division of the whole set into non-empty, disjoint subsets and to each of these subsets are given *names*. These names cannot be just arbitrary ones because they are meant to play some role for the semantic part of lexical fields.

I first want to give you an illustration before I explain this role in general terms. For instance, on page 305 you will find *zuletzt* as a member of the subset 'Final point in a temporal ordering within a frame fixed by the context' with regard to dimension 1. What does that mean? Well, I understand it in the following way: If we paraphrased the meaning of *zuletzt* under the semantic aspect 'Temporal ordering of events etc.', then under normal circumstances we would get among other things something like 'Final point in a temporal ordering within a frame fixed by the context'. In other words, under normal circumstances the name 'Final point in a temporal ordering within a frame fixed by the context' gives a *necessary* part of the meaning of *zuletzt* under the semantic aspect 'Temporal ordering of events etc.'. In general, the names of subsets from dimensions play the following role: Under normal circumstances they are necessary parts of the meaning of any member of the corresponding subsets with regard to the given semantic aspect.

There are two *exceptions* to this characterization of the role of names of subsets from dimensions, and these are the names 'Neutral' and 'Neither nor'. A look on page 306 shows that *vorher* is a member of the subset 'Neutral' with regard to dimension 2. This tells us that a paraphrase of the meaning of *vorher* under the semantic aspect 'Temporal ordering of events etc.' allows under normal circumstances something like 'Not first point in a temporal ordering' or 'First point in a temporal ordering within a frame fixed by the context' as part of it. So, if a word is a member of a subset named 'Neutral' with regard to a dimension, then under normal circumstances the meaning of this word is open with regard to the names of the other subsets which are different to 'Neither nor' within this dimension. One could say that *no decision* is made with regard to the division caused by the dimension.

We find a different situation in the case of 'Neither nor'. Again compare with page 306 . For instance, under normal circumstances a paraphrase of the meaning of *zuerst* under the semantic aspect 'Temporal ordering of events etc.' would neither involve something which is in relation to something following determined by the context nor something which is in relation to something preceding determined by the context. That is the reason why *zuerst* is a member of the subset 'Neither nor' with regard to dimension 3. So, for any member of the subset 'Neither nor' with regard to a dimension the following holds: A paraphrase of the meaning of a member of such a subset with regard to a dimension under the given semantic aspect would tell us that under normal circumstances the meaning has *nothing* to do with the names of the other subsets of this dimension; i.e. the negation of each of the names of the other subsets with regard to this dimension would under normal circumstances be a necessary part of the meaning under the given semantic aspect.

In my opinion I have exhausted the role of dimensions for lexical fields with these remarks. There is nothing more to it. For instance, the assigned dimensions and names of the caused subsets do not constitute the meaning of the elements in the lexical field under the given semantic aspect. This constitution of the meaning is one of the assumptions which have traditionally been made in connection with dimensions for lexical fields. It does not work like that because to my mind it is impossible to describe the meaning of a word adequately by combining single components. For instance, the menaing of *hinterher* under the given semantic aspect 'Temporal ordering of events etc.' does not consist of the sum or product or whatever of 'Not first point in an ordering' and 'In relation to something preceding determined by the context' and perhaps another, finite number of names. Such a sum or product or whatever is not the meaning, at most it is under normal circumstances a necessary part of it.

My interpretation of the role of dimensions for lexical fields differs from the traditional one especially in the case of a paradigm with *referential* words. In this case one normally assumes that the names of subsets with regard to dimensions state necessary properties for objects being the referents of the words. In my opinion this does not hold in general either. At best it holds for the so-called *stereotypes*. By the way this point was the very reason why I have always added the phrase 'under normal circumstances' for the characterization of the role of dimensions and names of the corresponding subsets.

Semantic relations

Realizing all this means drawing the following conclusions: Instead of expecting too much from our dimensions, which is exactly what has traditionally been done, we had better look around for something else. Again, not in the sense of an excluding alternative but in the sense of a fruitful addition.

In fact if I go right back to the beginning of my lecture and take the holistic view seriously, then I do not have to look very far. It seems to be reasonable to make use of *semantic relations* between words for my concept of lexical field. By 'semantic relations' I mean the ones which have been advocated, for instance, by Lyons.[3]

The *hyponomy-relation* and the *converse-relation* turned out to be especially helpful in that context, because words of a fragment which are related to each other by one of these relations under the given semantic aspect should presumably be members of the same lexical field. In addition, differences in meaning for elements of a lexical field may be easier stateable by means of these relations than by trying to introduce more and more dimensions which could not in any way capture information we get from relations between words. Concerning the scope of these semantic relations we must not forget that they are also relativized to the given semantic aspect of the lexical field and that they derive in this context from natural semantics.

---

3 Compare chapter 9 in Lyons (1977).

Informal definition of lexical fields

The role of semantic relations for lexical fields will be clearer if we look at my informal definition of the notion 'lexical field'. I shall give more explanations when the points come up.

Let F be a fragment of a natural language and α be a class of basic words in F. α is a *lexical field* in F iff α is a word paradigm in F such that conditions 1-5 hold. These conditions are:

1. α is a class of at least two elements.

We need this condition. Otherwise the empty set would trivially be a lexical field in F and classes with one element would not be excluded either. For the latter case one would only need the name of a very special object which does not have any other names and no hyponyms either. Now, classes with only one element have traditionally been excluded as lexical fields and in this case I am following the tradition.

2. There is a syntactic category C such that all members of α belong to this category C. This condition is self explanatory.

3. There is a semantic aspect A and there are finitely many dimensions $D_1,\ldots,D_n$ such that all members of α are relative to this aspect A and each dimension D_i (i=1,...n) causes a division of α into finitely many non-empty disjoint subsets named N_{im_i} (i=1,...,n). According to this condition one has to choose a semantic aspect and some dimensions with corresponding subnames for the subsets created by the dimensions. For instance, you can see on pages 300, 305, 306 which aspect and subnames for the three dimensions I have chosen for the ordering adverbs. I have explained before which role these dimensions and subnames play from the semantic point of view.

4. If $x \in \alpha$ and there is a basic word y in F belonging to category C which is converse to x with regard to aspect A, then $y \in \alpha$ and if $x \in \alpha$ and there is a basic word y in F belonging to category C which is a hyponym of x, then $y \in \alpha$ and if $x \in \alpha$ and there is a basic word y in F belonging to category C and relative to aspect A such that x is a hyponym of y, then $y \in \alpha$.

This condition practically splits up into three subconditions. It is a very important one because it solves a problem which has not been solved by any other notion of lexical fields I know of, namely the question of *completeness*. Here it is interpreted as a *closure condition* with regard to semantic relations. The idea behind it is that elements which fit into the general frame of the field should belong to the field themselves.

5. For all x,y ε α not differentiated by any of the subnames and not hyponyms of each other there is a z ε α such that z is a hyponym or converse to exactly one of the two words x,y or there is a z ε α such that exactly one of the two words x,y is a hyponym of z.
This condition guarantees that members of a lexical field are differentiated up to hyponymy in both directions. The point is that this condition does not necessarily have to be met by introducing further dimensions. This, to my mind, largely unnecessary introduction of further dimensions is a feature which has been shared by all major theories about lexical fields.

Constructing a paradigm

So much for lexical fields in general. It is about time to come back to my *ordering adverbs*.

Let me start with a suitable *verbal context*. I propose: *X hat er sich für ein Buch entschieden* [X he decided on a book].[4] The designated element X indicates the place where the substitution has to take place. In order to save time I shall not consider the general paradigm in F defined by this context but only the subparadigm in F with regard to a specified syntactic category and a given semantic aspect. In other words, I shall immediately try to take parts of condition 2 and 3 for lexical fields into account. Concerning the syntactic category I am aiming at basic words which can have the category t/t; i.e. basic words which can function as *sentence modifiers*. As *semantic aspect* I take 'Temporal ordering of events, states or actions within any interval'. Accepting these restrictions, we end up with the following class of basic words in F:

---

4 All translations from German to English are meant to give a rough idea of the meaning of the German sentence.

α = {*zuerst, darauf, danach, später, zuletzt, erst, dann, schließlich, davor, vorher, zuvor, daraufhin, hintennach, hintendrein, hinterdrein, hinterher, hierauf, letztlich, nachher, endlich, sodann, zuallererst, zuallerletzt, zunächst, hieraufhin*}.

Because of the chosen syntactic category words like *schnell* are excluded. We can also rule out words like *möglicherweise*, *weshalb*, *heute* and *vormittags* with the aid of the semantic aspect. *Vormittags* does not fall under it because it cannot be used within arbitrary time intervals but only for those restricted to days. The aspect determines which reading we should intuitively concentrate on in the case of ambiguous phrases. For instance, in the case of *zuletzt* and *zuallerletzt* it rules out an understanding of them as 'last time in the past' and in the case of *letztlich* an understanding of it as 'recently'.

Questions about elements of the paradigm

In the following I want to find out what sort of dimensions and subnames would be helpful for our class α. By using one's own intuition and consulting several dictionaries[5] one realizes that answers to the following *questions* are relevant for this problem:

1. Is there any event, state or action to follow the one mentioned under the ordering adverb in question?
Obviously, there will almost certainly be some event, state or action which follows the mentioned one but that is not the question. What I am interested in with this question is whether by using the ordering adverb in question one has to have *several* things in mind which one temporally orders by means of the adverb?

2. Can any event, state or action follow the one mentioned under the ordering adverb in question within a frame fixed by the context?
It is really important to restrict the sequence to a frame fixed by the context, otherwise this would trivially hold.[6]

---

5 The dictionaries are listed on page 314.

6 Sentence 4) will illustrate this.

3. Is there any event, state or action to precede the one mentioned under the ordering adverb in question?
This question is to be understood in the same way as question 1.

4. Can any event, state or action precede the one mentioned under the ordering adverb in question within a frame fixed by the context?
Again it is important to restrict the sequence to a frame fixed by the context.

5. Can any event, state or action come in between two other events, states or actions which are temporally ordered by the ordering adverb in question?
Obviously, in almost all cases something will come in between. So my question is whether one has to have in mind a *connection* between the two events, states or actions which are temporally ordered by the ordering adverb in question?

Now I do not have the time and probably you would not have the patience anyway to go through the whole set α for all five questions. Hence I shall try to discuss some *typical* examples for each question.

Question 1 especially concerns those adverbs which intuitively mark the first element in a sequence or the beginning of a process which can be temporally divided into several stages. Such an ordering adverb, for instance, is *zuerst*. Compare sentences (1) - (3):

(1) *Ich bin der einzige Gast, deshalb komme ich zuerst.*
[I am the only guest. That is why I arrive first.]

(2) *Sie serviert das Dessert zuerst, weil es weiter nichts gibt.*
[She is serving the dessert first, because there is nothing else.]

(3) *Diese Bedingung erscheint bei ihm zuerst, weil es keine weitere gibt.*
[This condition occurs first in his writing, because there is no other condition.]

The reading 'historically first' for *zuerst* is clearly ruled out for sentence (3) by the subordinate clause. You must not be surprised to find sentences which first of all involve ordinary ranking. Because this happens in time there is always a temporal interpretation too. In most cases both are inseparable. These three sentences show that in case of *zuerst* nothing has to follow in the sense of question 1.

Question 2 especially concerns those adverbs which intuitively mark the last element in a sequence or the end of a process which can be temporally devided into several stages. As examples I pick out *zuletzt* and *endlich*. Compare sentences (4) - (6):

(4) *Zuletzt tranken wir Kaffee, dann machten wir uns auf den Heimweg.* [The last thing we did was to drink coffee. Then we set off for home.]

(5)* *Zuletzt gab es Eiskrem, danach folgte noch ein Stück Kuchen in der Speisenfolge.* [The last course was ice-cream. This was followed by a piece of cake as part of the menu.]

This sentence would only be acceptable with an explicit *correction* of the first part. For instance, something like sentence (5'):

(5') *Zuletzt gab es Eiskrem, - halt nein - danach folgte noch ein Stück Kuchen in der Speisenfolge.* [The last course was ice-cream, - no, hold on - this was followed by a piece of cake as part of the menu.]

(6) *Endlich waren wir beim Dessert angelangt und dann kam auch noch ein Stück Kuchen.* [At last we got through to the dessert and then we still had to face a piece of cake.]

Sentence (4) shows the importance of restricting the question to a frame fixed by the context. Comparing sentences (5) and (5') suggests that for *zuletzt* nothing can follow *within* a frame fixed by the context. In contrast to that as sentence (6) shows something can follow for *endlich*.

Question 3 also concerns those adverbs which intuitively mark the last element in a sequence or the end of a process which can be temporally divided into several stages. As before I treat *zuletzt* and secondly *schließlich*. Compare sentences or sentence sequences (7) - (10):

(7)* *Ich bin der einzige Gast, deshalb komme ich zuletzt.* [I am the only guest. That is why I arrive last.]

(8)* *Ich serviere das Dessert zuletzt, weil es zuvor nichts gibt.* [I am serving the dessert last, because there is nothing to serve before it.]

(9)* *Ich serviere das Dessert zuletzt, weil es sonst nichts zu essen gibt.* [I am serving the dessert last, because there is nothing else to eat.]

(10) *Schließlich aßen wir ein Dessert. Ansonsten gab es nichts zu essen.* [Finally we ate a dessert. Apart from that we had nothing to eat.]

Sentences (7) - (9) suggest that for *zuletzt* something has to precede. The sequence in (1) is certainly alright. Nevertheless, one can only understand it in the sense that something other than eating came before, for instance, a talk or movie film. In other words, something seems to have to precede in the case of *schließlich* too.

With question 4 I have to go back to those adverbs which intuitively mark the first element in a sequence or the beginning of a process which can be temporally divided into several stages. Again I look at *zuerst* and also at *zunächst*. Compare sentences (11) - (13):

(11)* *Die Suppe kam zuerst, davor hatten wir in der Speisenfolge noch eine Melone gegessen.*
[Soup came first. Before that we had eaten a melon as part of the menu.]

(12) *Zuerst fand ich ihn unsympathisch, davor hatte ich ihn schon einmal sympathisch gefunden.*
[At first I did not take to him. Earlier on I once found him quite pleasant.]

(13)* *Zunächst kam die Suppe, davor hatten wir noch Pastete in der Speisenfolge.*
[To begin with we had soup. Before that the menu also included pâté.]

Sentence (11) suggests that for *zuerst* nothing can precede. The same holds for *zunächst*. Sentence (12) is not a counterexample to my claim about *zuerst*. The occasion before is considered to fall outside the frame fixed by the context. One must not be confused by a suggested reading for *zunächst* in Grimm's dictionary.[7] There it suggests that *zunächst* is occasionally used for an event, state or action which follows an event, state or action which has already been referred to as the first one. Examples for this case are sentence (14) and sentence sequence (15):

(14) *Dann nach all der glücklichen Zeit wird zunächst ein Übel kommen.*
[Then after all this happy time the first thing to come will be evil.]

(15) *Du bist ein feiner Kerl. Du sorgst wohl zuerst für dich, dann gilt aber, daß du zunächst für mich sorgst.*
[You are a fine lad. Admittedly, first you make sure you look after yourself, but then your first care is for me.]

In my opinion these sentences are no counterexamples to my claim about *zunächst*. In both cases a new beginning is marked by the use of *dann*. From then

7 Grimm (1954), columns 546-547.

on the event, state or action mentioned under *zunächst* is the first one.

With regard to question 5 one had better concentrate on those adverbs which either explicitly or implicitly refer to another event, state or action to the one mentioned under them. I shall look at *davor, danach, daraufhin* and *nachher*. Compare sentences or sentence sequences (16) - (22):

(16) *Es gab Hagel am Nachmittag. Davor hatten wir einen kräftigen Schauer, aber dazwischen schien die Sonne für einige Zeit.*
[We had a hailstorm in the afternoon. There was a heavy shower before that, but in between the sun was shining for a while.]

By the way if there is some stress on *da-* in the word *dazwischen*, then *dazwischen* seems to refer to an interval within the interval which is determined by the immediately preceding mentioned event, state or action. In (16) this would be the reading with the sun shining in between the shower. Obviously, I do not intend this reading. I want the reading where the sun is shining between the hail and the shower.

(17) *Beim 100m Lauf gab es ein knappes Finish. Unser Läufer mußte sich mit dem vierten Platz begnügen. Wie erwartet, davor kam der Engländer als erster und dazwischen schoben sich noch beide Polen.*
[There was a photo finish in the 100m event. Our competitor had to put up with fourth place. The English runner, as expected, came in before him in first place and in between there were still the two Polish runners.]

(18) *Diesen Sommer fahren wir in die Französischen Alpen, danach halten wir uns einen Tag in Paris auf und dazwischen wollen wir noch einen Abstecher nach Nizza machen.*
[This summer we are going to the French Alpes. Afterwards we shall spend a day in Paris and in between we still want to make a flying visit to Nice.]

Because it is practically impossible to dash down to Nice while you are staying in Paris for one day, this sentence would be odd with some stress on *da-* in the word *dazwischen*. The reason for that was explained after sentence sequence (16).

(19) *Sieger im Skimarathon war ein Schwede, danach kam ein Finne und dazwischen waren noch zwei Polen.*
[Winner in the skimarathon was a Swede. After him a Fin and between them there were still two Poles.]

(20) *Er wurde angezeigt und daraufhin verhaftet. Dazwischen war er noch einige Tage beobachtet worden.*
[He was reported and on the strength of that he got arrested. In between he had been kept under observation for a couple of days.]

(21)* *Es begann zu regnen, daraufhin drängten alle in die Kneipe und dazwischen hatte die Sonne geschienen.*
[It started raining, which made everyone run into the pub and in between the sun had been shining.]

(22) *Ich sauge jetzt den Teppich. Nachher staube ich ab, nachdem ich eine Tasse Kaffee getrunken habe.*
[I am hoovering the carpet. Later on I shall do the dusting after I have had a cup of coffee.]

In the case of *davor*, *danach* and *nachher* something can come in between. Wahrig suggests for *danach* an immediate following if it concerns ranking.[8] I have my doubts about that as sentence (19) shows. One seems to establish a fairly *strong* connection between two events, states or actions by using *daraufhin*. That is what we can learn from sentence (21). Nevertheless it is possible that something can come in between as sentence (20) proves. In this example one feels that the connection between the two actions (reporting and arresting) is not destroyed by the action in between. This is exactly the case in sentence (21), so it is not surprising that I find it strange.

Dimensions for the paradigm

I would propose the following dimensions with the experience of this survey:

Dimension $\underline{D_1}$:

(N_{11}) Not final point in a temporal ordering = {*davor, vorher, zuvor*}

(N_{12}) Final point in a temporal ordering within a frame fixed by the context = {*zuletzt, zuallerletzt, letztlich*}

(N_{13}) Neutral = {*erst, zuerst, zuallererst, zunächst, schließlich, endlich, darauf, danach, später, dann, hinterher, hierauf, daraufhin, hintennach, hintendrein, hinterdrein, nachher, sodann, hieraufhin*}

---

8 Wahrig (1978), p.191.

Dimension $\underline{D}_2$:

(N_{21}) Not first point in a temporal ordering =
{*darauf, danach, später, dann, schließlich, hinterher, hierauf, endlich, letztlich, zuletzt, zuallerletzt, daraufhin, hintennach, hintendrein, hinterdrein, nachher, sodann, hieraufhin*}

(N_{22}) First point in a temporal ordering within a frame fixed by the context =
{*zuerst, zuallererst, erst, zunächst*}

(N_{23}) Neutral =
{*davor, vorher, zuvor*}

By dimension D_1 in N_{12} I also capture the indexicality of *zuletzt, zuallerletzt* and *letztlich* and by dimension D_2 in N_{22} I capture the indexicality of *zuerst, zuallererst, erst* and *zunächst*. However, one cannot tell from these two dimensions that the words *darauf, danach, davor, hierauf, daraufhin* and *hieraufhin* are explicitly indexical as well as that the words *vorher, zuvor, später, dann, hinterher, hintennach, hintendrein, hinterdrein, nachher* and *sodann* are in a similar way implicitly indexical. This will be captured by my third dimension.

Dimension $\underline{D}_3$:

(N_{31}) In relation to something following determined by the context =
{*davor, vorher, zuvor*}

(N_{32}) In relation to something preceding determined by the context =
{*darauf,danach, später, dann, hinterher, hierauf, daraufhin, hintennach, hintendrein, hinterdrein, nachher, sodann, hieraufhin*}

(N_{33}) Neither nor =
{*zuletzt, zuallerletzt, letztlich, schließlich, endlich, erst, zuerst, zuallererst, zunächst*}

Schließlich and *endlich* are the only words in my paradigm which are not indexical, at least not in the sense as *zuerst* or *danach* are.

In my opinion it does not seem to be sensible to introduce further dimensions, not to mention the fact that there are no obvious ones anyway. If you remember my general discussion about the notion 'lexical field', then you know that we had better ask ourselves how the words of the paradigm are related to each other by means of semantic relations.

Semantic relations within the paradigm

Concerning semantic relations we can restrict ourselves to the *hyponymy-relation* and the *converse-relation*. The antonymy-relation is not applicable to any of the words of my paradigm because none of them allows gradation and we shall see that the incompatibility-relation can be derived from the others to a certain extent.

You will find the results of this section below in form of diagram 1 and diagram 2. The arrow stands for the hyponymy-relation and the converse-relation is marked with 'conv'.

Diagram 1

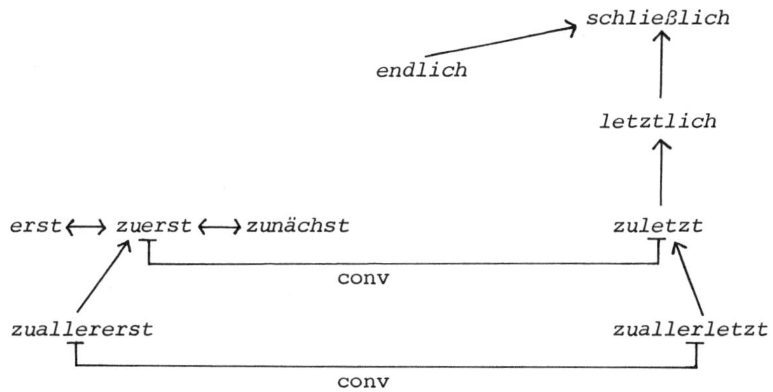

Diagram 2

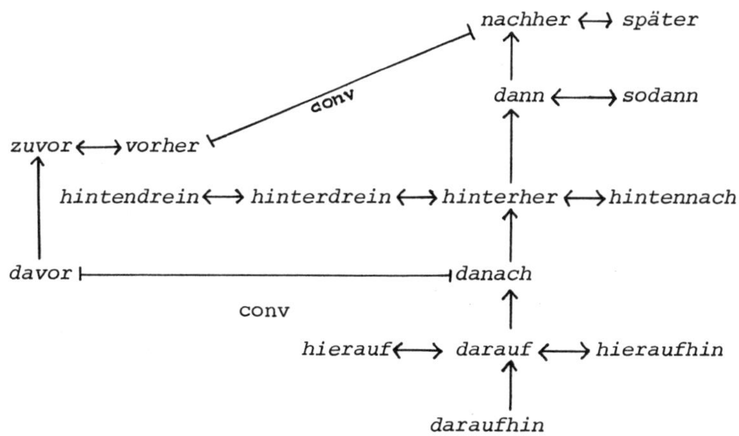

In the following I shall try to point out *general* lines of argument for the establishment of these diagrams. Again all relations are with regard to the given semantic aspect.

Normally the existence of a hyponymy-relation between two words is argued on the level of natural semantics with regard to situations in which we can appropriately apply the words. Examples are:

First the relation between *zuerst* and *zuallererst*.

Any event, state or action which is mentioned under *zuallererst* can be mentioned under *zuerst* too, in other words we certainly have a hyponymy-relation between *zuallererst* and *zuerst*. I hesitate to establish a hyponymy-relation in the other direction. Roughly, *zuallererst* seems to be more appropriate for marking the beginning of something with regard to the *whole* considered frame whereas *zuerst* seems to be more flexible in the sense that it can also mark the beginning with regard to any *part* of the whole considered frame. Compare two reports (23) and (24) on a discussion:

(23) *A eröffnet die Diskussion und behauptete X. B plädierte dagegen für Y. Zuerst fing dann C mit X an, schließlich schwächte er es wieder ab.*
[A opened the discussion and spoke in favour of X. B argued instead for Y. C started then first with X, finally his argument for X was weakened.]

(24)? *A eröffnete die Diskussion und behauptete X. B plädierte dagegen für Y. Zuallererst fing dann C mit X an, schließlich schwächte er es wieder ab.*
[A opened the discussion and spoke in favour of X. B argued instead for Y. First of all C then started with X, finally his argument for X was weakened.]

A's maintaining of X could be mentioned under *zuallererst* without any doubts, whereas I am not sure about C's part in the discussion.

Secondly the relation between *schließlich* and *endlich*. *Schließlich* is not understood in the sense of marking a strict end. On the contrary, it is to be seen as marking the end of a process which has gone on for a while. The same holds for *endlich*. In addition though, the relevant process has been subjectively experienced as a *long* one. So there is no question that *endlich* is a hyponym of *schließlich*. However, because of this additional aspect of *endlich*, I do not think that the hyponymy-relation holds the other way around. In this context I am thinking of a process with a time interval which is even subjectively experienced as a fairly short one. Compare the two sentence sequences (25) and (26):

(25) *Mir wurde eine Mathematikaufgabe vorgelegt. Schließlich löste ich sie in auch für mich unglaublich kurzer Zeit.*
[I had a maths problem to do. Finally I solved it in what seemed to me an incredibly short time.]

(26)? *Mir wurde eine Mathematikaufgabe vorgelegt. Endlich löste ich sie in auch für mich unglaublich kurzer Zeit.*
[I had a maths problem to do. At last I solved it in what seemed to me an incredibly short time.]

Thirdly, does *endlich* stand in the hyponymy-relation to *zuletzt*? We can immediately deny that because, as we have already learned, the event, state or action mentioned under *endlich* is not necessarily the last one within the frame ficed by the context, whereas it is necessarily the last one for *zuletzt*. *Zuletzt* is not a hyponym of *endlich* either. This can be explained by the subjective experience of a long process in the case of *endlich*. Compare sentences (27) and (28):

(27) *Zuletzt kamen 2 Clowns und die ganze Zirkusvorstellung war eigentlich wie im Fluge vorübergegangen.*
[Last came 2 clowns and the whole circus performance was practically gone in a flash.]

(28)* *Endlich kamen 2 Clowns und die ganze Zirkusvorstellung war eigentlich wie im Fluge vorübergegangen.*
[At last 2 clowns appeared and the whole circus performance was practically gone in a flash.]

These data are the reason why I refrain from following Grimm's dictionary which suggests for *endlich* a relation to *zuletzt*.[9]

In the case of words which intuitively involve the ordering of *two* events, states or actions I can use further arguments in support of the hyponymy-relation. One of them is that those words which are explicitly indexical, the so-called pronominal adverbs, somehow establish a stronger connection between two events, states or actions than those which are implicitly indexical. Examples for that are: First, *davor* is a hyponym of *vorher* as well as of *zuvor*. Secondly, *danach* is a hyponym of *hinterher*.

Another argument concerns the possibility of *local* or spatial readings. Local associations carry over to the ranking reading and this, as we know, also allows a temporal reading. Hence, if a word allows local associations because of its morphological form, then it establishes a stronger connection between two events, states or actions than any word which is not a pronominal

9 Grimm (1862), column 464.

adverb. As examples I can quote:
First, the hyponymy-relation between *hinterher* and *dann*. Secondly, the hyponymy-relation between *darauf* and *danach*. In this case one has to relate two pronominal adverbs to each other and the stronger local associations for *darauf* were the decisive ones. By the way, all the dictionaries I consulted agreed with my direction, namely that *darauf* is a hyponym of *danach*.

Of course an intuitive, detailed *individual* survey can also help for the hyponymy-relation. One finds the strongest influence of this argument in the case of *daraufhin*. If you can remember, when I considered the five questions in order to arrive at sensible dimensions for my lexical field we found out about *daraufhin* that it only allows something to come in between the two events, states or actions in special cases. In other words, it seems to establish the strongest connection of all comparable words. Therefore, *daraufhin* is even a hyponym of *darauf*.

The hyponymy-relation in both directions can only be a *necessary* condition for the synonymy-relation. In my opinion for synonymy one has to take into account conditions of *usage* too. I can illustrate this point by means of *vorher* and *zuvor*. They are not differentiated by means of the hyponymy-relation. Nevertheless, I would not call them synonymous to each other. For instance, both behave differently if they are used as predicate modifier. Compare sentences (29) - (32):

(29) *Hans kam Otto zuvor.*
[John was one step ahead of Otto.]

(30) *Hans kam Otto vorher.*
[John was walking in front of Otto.]

This sentence needs some stress on *-her*.

(31)* *Hans lief Otto zuvor.*
[John was walking in front of Otto.]

This sentence confirms the more idiomatic connection with *kommen* for *zuvor* in this verbal context. In contrast to that:

(32) *Hans lief Otto vorher.*
[John was walking in front of Otto.]

This sentence needs some stress on *-her*.

Now I shall turn to the converse-relation. In order to establish a converse-relation between two words we have to show on the level of natural semantics that both words involve asymmetric relations which are reverse to each other. I can illustrate this in the case of the words *zuerst* and *zuletzt*. We have seen that for *zuerst* nothing can precede within a frame whereas for *zuletzt* nothing can follow within a frame. Preceding and following are both asymmetric and reverse to each other. Of course, the same roughly holds for *erst* and *zunächst* in comparison to *zuletzt*. In support of *zuerst* as a converse of *zuletzt* one claims the morphological form. The same argument holds for *vorher* and *nachher*.

If we have two words which are converse to each other, then we often find that one of them is the *marked* element. This is actually the case for *zuerst* and *zuletzt*. As sentence (3) clearly showed *zuerst* can be used for the first element as well as the last element if there is a situation where both fall together. *Zuletzt*, however, is excluded for this situation, as sentence (7) showed. Again the combination of *zuerst* with *zuletzt*, in this order, is possible, compare sentence (33):

(33) *Ich bin der einzige Gast, deshalb komme ich zuerst und zuletzt.*
[I am the only guest. That is why I arrive first and last.]

This supports the thesis that *zuletzt* is the marked element of the set {*zuerst*, *zuletzt*}. If one asks in which sense *zuletzt* is marked, then accepting Lyon's distinction of formally marked, distributionally marked and semantically marked it comes closest to the sense of distributional marking.[10]

Incompatibility is another relation we can look into. However, the opposition in both diagrams is already expressed by means of the converse-relation, so we should not expect a lot of new information from this relation. Two words are incompatible to each other on the level of natural semantics if both words appropriately applied to the same situations always contradict each other. For the incompatibility-relation we can state the following two *rules* which relate it to the converse- and hyponymy-relation:

Rule 1 If {x,y} is a set of expressions which are converse to each other with regard to a semantic aspect, then {x,y} is a set of expressions which are incompatible to each other with regard to this semantic aspect.

Rule 2 If {x,y} is a set of expressions which are incompatible to each other with regard to a semantic aspect and z is a hyponym of x with regard to this semantic aspect, then {z,y} is a set of expressions which are incompatible to each other with regard to this semantic aspect.

Both rules should be pretty obvious, so I shall not spend any time in explaining them.

---

10 Compare paragraph 9.7 in Lyons (1977).

Using rule 1 we find the following sets of incompatible expressions: {*davor, danach*} , {*vorher, nachher*}, {*zuerst, zuletzt*} and {*zuallererst, zuallerletzt*}. Using rule 2 we find, for instance, the following sets of incompatible expressions: {*davor, nachher*}, {*davor, darauf*}, {*zuallererst, zuletzt*} and {*zuerst, letztlich*}.

Because of this relation to the hyponymy-relation and converse-relation I do not see any point in indicating the incompatibility-relation in the two diagrams.

Final remarks

Until now, I have been talking about two distinct diagrams. It is about time to ask whether words from diagram 1 are related in any way to words from diagram 2? Well, if we have *two* events, states or actions ordered by words of diagram 1, then we can also order these two events, states or actions by using one word from diagram 2. That works because the event, state or action mentioned under a word from the set 'First point in an ordering within a frame fixed by the context' necessarily comes before the event, state or action mentioned under a word from the set 'Final point in an ordering within a frame fixed by the context'. Now one of these two events, states or actions can be chosen to be the referent determined by the context. If it is the first event, state or action, then we can choose one of the words from the set 'In relation to something preceding determined by the context' and if it is the second one we can choose one of the words from the set 'In relation to something following determined by the context'. Only one word from diagram 2 can be involved, becuase in any sequence of words from each of the two sets 'In relation to something preceding determined by the context' and 'In relation to something following determined by the context' at least three events, states or actions would be involved. Compare sentence sequence (34):

(34) *Davor schien die Sonne. Danach regnete es.*
[The sun was shining before. Afterwards it was raining.]

Obviously there are sometimes restrictions on the appropriate choice. This is explained by the different conditions of usage for the words. Also one may have to change the order of sentences. Compare sentence sequences (35) - (40):

(35) *Erst regnete es. Zuletzt schien die Sonne.*
[It was raining first. In the end the sun was shining.]

(36) *Es regnete. Darauf schien die Sonne.*
[It was raining. After that the sun was shining.]

(37) *Es regnete. Dann schien die Sonne.*
[It was raining. Then the sun was shining.]

(38) *Es regnete. Später schien die Sonne.*
[It was raining. Later the sun was shining.]

(39) *Es schien die Sonne. Vorher regnete es.*
[The sun was shining. It had been raining before.]

(40) *Es schien die Sonne. Davor regnete es.*
[The sun was shining. Before that it had been raining.]

Normally one hesitates to use *daraufhin* because of its causal background.

We cannot expect a similar relation between words from diagram 2 and words from diagram 1. Words from diagram 2 involve just two events, states or actions whereas words from diagram 1 mark the beginning or end of any sequence of events, states or actions.

If you look at my two diagrams you must not forget that some differences between words are already explained by means of my dimensions. This remark is especially relevant to the example of the words *endlich* and *zuletzt*. Both words are hyponyms of a third one, namely *schließlich*, but they are not related to each other by means of semantic relations. Still, I can differentiate them with regard to dimension 1 because they belong to different subsets.

It is clear that my conditions 1 - 5 for a lexical field in F are met by my paradigm α in F. This especially means that condition 5 is satisfied, in other words, the elements of α are differentiated up to hyponymy in both directions by means of my dimensions or the hyponymy- and converse-relation.

A common label like 'Ordering adverbs' for this set is justified because we have seen that the elements of this set built a structure which can be described as a lexical field.

I hope it is also clear by now how helpful such investigations, in the sense of a holistic view, can be for establishing an interpretation of our fragment F on a *formal* level. Obviously, the meaning of an element of α on the level of formal semantics will have to be formulated in such a way that a paraphrase of such a meaning would incorporate the information the subnames of my dimensions give for the element in question in the sense I explained before. Also we should be able to derive in a formal way the counterparts of the relations the element undergoes to other elements in α according to diagram 1 or 2 from the menaing of the element. If we can guarantee this for our formal semantics of fragment F, then in my opinion we are taking a major step towards solving the problem of finding good approximations for our intuitive or natural semantics of fragment F.

Literature

Lieb, H.-H., 1977, Outline of integrational linguistics. Preliminary version. *Linguistische Arbeiten und Berichte* Heft 9. Freie Universität Berlin FB 16, Berlin (West).

Lutzeier, P., 1978, Semantic fields and other problems in word semantics reconsidered. *Amsterdam papers in formal grammar* Vol.II. Centrale interfaculteit Universiteit van Amsterdam, Amsterdam; pp.235-253.

Lyons, J., 1977, Semantics Vol.1. Cambridge University Press, Cambridge.

Matthews, P.H., 1965a, Some concepts in word- and - paradigm morphology. *Foundations of language* 1, pp.268-289.

Matthews, P.H., 1965b, The inflectional component of a word- and - paradigm grammar. *Journal of linguistics* 1, pp.139-171.

Dictionaries

Drosdowski, G. u. Mitarbeiter, 1976-?, Duden. Das große Wörterbuch der deutschen Sprache in 6 Bänden. Bibliographisches Institut Mannheim.

Grebe, P. u. Mitarbeiter, 1970, Duden. Band 10: Das Bedeutungswörterbuch. Bibliographisches Institut Mannheim.

Grimm, J./ Grimm, W., 1860, Deutsches Wörterbuch 2. Band.

Grimm, J./ Grimm, W., 1862, Deutsches Wörterbuch 3. Band.

Grimm, J./ Grimm, W., 1877, Deutsches Wörterbuch 4. Band, zweite Abteilung (bearbeitet von M.Heyne).

Grimm, J./ Grimm, W., 1885, Deutsches Wörterbuch 6. Band (bearbeitet von M.Heyne).

Grimm, J./ Grimm, W., 1889, Deutsches Wörterbuch 7. Band (bearbeitet von M.von Lexer).

Grimm, J./ Grimm, W., 1899, Deutsches Wörterbuch 9. Band (bearbeitet von M.Heyne).

Grimm, J./ Grimm, W., 1905, Deutsches Wörterbuch 10. Band, erste Abteilung (bearbeitet von M.Heyne).

Grimm, J./ Grimm, W., 1951, Deutsches Wörterbuch 12. Band, zweite Abteilung (bearbeitet von R.Meiszner).

Grimm, J./ Grimm, W., 1954, Deutsches Wörterbuch 16. Band (bearbeitet von G.Rosenhagen). All in: Verlag von S.Hirzel Leipzig.

Paul, H., 1976, Deutsches Wörterbuch (bearbeitet von W.Betz). Max Niemeyer Tübingen 7. Auflage.

Wahrig, G. u. Mitarbeiter, 1978, dtv - Wörterbuch der deutschen Sprache. Deutscher Taschenbuch Verlag München.

SIMPLE PRESENT TENSE*

Asa Kasher and Ruth Manor, Tel-Aviv University

Probably, one of the tenses most neglected by tense logicians and linguists is the simple present as used in descriptions of regularities, habits and states. It seems to us, however, that the present is philosophically, logically and linguistically as fascinating as the past and the future, even when just simple tenses are under consideration. Presently, we would like to make some observations and put forward certain proposals for an analysis of semantic (and pragmatic) contributions of simple present tense forms.

I. Our starting point is a class of arguments, some of which we consider valid and some invalid. Our problem is to explain all these given judgments, by analysing some of their semantic constituents.[1]

(1) The moon moves around the earth.
∴ The moon moves.

(2) Charles drinks wine.
∴ Charles is drinking wine.

(3) I smoke.
∴ I will smoke.

(4) Trevor jumps two meters high.
∴ Trevor has jumped two meters high more than once.

(5) He reminds me of my father.
∴ Always, when I see him I am reminded of my father.
∴ Usually, when I see him I am reminded of my father.
∴ Often, when I see him I am reminded of my father.
∴ Sometimes, when I see him I am reminded of my father.

(6) John speaks Spanish.
John dyes his hair.
∴ John speaks Spanish and dyes his hair.

---

* This is a completely revised version of the paper read in the conference. We thank many of the participants for making helpful remarks on the previous version of our paper.

1 We ignore semantic and pragmatic problems which may arise in a more complete analysis of the following arguments, e.g. adverbial problems.

(7) Russell serves his wife.
Russell serves his country.
∴ Russell serves his wife and his country.

(8) Henry tells me stories and jokes.
∴ Henry tells me stories.

(9) We often ride our horses.
We often read the Bible.
∴ We often ride our horses or read the Bible.
∴ We often ride our horses and read the Bible.

(10) He does not smoke.
∴ It is not the case that he smokes.
∴ He has never smoked a cigar.

These examples represent the spectrum of simple present tense phenomena that will interest us in the sequel. We turn now to a discussion of certain properties shared by propositions which serve as premisses or conclusions in such arguments. Later we shall turn to the arguments themselves.

II. Obviously, sentences such as the premisses of (1) - (10) are used for asserting that certain regularities hold for some periods of time.[2]

What is the time period that a serious and literal utterance of, say, (2) involves? We would like to show now that though this question seemingly lends itself to a simple answer, interesting but rather difficult problems arise when a second glance is attempted.

It seems natural to assume that the time period that an appropriate utterance of (2) involves is some *interval* around the time of utterance, but some difficulties arise immediately. First, an interval has end-points,[3] but there is no natural way to define a starting point and an end point for the period of time during which Charles drinks wine. The first few events of his drinking wine should, probably, be left outside that interval; a couple of drinks does not mark a regularity of drinking. Clearly, it would not make much sense to propose the sixth or the seventh, or for that matter the *n*th drinking event as the starting event of the regularity of wine drinking on the part of Charles. Indeed, there is a further dimension of vagueness lurking in these drinking events. Assuming that one of the events is a starting-*event* of the drinking habit, we are still short of being able to define a starting-*point* of the time period during which Charles drinks wine. Which time point of that event, if any, should be considered the starting point of the regularity

---

2 We limit our attention to sentences describing regularities extending over periods of time. There are, of course, other uses of the simple present. For example, sentences such as: "A lion is an animal." and "This doll walks." may be used to describe properties of the subject without having any specific temporal reference.

3 It does not matter now whether the interval is open or closed.

period? Should it be a time point defined by an act of pouring wine into a glass, an act of swallowing a certain amount of wine, or any intermediate event? We know no answer to this question which is not more than a bit arbitrary.

Moreover, even if a plausible argument is advanced to the effect that a certain act serves as a starting stage the beginning of which is also the beginning of the regularity period, a further deeply rooted dimension of vagueness is forthcoming: Is there any justification for ascribing exactly one interval -- one class of *real numbers* -- to an act of a human agent such as pouring wine into a glass? Is there any way of showing that a pouring event started at 10 o'clock rather than at 10^{-100} seconds past 10? Indeed, if exactly one class of real numbers should be ascribed to the pouring event under consideration such minute and pointless distinctions cannot be avoided.

If the time period involved in an utterance of (2) is an interval, problems of the end-point also abound. Beside the end-point problems which are counterparts of the starting-point problems above mentioned, there is another end-point problem which reflects rather a past/future assymmetry. My assertion that Charles drinks wine rests on what I *know* (or at least firmly believe) about the past behaviour of Charles and on what I *expect* the future behaviour of Charles to be. There does not seem to be a way to justify an expectation about a habit of a person that it terminates on a certain time-point, rather than 10^{-100} seconds earlier.[4]

A second cluster of problems which arise from an ascription of a time interval to an utterance of (2) is related to the density of the class of time points in such an interval. When I assert that Charles drinks wine do I assert something about time-points which are not included in his drinking events? There is, indeed, a natural inclination to answer this question in the affirmative: the regularity shown in the appearance of wine drinking events in Charles' behaviour is due to a habit on his part, to a trait of character he entertains even when he is not engaged in drinking. Dispositions and inclinations exist not only when they are clearly manifested. The first premiss of (6) is an extreme example: an assertion that John speaks Spanish says a lot even about John's most silent moments.

However, arguments of that form rest on a confusion of assertion and natural explanation. Uses of present simple tense, such as the premisses of (1) - (10), assert the existence of regularities, many of which are usually

---

4 On the role of justification in the present context, see M.Dummett, *Frege, Philosophy of Language*, e.g. p.622.

explained by resort to underlying laws (e.g., (1)), habits (e.g., (2) and (6)) etc., but the latter explanations are not part of what is asserted when the former sentences are appropriately used. To see it clearly, consider the following examples:

(11) As it happens, I meet him in the street almost every day.

(12) He wins the local lottery very often.

When these sentences are used under appropriate circumstances, what is asserted is clearly just the existence of a regularity. No allusion is made to some underlying explanation of the asserted regularities. On the contrary, the nature of the examples suggests that it is just by pure chance that I meet him in the street almost every day and that he wins the local lottery that often. When mere regularities are thus asserted to exist, nothing is implied about time points which are not included in the events which form the regular course of events. Nothing, then, is implied by an assertion that I meet him, as it happens, every day, neither about his or my properties nor about our relationships outside the time period of any of our meetings in the street almost daily. Thus, even if a time interval (t_1,t_2) could be naturally ascribed to a certain utterance of (11), nothing is implied by what is thus asserted, concerning properties and relationships entertained by the subjects of the assertion during several closed parts of that interval.[5]

Finally, notice that the premiss of (1) does not imply that time has a first point, nor does it imply that time has a last point. Thus, the time period of the regular turning of the moon around the earth, as asserted by utterring premiss (1), might lack both an upper bound and a lower bound. Strictly speaking, no interval of real numbers can be ascribed to that assertion.[6]

Later we shall make a suggestion as to what should replace the apparently natural time interval in the specification of the semantic and pragmatic contributions of the simple present.

---

5 The same would hold for the sentence "I meet him in the street almost every day."

6 Alternatively, one may claim that every interval can be ascribed to that assertion, namely, that there is no interval which is not within the period of the moons' turning around the earth.

III. We turn now from the problems of the time period of asserted regularities to those of the latter regularities themselves.

A simple observation about regularities, such as Charles's drinking of wine, during a certain period of time, is that the course of events which forms a regularity is spread along the whole time period in a characteristic way: if we divide that time period into segments, *in an appropriate way*, we find a representative of the course of events in each of these segments. (Strictly speaking, a regularity could be formed by a course of states or a course of processes rather than a course of events, but we use "event" in the sequel as a covering term for all such courses.).

What is an appropriate division of the relevant time period into segments?

Notice first that what is an appropriate division of a time period when one regularity is asserted could be clearly inappropriate when another regularity is asserted. In case of the premisses of (1) or (6) every division is appropriate, because the movement of the moon around the earth is continuous, as is also, in a somewhat different sense, John's knowledge of Spanish. The premisses of (2), (3) and other examples do not lend themselves to the same pattern of division of the related time period. Not every division of the time period ascribed to an assertion that Charles drinks wine will have in each of its segments a representative of the course of drinking events which forms that regularity; even Charles does not drink when he is asleep, but a division of his life into segments is possible which has all his sleeping periods among the resulting segments.

Moreover, what is an appropriate division of the time period ascribed to one assertion that Charles drinks wine might be inappropriate for another case of assertion that Charles drinks wine, even if exactly the same time period is ascribed to both assertions. To see that, consider the drinking habits of two communities, *A* and *B*. In community *A* wine is extensively consumed, being commonly served with every meal, on a par with beer and water which are as extensively used. In community *B* wine is a prohibited commodity, but some members of *B* manage to brew some wine and spirits, which they drink during their new year festivities only. Consider, now, Charles, who is a member of *B*, one of those who manage to enjoy drinking some wine almost every year during the new year family reunion. When a member of community *B* asserts, on the background of the drinking regularities prevailing in his community, that Charles drinks wine, a division of the time period ascribed to the assertion might consist of year long segments. Such a division would be inappropriate for an assertion that Charles drinks wine, if performed on the background of the drinking regularities of community *A*, because in that community one who manages to drink some wine just once every year could not be correctly described to be one who drinks wine. Assertions made on the background of community *A* to the effect that Charles drinks wine involve much more delicate divisions of the related time period, indeed.[7]

We shall show later how to capture that type of indexicality induced by the simple present tense.

The last problem we would like to mention presently is that of describing the atomic constituents of a regularity. If Charles drinks wine, then he participates in a course of drinking events each of which is streched over some time period. Our problem is how to describe what happens during such a time period. Put differently, our problem is whether the simple present tense should be analysed in terms of another, more basic tense or rather in terms of some tenseless constituents.

Obviously, a single case of Charles' drinking wine could not be adequately described by *Charles drinks wine during the time period of the single drinking event*. Nor could it be adequately described by *Charles is drinking wine during the time period of the single drinking event*, because the progressive tense is used for marking a proper part of the time period of the described act or process.[8] To be sure, one could use past or future forms for describing a single drinking event from appropriate time points after or before the event, but we see no reason for considering past and future forms as more basic than present forms.[9]

Such considerations make the use of tenseless predicates or relations seem indispensable. Notice, however, that just tenseless forms won't do, because the time of utterance plays a requisite role in any analysis of the simple present.

Our analysis of the simple present tense is, then, required to solve the

---

7 We admit that the concept of assertion made on a background is in want of analysis, but we cannot indulge in that here.

8 See M.J. Cresswell, "Adverbs of space and time", in: *Formal Semantics and Pragmatics for Natural Languages* (F. Guenther and S.J. Schmidt, eds.) p.175, for an example of that observation.

9 In the lecture we did suggest an analysis of the single events described in terms of the simple past tense. This was changed due to the reasons mentioned as well as the following: the single activity may not even be describable by the same verb used to describe the regularity. Z. Vendler, *Linguistics in Philosophy* (p.109) gives the example of a ruler who rules the contry by performing single activities which are of various natures.

problems we have mentioned concerning presentations of the related periods and regularities and to explain why arguments such as (1) and (3) are valid, why arguments such as (2) are invalid and why (7) and similar arguments might seem of some oddity.

IV. Our approach is essentially pragmatic. Although some semantic ingredients can be extracted from it,[10] our approach is of a pragmatic nature because it rests on use conditions of sentences, rather than on truth conditions and the like. Indeed, any analysis of the simple present tense should specify the related contextual elements, which are more complicated and much more interesting than the common indexical features of sentences, but our approach is pragmatic in a deeper sense than that involved in indexical semantics and its extensions.[11]

When one of us says, literaly and sincerely, "Charles drinks wine", he or she is *committed to specify, on demand,* first, *a time period,* during which Charles is taken to show a regularity of drinking wine, and secondly, *a type of regularity,* which takes the form of appropriate conditions for divisions of the specified time period into segments; almost each of the latter segments is taken by the speaker to include a representative of Charles' drinking regularity.

How does one specify a time period? The common, *semantic* answer to this question is simple, *viz.* by specifying the end-points of an interval, or the like, but as we have argued in section II of the present paper, that answer is incorrect.

The pragmatic approach lends an answer of a completely different nature.

Use conditions are appropriate for using the sentence sincerely and literally. Now, a major ingredient of an appropriate context of utterance is what the speaker is committed to do, or to show that he is able to do, upon demand, when asserting or doing something similar with words in that context. The speaker is committed to provide a suitable justification of whatever he asserts or implies to know,[12] and it is that commitment which is at the roots of our approach.

---

10 See "Intervals with vague edges", forthcoming.

11 For a programmatic background, see A. Kasher, "What is a theory of use?", in: *Meaning and Use* (A. Margalit, ed.), pp. 37-55.

12 See references in our fn 4 and fn 11 for general discussions of this point.

For the time period ascribed to an assertion to be known, information of two types is required: First, that certain time elements - say, points or intervals - belong to that time period, and secondly, that if some time elements belong to that period and fulfil certain conditions, then certain other time elements also belong to that time period.

For example, if Joan was smoking a cigarette when the midnight news were breadcasted, then one may be in a position to know that midnight was among the time elements of the time period of that smoking event, as well as that one minute past midnight also belongs to that time period. Moreover, one may be in a position to know that if two time points, which are one minute apart from each other, belong to the time period of the same smoking event of a single cigarette, then every time point between them, also belongs to the same time period of that smoking event.

Thus, the specification of a time period T, *within a pragmatic framework*, takes the form of what looks like an axiomatic system of some type. The axioms are propositions about set-theoretical relations which hold between *certain* time elements and the time period T, and the non-logical derivation rules of the system correspond to general propositions about time elements belonging to T given that some related elements belong to it and fulfill certain conditions. The time period T is accordingly characterized through the set of theorems of the axiomatic system having the form 'time element t belongs to time period T'.[13]

We define now, a *vague interval* to be a time period T as characterized by an axiomatic system X of that type,[14] which fulfills the following conditions:

(i) there are different time elements t' and t'' for which it is provable in X that t' belongs to T and that t'' belongs to T;

(ii) for every two elements t' and t'' for which it is provable in X that t' and that t'' both belong to T, it is also provable in X that all the elements between t' and t'' belong to T;

(iii) there are two time points s' and s'' for which it is provable in X that s' and every earlier time element do not belong to T and that s' and every later time element do not belong to T;

(iv) for any two elements t and s for which it is provable in X that t belongs to T and that s does not belong to T, there are different elements r' and r'' between t and s such that for any element r between

---

13 Whenever an axiomatic system is used in explaining human capacities the problem of the logical power of the human agent arises, but we abstract from such problems now.

14 We leave our definition and arguments at that informal level. A precise presentation is left for another occasion. See fn 10.

r' and r'' it is neither provable in X that r belongs to T nor that r does not belong to T.

Notice that knowledge about T may pertain to points outside the interval (t',t''), but indeed within (s',s'').

The only restriction that our definition imposes on the system X, with respect to theorems of the form 't belongs to T' or 't does not belong to T', where t belongs to either (s',t') or (t'',s''), is our condition (iv).

Now we define an *e-vague interval*, for a positive real number *e* , to be a vague interval which satisfies also the following condition:

(v) There are no two time points t and s which are less than *e* apart from each other, for which it is provable in X that t belongs to T and that s does not belong to it.

Condition (v) - which is clearly stronger than (iv) - imposes on the axiomatic system an indeterminacy of a special type. It is not the case that for every time point t in the "edge intervals" (s',t') and (t'',s''), it is either provable in X that t belongs to T or provable in X that t does not belong to T. In a sense, the minimal extent of indeterminacy occurs when there is a point t, say in the edge interval (t'',s''), such that it is provable in X that t belongs to T and that the time point t+*e* does not belong to T. According to condition (v) it is provable in X neither that a time point s belongs to T nor that it does not belong to T, if it is in the interval (t,t+*e*).

An important observation about condition (v) is that the definition does not imply that *e* is maximal. Given that a system X defines an e-vague interval, it is still possible indeed that X defines, say, a 2e-vague interval, provided that s' and t' are at least 2e apart and so are s'' and t''. Consequently, when one specifies a time period the edges of which are vague, by putting forward a system X which defines an *e*-vague interval, for some real number *e*, one may keep the extent of edge-vagueness also vague by not claiming that the real number *e* plays any particular role which any other real number *e*' could not have played. This second order vagueness of time specifications is not captured by the common semantic approaches. A fuzzy interval, for example, would have an explicitly determined point where a change in the characteristic function of the interval takes place, from 0 to positive values in the open interval (0,1) or from values in that interval to 1. This strictly determined extent of vagueness seems to be a major shortcoming of some explications of vagueness which the present definition seems to have overcome.

We are now in a position to state one of our major claims concerning uses of the simple present tense here under discussion: when a speaker makes an assertion by utterance of any of the premisses of arguments (1) - (10) or syntactically similar sentences, the speaker is committed to specifying, on demand, an e-vague interval, for some real number e . That time period is ascribed to the asserted regularity, playing the role of the temporal domain of the regularity.

Each of the asserted regularities consists of a course of events. We claim that when a speaker is committed to specifying a time period for each of those events, it is again a commitment to specify an e'-vague interval, for some real number e', which is involved.

It is easily seen that the problems we mentioned in section II, concerning the common suggestion that the time periods ascribed to the asserted regularities and to the events of which they consist are intervals, do not arise with respect to our vague intervals.

Another claim that we would like to make pertains to the temporal properties of the asserted regularities. We saw in section III that the assertion of a regularity involves divisions of the ascribed time period into time segments, almost each of which includes a representative of the constitutive course of events. Furthermore, the class of relevant divisions depends on contingent knowledge of prevalent regulations, practices and the like. This ingredient of our analysis of the simple present tense is reflected in the following use conditions of sentences such as the ones employed in assertions of the premisses of (1) - (10) and of similar sentences: the speaker is committed to specifying a real number d and to defending the claim that divisions of the ascribed time period into segments of the form $(t,t+d)$ will be such that almost each of the segments of a division includes a representative of the constitutive course of events. Indeed, d could be a day, a year, a second or what have you, and usually there will be real numbers that would differ from d, but serve the same purpose as well as d does.

Such an implicit allusion to major ingredients of what the speaker means is of some general interest because it reflects a philosophically as well as linguistically interesting gap between meaning and understanding. The addressee may form hypotheses concerning which α and which e the speaker has in mind, but he is in no position to know which α and which e the speaker is committed to specify, from what the speaker says. It has been argued elsewhere that this is not an isolated phenomenon but we cannot go into that now.[15]

V. We conclude the present preliminary study of the simple present tense by trying to explain our intuitions concerning arguments (1) - (10). Formal representations of the arguments will not be attempted here.

Argument (1) is valid:

The moon moves around the earth.

Therefore: The moon moves.

The time period that a speaker has in mind when asserting the premiss of the argument would indeed fit an assertion of the conclusion on the same context of utterance.[16] The same would hold for the appropriate divisions of that time interval into segments which include representatives of the constitutive course of movement events. A full account of the argument which will demonstrate its validity should not ignore some formal properties of the adverbial "around the earth" but this is of minor importance for our present purposes.

Argument (2) is, of course, invalid:

Charles drinks wine.

Therefore: Charles is drinking wine.

An assertion of the premiss rests on a commitment to specify divisions of the ascribed time period into segments, almost each of which includes a constituent of the regularity. Under ordinary circumstances, such a segment will be much longer than a common drinking event. Ample parts of such segments do not include drinking events of the asserted regularity. Thus, when the premiss is asserted during such parts, it may be true, without the conclusion being thereby true at the same time.

Argument (3) is invalid, though it holds, so to speak, a grain of validity:

I smoke.

Therefore: I will smoke.

---

15 For a detailed discussion of other examples, see A. Kasher, "Three kinds of linguistic commitment", in the present volume, and "Meaning and Understanding", forthcoming.

16 Note that if the speaker intends "all times" as the period ascribed to the premise, then the period does not correspond a vague interval, since (iii) is violated. Yet, if we interpret "all times" as referring to all time periods (or vague intervals), then argument (1) is valid, since "all periods" fit the assertion of the conclusion as well.

As far as logical possibility is under consideration, the moments of the assertion of the premiss might be the last moments of my life, in which case the conclusion will be found out to be false. However, if I am correct in asserting that I smoke, then I am correct in asserting that I am an active agent in an ongoing regularity. My assertion implies that I intend to continue smoking, since it is an assertion about a present regularity, i.e. a regularity which is at least expected to extend into the future. In a sense, argument (3) is an enthymeme of some pragmatic nature.

Argument (4) seems to us to be valid:

Trevor jumps two meters high.

Therefore: Trevor has jumped two meters high more than once.

The assertion of the premiss is an assertion that a certain regularity is in effect, and a regularity in matters of jumping about two meters high seems to require more than one occasion of such an achievement: no regularity without a plurality.

Under (5) we have four different arguments, all of one and the same premiss: "He reminds me of my father." Assume that I assert the premiss. Where the conclusion is that *always*, when I see him I am reminded of my father, the argument seems to be invalid. For a regularity to hold along a series of seening events it is sufficient that almost all of those events are typical constituents of the regularity. Conclusions such as "Usually, when I see him I am reminded of my father", "Often, when I see him I am reminded of my father" and "Sometimes, when I see him I am reminded of my father" do, indeed, follow from that premiss, though an assertion of the last one in a context of utterance which is appropriate for asserting the premiss would violate a conversational maxim.

Arguments (6) and (7) have conclusions which are formed from the respective premisses by coordination processes.

John speaks Spanish.
John dyes his hair.

Therefore: John speaks Spanish and dyes his hair.

and

Russell serves his wife.
Russell serves his country.

Therefore: Russell serves his wife and his country.

Since the conclusions of these arguments are in the simple present tense, their assertions require commitments to specify appropriate time periods and appropriate divisions of them into segments. Since the *logical* form as "p,q, therefore: p and q" is valid, what we have to find out in order to explain our intuitions about arguments (6) and (7) is just whether such specifications of a time period and appropriate divisions for an asserted conclusion could be formulated in terms of the corresponding specifications for the asserted premisses.

While there does not seem to arise a problem in ascribing a time period to each of those conclusions, a problem does arise with respect to the appropriate division of the time period into segments. For example, under ordinary circumstances, the regularity which is asserted to exist by saying that Russell serves his wife consists of events much closer to each other in time than the constitutive events of the regularity described by saying that Russell serves his country. To be sure, under such circumstances the conclusion itself -- "Russell serves his wife and his country" -- sounds odd, but other, less ordinary situations are imaginable in which no such oddity is induced by the coordination. The lesson to be drawn from such examples seems to us to be that the contingent ingredients of the assertions involved should be spelled out before validity judgments are formed.

Argument (8) shows another problem of coordination:

Henry tells me stories and jokes.
Therefore: Henry tells me stories.

A regularity of telling stories and jokes cannot consist of an abundance of joke telling events and just a few story telling events. Both jokes and stories should be well represented in the series of the constitutive tale telling events, to an extent that would substantiate a claim that a regularity of story telling events also exists.

Assume now that the premisses of argument (9) can be coordinated:

We often ride horses.
We often read the Bible.

It follows clearly from these premisses that a course of events exists which form a regularity, each of the constituents of which is a horse riding event or a Bible reading event. However, the conclusion "We often ride our horses and read the Bible" seems to be ambiguous and the argument is valid on just one of the readings of the conclusion. If the latter is transformationally related to "We often ride our horses and often read the Bible", then the argument is obviously valid. However, if the reading of the conclusion is not related to the premisses in the same way, the argument is invalid, because its conclusion amounts to an assertion that a course of events exists whose constituents are each a horse riding and Bible reading event. The existence of such an unusual course of events is not warranted by the existence of two independent regularities of horse riding and of Bible reading.

Argument (10a) is clearly valid:

> He does not smoke.
> Therefore: It is not the case that he smokes.

Both the premiss and the conclusion describe the same state of affairs: there is no smoking regularity on his part. On the other hand, argument (10b) is not valid:

> He does not smoke.
> Therefore: He has never smoked a cigar.

Although no regularity of smoking exists on his part, he may have smoken one or two cigars. One or two smoking events do not form a regularity, indeed.[17]

17 In the lecture we have discurred some features of the relations between vague-intervals, e.g., weet, inclusion and temporal precedence, which are involved in some of the arguments we have discussed. We deter the presentation of our results to another occasion.

PRE-TENSE

Helmut Schnelle, Ruhr-Universität Bochum

0. Introduction

The analysis of time and tense leads to a fundamental problem of semantics: What is the nature of the conceptual scheme appropriate for our analysis of language understanding? In particular: Is the conceptual scheme that turned out to be useful in natural science adequate for a semantics of ordinary language? I think that this question has not been seriously discussed so far. Linguists tend to rely either on their analysis of language forms supplemented by some ad-hoc notions of semantics or else on some conceptual scheme imported from philosophy, logic, and natural science without further discussion.

The sometimes heated debates among linguists whether tense primarily expresses time or refers to time, and in which sense show that purely linguistic data do not shed much light on the semantic side of the problem. The analysis of the forms of verb inflection or verb constructions provide only for a system of subclassification which, if directly interpreted, indicates at most a corresponding system of modifications of the contents of verbs. But time is more than subclassifications of meanings, even more than the subclassifications into past, present, and future. As long as the contents of the notions "past", "present" and "future" keep to be taken in their intuitive, unanalysed meaning nothing is gained. In their attempt to give the linguistic subclassifications a precize meaning some linguists have turned to notions developed by physics and clarified by philosophy of science. This step seems justified in so far as these notions have even become common knowledge in our cultural context. Jespersen is one of the first to start the analysis of time and tense along this line. The force of his cultural context is obvious when he writes "there can be no doubt that we are obliged (by the essence of time itself, or at any rate by a necessity of our thinking) to figure ourselves time as something having one dimension only, thus capable of being represented on a straight line." (Jespersen, 1924, p.256). The logicians Reichenbach and Montague followed his analysis of tense. At present, analysis of tense in the framework of Montague grammar provide for the most advanced tense analysis of this type. In spite of the apparent success of the approach I do not believe that it is appropriate. There are two essential aspects of the underlying notion of time that make it inappropriate as a semantic concept of ordinary language. The first aspect is that it is a spatialized notion of time[1], the second aspect is that it does not provide a proper basis for temporal deixis. Let me discuss these aspects.

As correctly incicated by Jespersen, the classical notion of time is the bothways infinite and everywhere dense sequence of time that may be represented by a straight line. This representation relates it to the notion of one dimension in space. Four-dimensional space-time is therefore a natural consequence of this spatialized notion of time just as interpreting the laws of movement as constraints over possible configurations in four-dimensional space-time.

The success of this spatialized notion of time as the basic element of physics is not a sufficient reason for importing it to the semantics for ordinary language. In fact, the difficulty of conceptual analysis and the long-lasting struggle to arrive at the pure notion rather indicates that semantic factors of primary importance impeded conceptual progress and affected the classical notion by aspects that are quite alien to it. Rigorous conceptual analysis has shown - and this is certainly a major achievement for natural science - that in particular "nowness and temporal becoming are [not] entitled to a place within physical theory, ..." and that "the temporal relations of earlier than, later than, and of simultaneity do ... obtain among physical events in their own right in the sense familiar from the theory of relativity" (Grünbaum, 1969, p.172). Grünbaum, who thus summarizes the achievements, continues: "if the 'meaning' of time is held to comprize becoming or passage, then one of the features of time is mind-dependent." (ibid). But he adds: "in characterizing becoming as mind-dependent, I allow fully that the mental events on which it depends themselves require a biochemical physical base or possibly a physical basis involving cybernetic hardware." (ibid, p.173). I contrast to this, Grünbaum states in the introduction to his discussion: "Our commonplace use of tenses codifies our experience that any particular present is superseded by another whose event-

1 See e.g. Quine, 1960, p.172, Čapek, 1976, pp.XXVII,LI,503, a.o.

content thereby 'comes into being'. It is this occurring *now* or coming into being which is called 'becoming' or 'passage'. Thus by involving reference to present occurrence, becoming involves more than mere occurrence at various serially ordered clock times" (ibid, p.147).

If we agree with these statements of Grünbaum, as I think we must, tenses are correlated with commonplace experience whereas the conceptual scheme of physics, empirically related to serial orderings of clock times only, does not involve the essential ingredients of this experience. In particular, it does not capture the features of nowness and becoming and, hence, the elements basic to temporal deixis. It is therefore inappropriate as a conceptual scheme for the semantics of ordinary language. How is tense analysis of Montague-grammarians to be judged in this context? It appears as if their concept involves at least an aspect of 'nowness' since they show how to treat temporal deixis. But in a closer analysis this is not so. What is involved in their concept is simply this: The meaning of a verb is a time function.[2] The value of a particular time coordinate separates the time function into three sections, the section before that value, the section at that value and the section after that value - since the relation of before (or after) is a primitive relation in the set of time moments. But every possible value of the time coordinate gives this result. Thus the analysis is not yet a correlate of 'nowness' since it is unclear how we are going to identify the appropriate coordinate value of 'now' at an actual now. We may interprete time moments as possible dates and clock readings. In this case, we could identify now the appropriate coordinate value by looking at our watch now - assuming that its time is correct - and by identifying the present moment with the time moment read from the watch. But what, if we do not have a watch available? It is obvious, that Montague does not give a meaning to 'nowness' nor does he give a meaning to the "flow of time" or to the notion of "becoming". Since, however, we are analysing human beings and not the laws of physics this is a shortcoming of Montague's analysis just as of any analysis, making use of the time line, i.e. the spatialized time concept only. The laws of physics may be stated as constraints over configurations but our experiences of time are not captured in this way.

What are we to try instead? Obviously, the semanticist is forced to leave his purely formal argumentation. He has to join the efforts of con-

---

2 neglecting the possible-world coordinate.

ceptual analysis of human experience. Various ways are open for such an analysis.[3]

I think that conceptual analysis should start from the assumption that no concept is acceptable as a meaning component of a linguistic utterance - or of a word or a sentence - unless it is clear how it is available to the human mind. In other words, the meanings of linguistic utterances are not determined unless with respect to the parameters of mental organization.

One attempt at explicating mental organization may follow Grünbaum's suggestion to relate mental events and the structure over mental events to a physical basis as it may be described by information flow models, especially by nerve nets. Indeed: Why not write grammars and models of the world directly as nerve nets, i.e. as formal parameters of a part of the brain?[4]

With respect to semantics this approach is a concrete step towards Bierwisch's recent proposal[5] to re-interprete the system of possible extensions and intensions in such a way "that it becomes the structure of the enternal, cognitive representation of the experience of a person p" (ibid,p.8).

Exploiting such models for a linguistically appropriate analysis of tense[6] leads to a program that can only be outlined: Temporal concepts relevant for the semantics of tense are only acceptable insofar as their representation is specified in terms of the organisation of mind. The analysis of mental organisation as provided by cybernetics indicates different degrees of complexity of the relevant models. Depending on these degrees we get different aspects of the notion of time and tense that may be ordered into a sequence of increasing complexity. At the present stage of my research I distinguish seven notions:[7]

---

3 e.g. phenomenological analysis of time, which will be discussed elsewhere.

4 Philosophically, this is closely related to Harman's or Quine's ideas as explained in Harman, 1973, ch.3, or to Quine, 1974, p.14.

5 Bierwisch, 1979.

6 in the wide sense, i.e. including aspect, mood and mode.

7 In addition to the following brief indication of the differences between these notions it may be helpful to correlate the notions with some work done so far on these notions: 1. Chafe 1973, 1974; Longuet-Higgins 1972; Isard 1974, 1975; Stephany 1978; 2. Work on classical aspects such as imperfective, perfective etc.; 3. Schnelle 1976, 1979; 4. Work on usual aspects not subsumed under 2; 5. Work on omnitemporal tense; 6. Jespersen 1924, Montague 1974.

1. Pre-Tense, 2. Simple Aspectual Tense, 3. Simple Narrative Tense, 4. Complex Narrative Tense, 5. Complex Aspectual Tense, 6. General Tense, 7. Time-Line Tense.

The uses of language characteristic for these stages are:
1. Use with reference to actual or actually still relevant situations only, 2. Use with reference to simple differences of activities and achievements, 3. Simple narrative use, 4. Complex narrative use, 5. Duration specifying use, 6. Generalized statements use, 7. Use with reference to dates and clock-readings (from an infinite set of such dates and clockreadings). There is a certain development in this sequence correlating roughly to ontogenetic development. The expressional means, whose emergence marks the acquisition of the respective stages, are:
1. Simple verb modification, 2. Simple verb modifications related to achievements, 3. Text arrangements and simple text connectors (such as "and then"), 4. Temporal conjunctions and prepositions, 5. Duration specifying adverbs and aspectual verb modifications related to durations, 6. Generalizing adverbs (e.g. "always"), 7. Temporal moments or intervals naming expressions from an infinite set of such expressions.
It seems that the various means of referring to temporal aspects interact in such a way that means of later stages serve to provide for futher differentiation of the means provided by earlier stages. If so, we might expect to find uses of simpler stages even in fully developed languages.

1. *Linguistic considerations on pre-tense*

Let us first discuss some linguistic aspects connected with the notion of pre-tense. In 1973 Chafe argued for a new analysis of tense that would provide new perspectives - in his view - to a "psychosemantics". In considering the role of temporal adverbs in sentences with the past tense, he is lead to the view that sentences without a temporal adverb usually have a special status. He invites consideration of sentences such as

(1) Steve fell in the swimming pool.

and distinguishes three cases:
1. The sentence is uttered in a co-text explicitly containing a time reference (and there is no indication that this time reference is no longer relevant).

2. The sentence is uttered in a context where the speaker may assume that at that moment he and his adressee are focussed on the same time reference.

3. The sentence is uttered "'out of the blue' with no time reference previously established." (Chafe, 1973, p.264).

In the first case either the sentence itself contains an adverbial containing the time reference or a preceding sentence contained an adverbial that is still relevant for the sentence under consideration.[8] Sentences of this type have been studied in detail, in particular with respect to the interaction of tense and possible adverbials.[9] Sentences not containing adverbials nor any adverbial in their context are usually explained elliptically: Utterances of such sentences do not need to express time reference by adverbials, since what the adverbial could refer to is already in the mind of the speaker and hearer.

Time reference is, in these cases, of three different kinds: 'at/on'-adverbial containing an expression referring to dates or clockreandings, temporal preposition or conjunction adverbial containing an expression referring to an event, adverbials (such as yesterday, tomorrow) referring to a time-interval implicitly related to the interval determined by the utterance.[10] A sentence like (1) might also occur in a fiction story without being located by an adverbial. In this case, the location of the occurrence of the sentence in the story itself is sufficient for locating the event referred to relative to other events mentioned in the story. In this case, the past tense of the sentence is determined with respect to the narrative structure only, in other words it is a case of tense description of stage 3 mentioned above.[11]

Let us now turn to case 3 discussed by Chafe. Chafe writes that an utterance of a sentence according to case 3 "communicates, among other things, the information that the event in question is still fresh in the speaker's mind. Typically, it is something perceived by the speaker recently enough so

---

8 on the transfer of relevance of preceding adverbials see Languet-Higgins, 1972, Isard, 1974, 1975 and C.Smith, this volume.

9 See already Wunderlich, 1970. Compare also Seiler, 1977, pp.432 - 448.

10 the relation to stages 7, 4 and 5 mentioned above should be obvious.

11 On tense analysis relative to this stage cf. Schnelle, 1976, 1979. Vgl. also Wunderlich, 1970, § 1.6.

that it has not yet left his consciousness" (ibid). Chafe introduces the notion of surface memory relating to information kept in a memory similar to a short term memory. He claims, that during the period when information is retained in surface memory, a speaker may appropriately communicate it (in a context where no time reference has been established) with a sentence containing no temporal adverb. Chafe emphasizes that "retention of something in surface memory is not a function of 'real' or 'physical' time, but rather of what might be called EXPERIENTAL TIME" (ibid, p.266).

It is crucial to see, that neither time moments nor time intervals, nor events nor discourse structure in a narrative context is needed for the analysis of past tense in this use: It is the presence in surface memory that is sufficient. This is the essential feature I want to capture in the notion of pre-tense; its analysis does not depend on something that could or should be referred to by some additional phrase or clause. It simply depends on the situation of use in a similar way as strict present depends on it.[12] The reference to surface memory is not only communicated by past but also by *non-generic perfective*, such as

(2) Steve has fallen in the SWIMMING pool.

As usual, Chafe adds, that in an utterance of a sentence like this, the speaker is not only concerned with Steve's falling in the pool, but also with some other event or state, obvious to speaker and hearer, such that this other event is a consequence of Steve's falling in the pool. The non-generic perfective is relatively complicated since it refers to two events (instead of one) and a relation between the two.
Chafe remarks that the interpretation of a generic perfective is different. "If sentence [(2)] carried the generic meaning, it would convey the idea that Steve has fallen in the pool on one or more occasions during his life, the times being left unspecified." (ibid, p.267).

Chafe's discussion is related to the ordinary use of tense in a fully developed language. However, it seems to me that this particular use of tense - the pre-tense - is the earliest one that can be acquired by a child. Some

---

12 Not quite. A sentence in strict present is an occasion sentence in Quine's terminology. An utterance of (1) in its use of Chafe's case 3 is an example of a standing sentence, according to Quine, 1960, p.36 though not of an eternal sentence.

13 where the reference to one of the events is implicit, just as the reference to the relation.

linguists seem to hold that the most primitive use of tense is strictly related to aspect or, rather, that aspect is more primitive than tense.[14] But the facts discussed only show that tense in the *developed* sense is later than aspect, whereas pre-tense is strictly correlated with aspect to the point of being indistinguishable from it by form only.

I shall now turn to the discussion of the theoretical framework of conceptual analysis. It will be introduced by a re-interpretation of Montague's notion of past, which is - linguistically - not past, but the generic perfective as characterized by Chafe.

2. *Montague's past re-interpreted as a quasi pre-tense*

As is well known from "The proper Treatment of Quantification in Ordinary English", a sentence in the past tense is translated into a sentence of intensional logic of the form $\ulcorner H\phi \urcorner$, if $\ulcorner \phi \urcorner$ is the translation of the same sentence in the present tense. Interpreting $\ulcorner \phi \urcorner$ and $\ulcorner H\phi \urcorner$ on the basis of some interpretation system $\mathfrak{A}$ consists in assigning an intension to $\ulcorner \phi \urcorner$ as well as to $\ulcorner H\phi \urcorner$. Montague proposes that functions over an infinite sequence of time moments[15] with values in $\{0,1\}$ may serve as appropriate intensions.

Let us assume that the function assigned to $\ulcorner \phi \urcorner$ is f_ϕ and that the function assigned to $\ulcorner H\phi \urcorner$ is $f_{H\phi}$. $f_{H\phi}$ depends on f_ϕ in the following way: $f_{H\phi}$ has the value 1 for some time moment t_1 if and only if there is some time moment t_2 earlier than t_1 ($t_2 < t_1$) for which f_ϕ assigns the value 1. Let us assume that the sequence of time moments is discrete. Let t' be the time moment which immediately follows t in the sequence. In this case, the condition relating $f_{H\phi}$ to f_ϕ is equivalent to the following statement:

(1) For all t

$$f_{H\phi}(t') = f_{H\phi}(t) \vee f_\phi(t)$$

and

(2) if there is some t_1 for which $f_{H\phi}(t_1) = 1$ there is some t_2 such that

$$f_{H\phi}(t_2) = 0 \text{ and } f_\phi(t_2) = 1$$

---

14 cf. U.Stephany, 1978, p.6, referring to supporting evidence of Ferreiro 1971, a.o.

15 contained in the system $\mathfrak{A}$; the sequence does not have initial nor terminal elements.

In words: The truth of ⌜Hφ⌝ at some time t depends on the disjunction of the truth of ⌜Hφ⌝ and the truth of ⌜φ⌝ at the preceding moment, and if fφ is true somewhere then at some moment ⌜φ⌝ is (already) true and ⌜Hφ⌝ is not (yet) true. This, together with (1) implies that t_2 marks an initial occurrence of φ .

In this case of a discrete sequence the relation (1) can be realized by a modular net as in figure 1.

Figure 1:

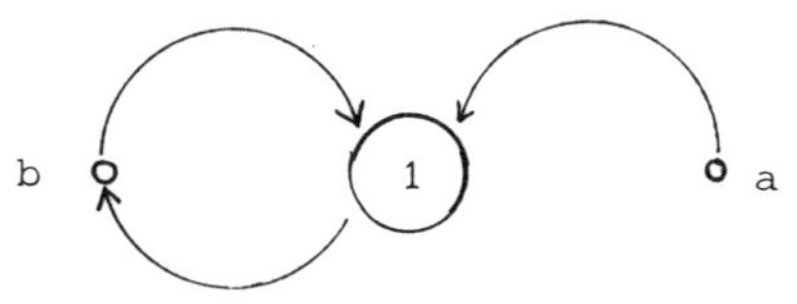

Modular nets are used as formal models of nerve nets[16]. For the moment it is sufficient to understand, that each point, a and b, may take a value 0 or 1 at some moment and that the operative element marked by 1 produces at its output a 1 immediately after *one* of its inputs has the value 1. Obviously, once b is marked 1 it will keep to be thus marked, due to the operative element. If a is marked 1 and b is marked 0, then b will be marked 1 the moment afterwards. It is easily seen that the network realizes the relation

(3) $f_b(t') = f_b(t) \vee f_a(t)$

This relation is the same relation as that required by (1) to hold between the values in the intensions for ⌜Hφ⌝ and ⌜φ⌝ . As a further condition we require, that a position marked by 1 loses this marking by itself, if it is not replaced by 1 due to an operative element whose output the position is.

Let us now assume that the position a gets the value 1 exactly when ⌜φ⌝ applies to a current state of the world. Let us supply our system in figure 1 with a world signal analyzer that determines, if confronted with a current state of the world, whether ⌜φ⌝ applies to it or not. Our world analyzer could also take account of other states of affairs. Let us simply assume that it also analyses the current state of the world with respect to the possible application of another sentence ⌜ψ⌝ . We get the situation of figure 2.

---

16 cf Arbib (1964), p.5. See also p.23 below where modular nets are shown to be special cases of operative networks.

Figure 2:

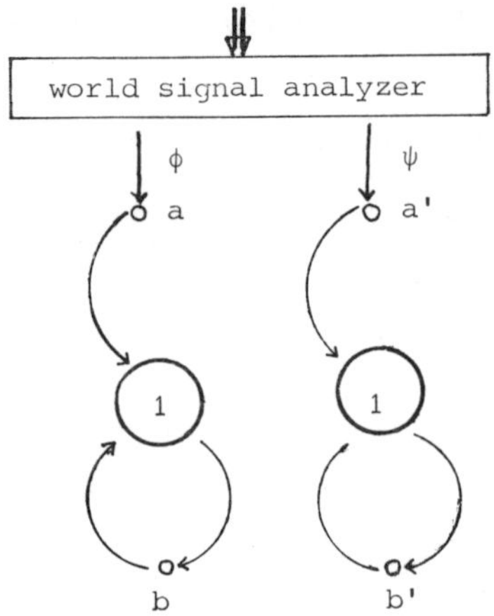

From figure 2 it is clear that position a gets only activated, i.e. can assume the value 1 only if the signal from the world is such that ⌜ϕ⌝ applies to the world according to the signal analyzer. But, as shown above, this is equivalent to condition (2). Therefore figure 2 captures the content of Montague's notion past.

Let us summarize what we have achieved so far by our re-interpretation: Montague's relation between two functions, - the function giving the intension of a sentence in the present -, has been transformed into a relation between the values of neighbouring positions for neighbouring times, i.e. a condition regulating the information flow between spatially contiguous positions. Moreover, the information flow may be interpreted as an internal information flow - internal to the interpreting system - or as an information flow memory. It is easy to show that our re-interpretation and Montague's proposals are not strictly equivalent or rather would only be equivalent if the world analyzer got always complete and correct information about the current world state. We shall not go into the details here, since the discussion serves only as an introduction to the analysis of pre-tense.

In order to show the structure of the system let us make the system more complete. Phonetics, Phonology, Morphology, Syntax and Lexicon provide the basis for a system that may be called a performance model of a sound signal analyzer.[17] It takes sound signals as inputs and determines whether a given

17 The details of this performance model cannot be discussed in this article. The syntax part corresponding to a constituent structure grammar has been developed in Schnelle 1964.

signal represents either $\ulcorner\phi\urcorner$, $\ulcorner\psi\urcorner$, $\ulcorner H\phi\urcorner$, or $\ulcorner H\psi\urcorner$. Let us select four positions *c*, *c'*, *d*, *d'*, taking values in {0,1} and let *c* correspond to $\ulcorner\phi\urcorner$, *c'* to $\ulcorner\psi\urcorner$, *d* to $\ulcorner H\phi\urcorner$, and *d'* to $\ulcorner H\psi\urcorner$. *c*'s having the value 1 indicates that the signal is taken as an utterance of $\ulcorner\phi\urcorner$. (The values at other positions are interpreted correspondingly).

The system of world (signal) analyzer supplemented by a memory and a sound signal analyzer may be supplemented by an agreement analyzer in order to provide the reaction that either what was uttered by means of the sound signal agrees with the actual or memorized experience or not. We get the system as represented in figure 3. [17a]

Figure 3:

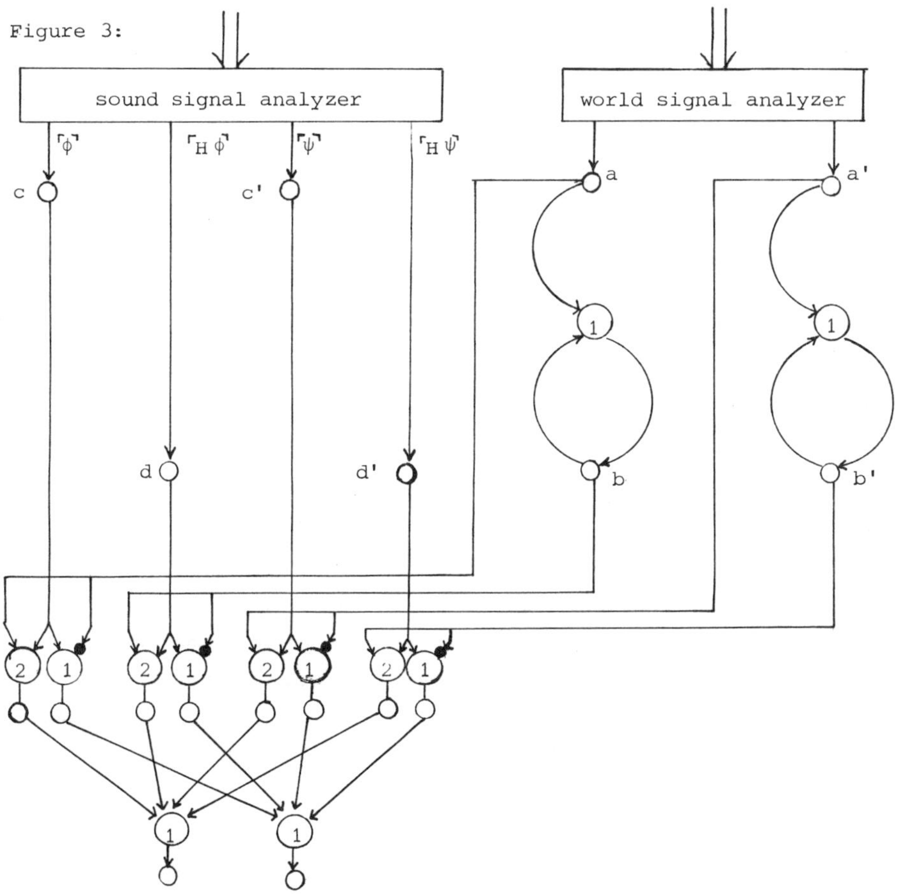

17a A black dot at an input of an oparative element indicates an inhibitory synapse. See Arbib 1964.

Now, if point f is activated (gets the value 1) the sentence just perceived does not agree with the content of the cognitive memory (i.e. is false for the network) and if point t is activated the sentence just perceived does agree with the cognitive memory (i.e. is true for the network).

So far our analysis tried to capture Montague's notion of past only, i.e. the linguistic notion of the generic perfective (cf. Chafe). The generic perfective is not a strict pre-tense category, but I would say that it is quasi pre-tense, since it can be analyzed by the same means as ordinary pre-tense.[18]

An ordinary pre-tense category is Chafe's simple past as referring to surface memory. It is easy to give it an interpretation in terms of a modular net. This interpretation is indicated in figure 4 which should be compared to figure 1.

Figure 4:

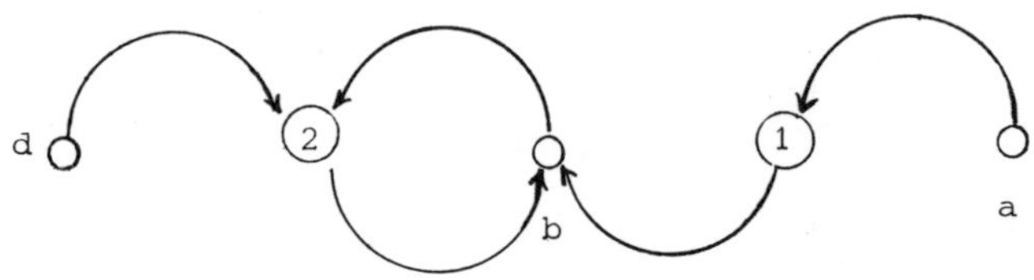

The formal neuron relating a and b is no longer operating as a logical disjunction with delay but only as a delay. This delay only provides for storage in the surface memory. In order to account for the variable duration the information is kept in surface memory,[19] there is a conjunction operator (marked 2) which keeps position b activated as long as some other information at c determined by the rest of the network is 1. It seems that in very young children, this interaction with the rest of the network (via c) is not yet much developed; in this case the duration of availability in surface memory is determined by the natural decay of information storage at a memory position like b.

18 The power of the analysis illustrated seems to be confined to such notions that can be analyzed by means of *one* temporal index only in Montague's framework. Analyses making use of multiple indexes seem to be outside the scope of pre-tense analyses.

19 Cf. Chafe, 1973, p.266: "Although what is in surface memory tends to be that which was perceived a relatively short time before, the length of this time seems to be determined by something like the salience of the information to the individual."

After this analysis of simple past in its "surface memory use" and of generic perfective, let us turn to an analysis of the category future.

3. *The analysis of immediate future*

In ordinary time-line analysis, future corresponds directly to past. If e_1, e_2 and e_3 are three events and t_1, t_2, and t_3 are the time moments at which they occur and t_2 is the time moment of now, then e_1 is in the past if and only if $t_1 < t_2$ and e_3 is in the future if and only if $t_3 > t_2$. The analysis of future tense with respect to the time-line is accordingly. In Montague's analysis, for instance, a sentence $\ulcorner H\phi \urcorner$ in the past tense is true at t_2 if and only if there is a time-moment t_1, with $t_1 < t_2$, for which $\ulcorner \phi \urcorner$ is true. As can be expected, the corresponding sentence $\ulcorner W\phi \urcorner$ in he future tense is true at t_2 if and only if there is a time-moment t_3, with $t_3 > t_2$, for which $\ulcorner \phi \urcorner$ is true.

In trying to re-interprete this analysis in terms of a modular or operative network, we remark that the category of immediate future is not at all similar to simple past with reference to surface memory. Pre-tense future is not symmetric to pre-tense past, but quite different.

In order to provide for an appropriate interpretation of immediate future, let us refer to two events that may be rendered by $\ulcorner \phi \urcorner$ and $\ulcorner \psi \urcorner$ as before (cf. figure 2). In addition to positions a and a', recording the presence of ϕ and ψ , and positions b and b', recording their presence in surface memory, we add positions c and c', for recording the immediate future of ϕ and ψ (see figure 5).

Figure 5:

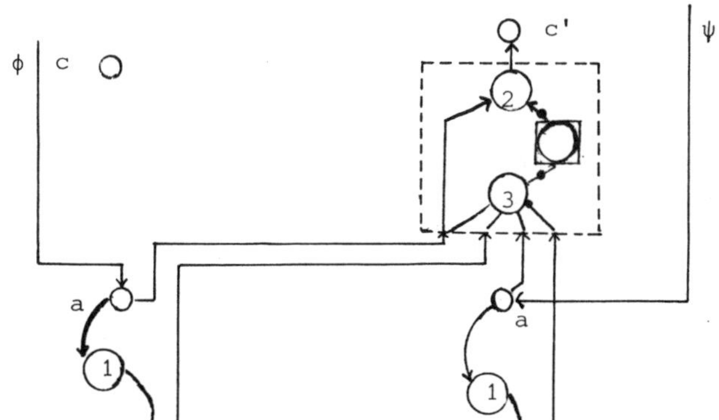

The problem is now how to relate say the future of ψ to the presence or immediate pasts of ϕ and ψ.

Let us first remark that future is, in its simplest meaning, a category of expectation. In order to analyse its function properly, let us first discuss a concrete example: A baby is lying in its cradle. During the nights mother sometimes enters the room, switches the light on, approaches the cradle and bows over the baby. For the baby two events come to be correlated: the sudden appearance of light during nights and the appearance of mummy's face. Let position a record the activation of the event of the appearance of light and position a' record the appearance of mummy's face. After several sequences of the appearance of light and the appearance of mummy's face the baby will develop the expectation that mummy's face is going to appear when light appears during night. This expectation is the semantic basis with respect to which later on a sentence like "Mummy's going to appear" could be judged, i.e. the basis of the judgement is the expectation of the future event to be judged.

Ordinarily, the expectation is discussed in the theory of inductive logic. The expectation of an organism X at a given moment may be rendered by $Cr(\psi)$, the degree of credence[20] about the hypothesis rendering the event. Stegmüller indicates (1973, p.394) that $Cr(\psi)$ may be considered to be a partial determination of the microstate of the organism's brain. He shows furthermore that a dynamic consideration of $Cr(\psi)$ relating to the following rationality criterion: The change of the degree of credence $(Cr_n(\psi) \rightarrow Cr_{n+1}(\psi))$ is only determined by the experience E_{n+1} made immediately preceding the change $(Cr_{n+1}(\psi) = Cr_n^{*}(\psi/E_{n+1})$; (Stegmüller, 1973, p.401. Dritte Rationalitätsbedingung). The degree of credence is related to the conditional probability. If we want to have only the value 0 or 1, we simply have to add a threshold, such that the system expects the event if and only if the degree of credence exceeds the threshold.

In the present context, we may leave open the precize nature of summation over experience and the threshold operation over the result of this summation. Many operations of this kind can be realized by an operative network. Let us represent the operative network required by —⊡— . Whatever its internal structure may be, its output is 1 exactly when its input causes the internal storage of the degree of credence to exceed the threshold.

---

20 Carnap 1971, p.10; Stegmüller 1973, p.313.

In considering the example discussed it is easy to realize what the evidence from experience, E_{n+1} is. It is determined by the (absolute or relative) number of cases of the perceptions of the following configuration: It is light and mummy's face appears and it has been light a moment ago but at that time mummy's face has not yet been there. Relative to this example it is easy now to represent the dependency of position c' 's activation, recording the future of ψ, on the activation of a, a', b, and b' as in figure 5.

The operative complex relating the activation of c' to the activations of a, b, and a' and the non-activation of b' consists of a formal neuron with weight 3 whose output, the position d' , is a credence recording position which is activated by several appearances of the configuration detected by the neuron with weight 3. After the activation of d' every activation of a - i.e. the appearance of light - will cause the activation of c', i.e. the expectation of mummy's face.

We see that the pre-tense category future relates to a configuration in the operative network that is quite different from the configuration for the immediate past. The latter means simply delay or short-term storage, the former means learning the sequential regularity and potentially causal relation between events.

These examples for an analysis of pre-tense and quasi pre-tense categories may be sufficiently clear for an illustration of its essential aspects in time analysis. After this discussion I shall procede to present the general framework for interpretations in operative (especially modular) networks.

4. *The essentials of pre-tense analysis*

As already indicated in the introduction it is essential for the notion of pre-tense that it refers to a feature of an information flow network instead of referring to the classical concept of time.

In this, it contrasts with Montague's analysis which is relative to some interpretation system $\mathfrak{A}$ containing the infinite time sequence, i.e. the set J of time moments and the ordering relation imposing a strict order onto J. Somebody who is able to interprete sentences and expressions in terms of $\mathfrak{A}$ should have command of the entities of $\mathfrak{A}$; in particular he should have command of the notion of an infinite sequence.

However, as our re-interpretation shows, the interpretation of sentences in the present and past need no depend on the command of a sequence of time moments. What our system has available instead is an operative element containing a delay for signals, or, more generally, a causal element relating an effect to its cause in such a way that the effect follows its cause. But this is the essential feature of every piece of physical space and every piece of matter: effect is not simultaneous to nor does it precede its cause but it follows its cause. Signal transmission through space is a special case. Sometimes it is said that signal transmission takes time. But this is another way of saying that effect follows cause. Furthermore, it is usually said that a piece of matter has the disposition or property to react in certain ways to certain conditions applied to it. The disposition to react is another aspect of the causal relation.

It turns out that, on this level of analysis, *the notions of time, causation and disposition are not independent* of each other. You may take one of these notions and define the others on its basis. Our re-interpretation can now be seen as a shift from time as the basic notion to disposition or causation as for pre-tense the notion of disposition is sufficient. The next notion in our sequence of seven tense notions - namely simple aspectual time - depends, as I shall show elsewhere, on certain differences in the notion of causation. On the other levels of tense analysis the fruitfulness of the shift from the infinite linear time notion to disposition and causation will be further corroborated.

So much for the general aspect of our re-interpretation. The application to brains is straight-forward: neurons are pieces of matter disposed for signal transmission. The effected signal follows the signals causing the effect. Signals travel with a certain speed along the axons. The firing of the next axon is, therefore, delayed with respect to the inputs at the synapses. This shows that, since disposition to react is an essential element of each part of an organism, in particular his neural parts, the notion of time is available as much as it only depends on disposition. This is particularly important for the early phases in the development of conceptual systems in organisms and in man, in particular. As can be shown, not all aspects relevant for the analysis of time can be related simply to disposition by reserving one storage position only for each tense form of each verb. Those aspects of tense that can be treated thus will be called pre-tense. It is easily understood that from narrative tense onward, an appropriate analysis of tense requires the storage of a text or an experience that may be rendered by a text. Therefore, more is needed than in our operative system. This is obvious since in texts the same verb in the same tense form may occur several times.

This concludes our introduction to the notion of pre-tense. We can now turn to the discussion of the general framework used for grammatical and interpretational analysis: The framework of operative networks.

5. *Operative network grammars and interpretations*

Just as in the preceding, introductory, sections it seems appropriate to present the general principles of universal grammar implied by our approach in comparison with Montague's conception of universal grammar.

A grammatical analysis according to *Montague's framework* is based on three interrelated systems each organized by a recursive combinatorics: the syntax $\mathcal{S}$ for an ordinary language L, the intensional logic calculus $\mathcal{I}$ and the reference hierarchy $\mathcal{M}$. Each of the three systems contains a set of entities - 1. the structured expressions of L, 2. the expression of IL-calculus, 3. the meanings.

The sets are recursively generated from finite subsets - 1. the basic expressions, 2. the basic constants, variables and auxiliary symbols of IL calculus, 3. the system $\mathcal{M}_o$ of basic meanings - by means of a finite system of combinatoric principles or rules of $\mathcal{S}$, $\mathcal{I}$, and $\mathcal{M}$, respectively.

The interrelation between the three systems is to provide a) an interpretation F assigning each expression of the I-calculus a meaning (or several meanings if it is ambiguous), b) a translation T assigning each expression of L one (or several) expressions of I-calculus as I-readings and c) - by combining a) and b) - an indirect interpretation assigning each expression of L a meaning (or several meanings, if ambiguous). The interrelations depend on the recursive combinatorial or rule governed structure of the three systems.

More formally a Montague grammar is given by the quintuple

$$G = (\mathcal{S}, \mathcal{I}, \mathcal{M}, T, F).$$

In the present context, $\mathcal{M}_o$, i.e. the system of basic meanings, is of special interest. It is the quintuple (according to PTQ)

$$\mathcal{M}_o = (E, I, J, \leq, V)$$

where E is the set of entities, I is the set of possible worlds, J is the set of time moments, $\leq$ is the strict ordering defined in J and V is the set of values {0,1}. It is easily seen that, since in Montague $\mathfrak{U} = (E,I,J,\leq)$ is called an interpretation, $\mathcal{M}_0$ is a pair of an interpretation and the set of values assignable as extensions to sentences. $\mathcal{M}_0$ is a collection of sets such that one set, namely J, is strictly ordered: $(J,\leq)$ is the basis of interpretations referring to time.

As a result of a grammatical analysis in terms of a Montague grammar, it assigns to each signal an expression, its translation(s) and interpretation.

If contextual information expressible in terms of $\mathcal{M}_0$ (or $\mathfrak{U}$) is given, an expression gets also (in addition to a meaning) an extension that can be determined on the basis of its interpretation and the contextual information. The possible extension of a sentence is 0 or 1 as already indicated.

In contrast to a grammar in Montague's framework, an *operative network grammar* is a special kind of an operative network. I shall first characterize the notion of an operative network in general and of a modular network in particular. The special application in grammar analysis will be given with reference to the general characterization.

Whereas a generative grammar as well as a Montague grammar is, mathematically speaking, a special case of universal algebra, an operative network is, mathematically speaking, a special case of topology or of its offspring the dynamic systems.

An operative network is a system that may be represented formally by the quadruple

$$O = (P,Q,V,D)$$

where P is a set of positions, Q is a topology over P, V is a set of values assignable to P. D is the essential part in the present context, but is not as easily characterized as the preceding notions. Formally, it is a function defined in P with values that are mappings of sets of tuples (vectors) over V into V. The role of P,Q,V, and D is, perhaps, best understood relative to the explicandum it is intended to explicate. The positions, i.e. the elements of the set P, are each related to their neighbouring positions by Q. In other words: Q determines for each position which other positions are its neighbours - in some sense of neighbourhood explained in topology -. V is a set of values that could appear - or that could be measured - at each position.[21] Now, the change of the value at each position is to depend on certain configurations of values at neighbouring positions. We shall say that the position is sensitive to the corresponding value configurations in its environment. The position reacts to the configuration by a certain change of its own values. Obviously, we shall have to specify for each possible change of values of a given position which value configurations of its neighbourhood would result in that change. This specification will be called the *sensitivity law* D_x for the reaction of the position x. D is the function that assigns to each position its sensitivity law. In physics, one mostly considers uniform spaces characterized by the fact that every position has the same topological condition and the same sensitivity law. Uniform fields of hydrodynamics or electrodynamics and iterative cellular automata[22] are cases in point. By contrast, we find in nerve nets extremely varying topological conditions as well as varying sensitivity laws for the different neurons to be taken as the positions of the network.[23]

It is easily seen in which sense a modular network - the formal model of a nerve net - is a special case of an operative network: Q can be specified by assigning to each position, i.e. element of P, a finite subset of P, the set of its immediate neighbours. Consequently, Q is a mapping of P into the power set over P. V is the binary set {0,1}. A position can react to a configuration it is sensitive to by the change (0,1), i.e. by firing. The change (1,0) comes "by itself", it is not caused by configurations in the environment. In consequence, a sensitivity law for a position having n immediate neighbours consists in a pair whose second member is 1 and whose first member is a set of n+1 place vectors over V. The last component of the vector is to indicate a value of the position itself, since the position is partially sensitive to its own previous state; this value should be 0, since

21 Since we assume only one set V, this corresponds to a physicalistic system relative to one observable measurable quantity only. Most physical systems must be described with respect to several observables. This shows that our networks are special cases of physical systems in this sense. Our analysis is closely related to the general theory of dynamical systems as presented e.g. in Kalman et al. 1969.

22 see e.g. Burks, 1977, p.562.

23 In order to make these networks work properly, the refractory period of neurons will have to be taken into account. I shall not go into this in the present article.

we are interested in the (0,1) reactions. Since there is only one second member in a sensitivity law, it is sufficient to give the first member, the set of sensitivity configurations only. Now, D is the function assigning to every n-place position an appropriate set of sensitivity configurations.

For ordinary modular nets the sensitivity configuration at a position (neuron) having n immediate neighbours is determined by n+2 integers, i.e. n+1 weights w_i $(0 \leq i \leq n)$ and a threshold t. The set of sensitivity configurations contains all vectors $(x_0, x_1, \ldots, x_n)$ such that the weighted sum of the components is equal or greater than t, i.e.

$$\sum_{i=0}^{n} w_i x_i \geq t$$

The modular nets used in our illustrations are still simpler. Their weights are either −1 or +1 (inhibitory or excitatory) and their thresholds are as marked by the natural number in the circle of the operative element.

Our *application of operative networks to simple language analysis* is as follows: The network contains two subnetworks, the operative syntax network and the operative referential network. The two subnetworks are embedded in a network relating both. They are further supplemented by input-output-networks.[24] In simple cases, the positions of the operative syntax network record the presence of simple sentences and word-forms and the positions of the operative referential network record, whether the conditions for the application of the corresponding simple sentences and word-forms to the world obtain. A recorded sentence is true for the network, in virtue of the wiring of the embedding network, if the conditions for the application obtain. As we have seen, there is a basic operative connection - i.e. specific assignments of D_x - between certain positions. This provides, as we have seen, for the differentiation of tense forms of the otherwise same sentences.

It is obvious that not all recursive properties of language can be captured by the model as described so far. However, two remarks are at order. First: There is an enormous amount of problems in language analysis that do not rely on any recursive device. If the time applied for learning a section of grammar of a foreign language could be taken as a rough measure of the

---

24 On a very simple level the structure of such a network is illustrated by figure 3 above and the discussion related to it. The input-output networks are the sound signal analyzer and the world signal analyzer.

amount of grammatical content to be managed, it seems that the non-recursive part is far more important than the recursive one. Second: It is quite easy to show how more recursive properties can be incorporated into models of information flow. The theory of Pushdown-store automata gives a first indication. I have given, along these lines, a general method to describe CF grammars.[25]

6. The notion of time in operative networks

Let us now turn to the question how the notion of time is determined in an operative network and whether, in contrast to the current notion of spatialized scientific time, the aspects of "now" and of the "becoming" are represented as well as that of temporal succession.

Formally, our framework is an abstract system of relations between positions and values at the positions. However, the indicated explicandum of the D-functions already hints at its relevance for our analysis of time: A D-function assigns to each position a sensitivity law that determines how, i.e. by which change of its values, it reacts to environmental configurations. The sensitivity law can also be interpreted as a disposition to react. Formally, it is a system of possible changes of values. Instead of an ordered set of temporal moments (or intervals) we have a system of potential changes or of dispositions as our basic concepts; the ordering of temporal moments is only indirectly determined by them.[26] But note that, so far, the ordering between recordings of events provided for by the sensitivity laws or dispositions is only local. In an operative network only local "time" relations are directly available.

This does not exclude, that we, as analysts, apply analytical concepts to the network and its operation, that may prove useful for us even if the system itself does not have them available. However, if we do so, it will be necessary to remember, that these analytical concepts are only analytical hocus-pocus and have no direct descriptive status; the system is described *as if* it had the conceptual apparatus of the analytical tool available. Let us discuss some analytical tools related to the time line and used by

---

25 See Schnelle, 1964: Cf. also on more general aspects Schnelle 1970.

26 This seems to be related to the arguments of H. Kamp, this volume, that change is prior to time intervals or time moments.

linguists in their analysis.

As a first element in our analytical hocus-pocus, we introduce the notion of a topological world line connected with each position. We may identify a position in an operative network with the substantial points of Minkowski's analysis and proceed as he does:

> "We fix our attention on the substantial point which is the world point x,y,z.t and imagine that we are able to recognize this substantial point at any other time. Let the variations dx,dy,dz of the space co-ordinates of this substantial point correspond to a time element dt. Then we obtain, as an image, so to speak, of the everlasting career of the substantial point, a curve in the world, a *world-line,* the points of which can be referred unequivocally to the parameter t from $-\infty$ to $+\infty$. The whole universe is seen to resolve itself into similar world-lines, and I would fain anticipate myself by saying that in my opinion physical laws might find their most perfect expression as reciprocal relations between these world-lines." *(Minkowski, 1976, p.340)*

I want to apply this idea as follwos: As I said, a position corresponds to a sentence or a verb-form or a nominal expression etc. that may apply in a situation or not. The application is recorded by the value 1 at the position. A "period" or stretch "during" which the expression applies and, correspondingly, ists position has value 1, will be called an event of the expression. If the exprssion is E_k we will call each such event an E_k-event. It follows that a "world-line" related to some position, i.e. applicable expression E_k which the expression applies and, correspondingly, its position is a sequence of E_k-events interrupted by intervals where E_k does not apply. Such a "world-line" could be represented as in figure 6, if we bear in mind that the lengths of intervals of the representation are without significance.

Figure 6: The "world-line" to E_k

E_k 

A simple E_k-event could be represented as in figure 7, i.e. by a simple interval.

Figure 7: An E_k-event

E_k 

As Minkowski writes, the laws should determine reciprocal relations between world-lines. In our case linguistic expressions should determine the reciprocal relations between the "world-lines", i.e. sequences of events. Usually they do this under the assumption that certain events from different "world-lines" are selected and related to each other. We get various relations of sequence, simultaneity etc., that could be represented, as e.g. in figure 8, by merging the time-lines to which the different events belong and marking the intervals appropriately.

Figure 8: The "merging" of events

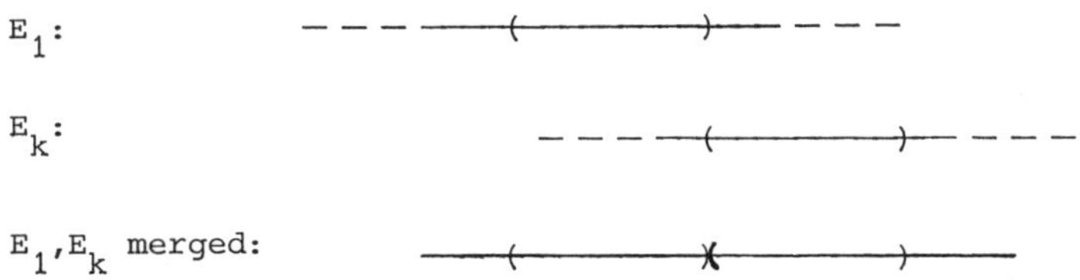

This manner of representation is widely used; there is no need to go into the details. We see that by succesive mergings we may obtain structures in which the events are temporally ordered with respect to each other and finally we might, perhaps, reach an ordering with respect to one time line only, i.e. one common temporal reference system for all events.

But this generalization leads us already far beyond the scope of pretense. Still within the scope we find, however, immediate témporal relations between *neighbouring* world-lines. Consider once again

(1) Steve fell in the swimming pool.

and

(3) Steve is falling in the swimming pool.

The structure of the network represented in figure 4 determines a relation as represented in figure 9.

Figure 9: The ordering for an event reported by (3) to the subsequent state (event) reported by (1)

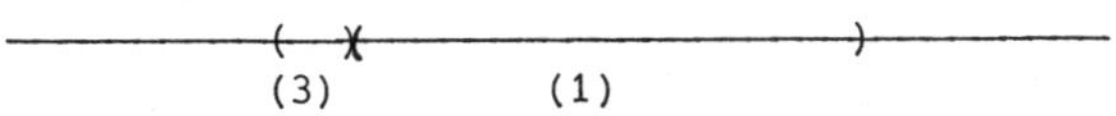

In other words: Figure 9 represents our analytical (hocus-pocus) characterization which the network itself determines as rendered in figure 4.

Let us now push our analytical hocus-pocus further to a point where a comparison with Montague's analysis (or the classical time analysis) is possible. Let us assume that it is possible for every two events - i.e. applications of sentences or word-forms - to determine whether they are simultaneous or not. This makes it possible to "merge" the time lines and to arrive at one time line only. In this case, we could determine for each "moment" and each sentence or word-form whether it applies at that "moment" or not. A particular configuration of application or non-application for the various expressions may be taken as a *global state* of the network. At each "moment", then, the network is in exactly one such state. Insofar as it makes sense to define a global state of the network, we may identify each given state as a "now" for the operative network. This "now" is directly related to a "becoming" by the fact that the system of dispositions for all points determines the coming about of the next state.

This analysis based on the existence of global states of a network can be related to Montague's analysis as follows: Consider the sequence of global states of the network actually brought about. Consider all alternatives to this sequence which might be obtained by substituting one or several of the global states by alternative states that might have come about if either the inputs or the disposition states of the network had been different. The set of sequences thus obtained may be taken as the correlates of Montague's possible worlds - the possible worlds for the network. The "positions" in the sequence may be taken as time moments of the temporal order. This reconstruction of Montague's concepts in terms of features of operative networks may serve as a basis of Bierwisch's proposal mentioned above.[27]

We may now ask, in which sense the analysis just given incorporates the notions of "now" and of "becoming". It is very simple: Take the global state of the network in each interpretative situation as the "now" and the network of dispositions for change as the "becoming". What the system has available at each "moment" is its "now" and its "becoming". The sequence of states and the time line underlying this sequence is merely implicitly determined by "now" and "becoming", though, in the hocus-pocus of the analyst, the sequence of states and the time line may be assumed to exist. But what are the aspects of the operative network that could be taken to represent "now" and "becoming", if the assumption of global simultaneity does not make sense? In this case, we would postulate a focal "awareness" operator selecting at each moment exactly one position that is in the system's special awareness focus.[28] In this case, "now" is represented by the environment state of the

---

27 Bierwisch, 1979, see p.5 above.

28 Conrad has given an indication how such a selection could operate. It would be determined by a supervising neuron. See Conrad, 1974,p.112.

selected position and "becoming" by the activation of the disposition or sensitivity law at the position. This concludes our explication of the notion of time in one of its most primitive aspects.

7. *Conclusion*

Let me emphasize that the present article only presents the first step in a theoretical analysis of time and tense which is planned to lead by at least seven steps of increasing complexity of internal notions of the interpreting system to the operationally most complex notion, the spatialized, linear, and dense concept of time and tense underlying the time line analysis. Our analysis is based on the methodological criterion, that the conceptual apparatus of semantic interpretations should include all aspects that seem to be important introspectively - at least insofar, as we can get a clear picture of them - and to eliminate all notions for which it does not seem probable that the interpreting organism does have them. "Now" and the impression of passing is an ingredient that seems to be fundamental in operative organisms since their actual state and the change of state always marks a now and the passing of now. It therefore seems desirable to incorporate it into the primitive system of interpretation. On the other hand, infinite sets of entities can only be manipulated indirectly on the basis of generalizing inductive insight. Insight of this status develops only relatively late in the human organism. It is only natural that the ability to manipulate correctly clock reading and dates is acquired only at the age of six to eight years. There can be no doubt, however, that children master more primitive notions of time and the corresponding interpretation of tense much earlier. In fact, the semantic basis for pre-tense is already acquired by the human organism long before it starts to speak its first word.

References

Arbib, M.A, 1964,"Brains, machines, and mathematics", New York, McGraw Hill

Bierwisch, M., 1979, "Utterance, meaning and mental states mimeo", Akad. d. Wiss. der DDR

Braitenberg, V., 1974, "On the representation of objects and their relations in the brain", pp.290-298 in: M. Conrad et al., 1974, s.d.

Burks, A.W., 1977, "Chance, cause, reason", Chicago, University of Chicago Press

Čapek, M. (Ed.), 1976, "The concepts of space and time", Dordrecht/Boston: Reidel

Carnap, R., 1971, *Studies in inductive logic and probability,* Vol. 1, Berkeley, University of California Press

Chafe, W.L., 1973, "Language and memory", *Language* 49, pp.261-281

Chafe, W.L., 1974, "Language and consciousness", *Language* 50, pp.111-133

Conrad, M., 1974, "Molecular data structures, pp.108-127, in M. Conrad et al. (Eds), *Physics and mathematics of the nervous system,* Berlin-Heidelberg-New York, Springer

Ferreiro, E., 1971, "Les relations temporelles dans le language de l'enfant", Geneva

Grünbaum, A., 1969, "The meaning of time", pp.147-177 in Anderson et al. (Eds), *Essays in honor of Carl G. Hempel,* Dordrecht-Holland, Reidel

Harman, G., 1973, "Thought", New Jersey, Princeton Univ. Press

Isard, St., 1974, "What would you have done if ...", *Theoretical Linguistics* 1, pp.233-256

Isard, St., 1975, "Changing the context", pp.287-296, in E.Keenan (Ed.), *Formal Semantics of Natural Language,* London, Cambridge Univ. Press

Jespersen, O., 1924, "The Philosophy of grammar", London, G.Allen and Unwin

Kalman, R.E., Falb, P.L., Arbib, M.A., 1969, "Topics in mathematical system theory", New York, McGraw Hill

Kintsch, W., 1974, "The representation of meaning in memory", New York, J. Wiley

Longuet-Higgins, 1972, "The algorithmic description of natural language", Proc. Royal Soc. London B, 182, pp.255-276

Minkowski, H., 1976, "The union of space and time", in M. Čapek (Ed.), *The concepts of space and time* (tr. from the original...)

Montague, R., 1974, "Formal philosophy", New Haven, Yale University Press

Quine, W.V., 1960, "Word and object", Cambridge, MIT-Press

Quine, W.V., 1974, "The roots of reference", La Salle Ill, Open Court

Schnelle, H., 1964, "Programmieren linguistischer Automaten", in K.Steinbuch, S.W.Wagner (Hrsg), *Neuere Ergebnisse der Kybernetik,* München, Oldenbourg

Schnelle, H., 1970, "Linguistics and automata theory", in B.Visantini (Ed.), *Linguaggi nella società e nella tecnica,* Milano, Edizioni di Comunità

Schnelle, H., 1976, "Circumstances and circumstantial expressions", in J. Groenendijk, M. Stokhoff (Eds), *Amsterdam Papers in Formal Grammar,* Vol. I, Amsterdam, Universiteit, Centrale Interfaculteit

Schnelle, H., 1979, "Circumstantial expressions", in A. Margalit (Ed.), *Meaning and use,* Dordrecht/Boston, Reidel

Seiler, H., 1977, "Sprache und Sprachen", München, Fink

Stegmüller, W., 1973, "Personelle und statistische Wahrscheinlichkeit", 1. Halbband, Berlin-Heidelberg-New York, Springer

Stephany, U., 1978, "The modality constituent", *Arbeitspapier* Nr. 36, Inst. f. Sprachwissenschaft Köln, Okt. 1978

Wunderlich, D., 1970, "Tempus und Zeitreferenz im Deutschen", München, M. Hueber

TEMPORAL STRUCTURES IN DISCOURSE

Carlota S. Smith, University of Texas

In conversation, narrative, and other discourse, temporal reference often extends over several sentences. I will be interested in this paper in how extended temporal reference is established and maintained. Most of the discussion focuses on English, but I develop a general account.

The study of extended temporal reference is essentially that of syntax beyond the sentence. The linguistic expressions of temporal reference are syntactically based: tense, verbal auxiliaries, adverbials. Yet they participate in semantic structures that have domains beyond the sentence, as well as within sentences simple and complex. Examining temporal reference within and across sentences leads to some new insights about the system of English, and about temporal systems in general.

In the first section of this paper I will discuss temporal reference across sentences, and will formulate principles to account for it that are based on the temporal expressions that occur in sentences. I will then make some remarks about the temporal system of English and other languages, and will give two arguments in favor of a Reichenbach-type approach to temporal reference. Finally, I will raise some semantic questions about the principles of section 1, and claim that syntactically-based principles cannot fully account for semantic structures beyond the sentence.

§1. The temporal structures that I am interested in extend over syntactically independent sentences. In such structures, temporal information from one sentence is part of the temporal interpretation of another. The sentences are dependent semantically, much as pronominal anaphors depend on antecedents.[1] I will refer to such temporally dependent combinations as extended temporal structures. Most of the sequences presented here have two or three sentences,

---

1 I discuss the relation between temporal and pronominal anaphora in Smith, in press. Recent studies of pronominal anaphora across sentences include Stenning (1978) and Webber (1978).

but in principle they are of indefinite length.

In this section I will present sequences of sentences with various temporal expressions and ask whether and when they can be interpreted as single temporal structures. I will then formulate principles that account for the interpretations. The account that I give is based on my analysis of temporal reference, Smith 1978. In this analysis surface structures are related to semantic representations by interpretive rules. The semantic representations and the general approach are deeply indebted to the work of Hans Reichenbach.

§1.1 Before discussing the way temporal interpretation can be extended beyond the sentence, I will outline some basic assumptions of my analysis for sentences. The assumptions are explained and justified in the paper cited above. I assume that sentences have one tense and one time adverbial each, the adverbial of unlimited complexity. I also assume that complete temporal reference in English is made by a combination of tense and a time adverbial.[2] The semantic unit for temporal reference, then, has a temporal expressions that is optional syntactically. Sentences without a time adverbial are vague, or incomplete, because they do not have enough information to establish temporal reference. In practice they are usually interpreted with information from context, linguistic or extra-linguistic.[3] Sentences are also incomplete temporally if they have certain combinations of tense and time adverbial that fail to establish temporal reference on their own (see the table in 2 below).

The temporal expressions of the language are assigned consistent relational values: simultaneity (=), posteriority (→), and anteriority (←). Their interpretations vary, depending on syntactic and semantic factors, an essential feature of a temporal system. In English past tense is assigned its traditional value of anteriority, and present tense the value of simultaneity.[4] They are interpreted as anterior to, or simultaneous with, the time of speech or another orientation point. Time adverbials are also assigned relational values, according to their normal interpretation. A representative sample of temporal order adverbials is given in 1; adverbials

---

2 Aspect will not be considered in this paper.

3 For discussion, see Smith (1977), and Partee (1973).

4 In this analysis Future is indicated by the combination of present tense and future adverbial; there is no future tense.

of frequency or duration are not included.

1. Relational value of English time ordering adverbials (not an exhausitve list)

	Anterior (←) Past	Simultaneous (=) Present	Posterior (→) Future
a.	last week 3 days ago yesterday	now this minute	next week in 3 days tomorrow
b.	previously earlier	at the same time	afterwards later
c.	on Tuesday[5]		on Tuesday
d.	before John left[6]		before John left

There are six possible combinations of tense and time adverbial, according to relational values. All six are grammatical in English. Four combinations establish a time independently, and two do not; sentences with these two need further information to be interpretable temporally. The combinations are given in 2, with examples of each.[7]

2. Combinations of tense and time adverbials

	Tense	Adverb	Time	
a.	present	present	Present[8]	
b.	present	future	Future	
c.	present	past	Past	
d.	past	present	-	(dependent)
e.	past	future	-	(dependent)
f.	past	past	Past	

a. Jane is swimming now
b. Jane is swimming tomorrow
c. Last week, Jane is swimming all alone when...

d. Molly understood the situation now
e. Molly was arriving in 5 days
f. Molly played soccer yesterday

---

5 This example indicates a class of adverbials in which prepositions *(at, on, before,* etc.*)* introduce NPs that refer to clock or calender times (hours, minutes, months, etc.) or to regularly scheduled events that can be taken as indicating a time (breakfast, school, etc.).
These adverbials have the relational value of posteriority or anteriority (not both at once).

6 This example indicates a class of adverbials in which prepositions introduce sentences. The sentences may be reduced (*John left before Mary*) if they contain material repeated in the main sentence.

7 I ignore here the conditional and irrealis interpretations of past tense.

8 I capitalize temporal interpretation -Past, Future- and use lower-case letters for temporal expressions -e.g. past tense.

These combinations are crucial for establishing temporal reference, within and across sentences.

§1.2 I can now introduce extended temporal structures. The examples of 3-5 are sequences in which the (b) sentences do not establish temporal reference on their own, but are fully interpretable temporally with information from the (a) sentences; they are extended temporal structures.

3. (a) I talked to Mary last night. (b) She was happy.
4. (a) I talked to Mary on Friday. (b) She was leaving in 3 days.
5. (a) John arrived at noon. (b) Mary came later.

3b has no time adverbial, and cannot be fully interpreted temporally. The other (b) sentences have time adverbials but cannot be fully interpreted either. The times established in the (a) sentences allow full interpretation of the (b) sentences.

I am assuming a sort of Gricean co-operative approach in these interpretations, according to which one looks for a related interpretation of sentences presented together.[9] I will ask whether sentences can function together as a temporal unit, and will not consider other interpretations that the sentences might have.

The relationship between 3-5, considering temporal interpretation only, can be described with a military metaphor. The (a) sentences can be said to have "captured" the (b) sentences, since the latter are dependent on the former for interpretation. Continuing the military metaphor, I will focus in this section on sentences that may become captives, that is, on dependent structures. There are three possibilities: a sentence may *demand* capture, it may be *available* for capture, it may be *protected* from capture.

Consider first sentences that demand capture. Such sentences do not establish temporal reference on their own; one class that demands capture is that of sentences with no time adverbials, as shown in 6-8. The (b) sentences are captured by the (a) sentences, the latter having time adverbials.

6. (a) Scott isn't at the class picnic. (b) He is working.
7. (a) Scott can't come to the picnic tomorrow. (b) He is working.
8. (a) John had a good time last night. (b) He went to the movies.

6 and 7 show that the present tense indicates Present or Future time, depending on the adverbial that completes the interpretation.

---

9 See Hobbs (1976) and Stenning (1978) for discussion.

Other sentences that demand capture have adverbials but do not establish temporal reference. Certain adverbials such as *previously, afterwards* never establish a time; they are intrinsically dependent on times given elsewhere. 9b exemplifies this type. There are also combinations of tense and adverbial that do not establish a time, as in 2d and 2e above, and 10b below.

9.(a) Isobel went riding at noon. (b) Sylvia rode later.
10.(a) I talked to Michael at noon. (b) He was going in an hour.

In these sequences the (a) sentences are the captors of the (b) sentences, giving the information that allows full interpretation. Thus 9b tells us that Sylvia rode later than some time, but doesn't fix this time; 10b tells us only that Michael planned the leave after some time. Thus the (a) sentences, giving times, provide what I will call "anchors" for the temporal expressions of the (b) sentences. In sum, sentences that demand capture do not establish temporal reference on their own; they depend on information given elsewhere. (Such information may be part of extra-linguistic context as well as in a neighboring sentence.).

In contrast, sentences that are *available* for capture do establish time reference in certain contexts, and in isolation. They are interpretable as anchored to the time of speech, ST. But they can also be captured to form part of an extended temporal structure. As captives, they anchor to a time established in another sentence. 11 and 12 exemplify this situation; the (b) sentences are available for capture, and are captured by the (a) sentences.

11.(a) Every Saturday afternoon Paul took the bus to town.
(b) He (Paul) went to the movies at 5pm.

12.(a) Next Sunday we're flying to Amsterdam.
(b) We visit friends on Monday.

11b alone indicates that Paul did something at a certain time -in the simplest interpretation, at 5 o'clock prior to the time of speech (ST). This is the simplest interpretation because it requires no information other than the given in the sentence.[10] In sequence with 11a, however, 11b has quite a different interpretation: it specifies an habitual activity of Paul's on Sunday afternoons. The frequency adverbial is captor, on the extended temporal interpretation of 11.

10 See Smith (1977) for discussion of the notion of the notion of simplicity in interpretation.

Similarly, the interpretation of 12b is different in sequence with 12a from its interpretation alone. On the simplest interpretation, 12b alone anchors to ST, indicating the Monday following ST. In sequence 12 it indicates the Monday after next Sunday: the time of 12a acts as anchor.

The sentences that are available for capture have time adverbials that may be taken as anchored to ST; they can, however, anchor to a time established in another sentence. To interpret a time adverbial it is essential to know how it anchors, and this knowledge will also allow prediction of the capturing possibilities of sentences in which the adverbial appears.

There are some sentences that cannot be captured: they are, in the terminology suggested above, protected from capture. Consider for instance sequences 13 and 14.

13. (a) Robert talked to Bonnie at noon.
    (b) He (Robert) arrived 3 hours ago.

14. (a) Every summer Robert travels.
    (b) He (Robert) is leaving for Greece tomorrow.

Although these sequences are not incoherent, they cannot be taken as extended temporal structures. In isolation 13b indicates a time 3 hours before ST, and it has the same interpretation in sequence with 13a. The situation is the same with sequence 14, in that the interpretation of 14b is unchanged by the preceding sentence. Both sentences of 13 and 14 are anchored to ST.

The same type of adverbial appears in the (b) sentences: *tomorrow* and *3 hours ago* are anchored to the moment of speech, making deictic reference. Deictics protect a sentence from capture because they anchor to ST and not to times established in other sentences (actually the situation is a little more complicated than this, as I will show directly).

Temporal adverbials can be classified according to how they anchor, and whether they can re-anchor. These are essential factors for interpretation. There are 3 main classes: Deictics, that anchor to ST only; Dependents, that anchor to a time given explicitly (the time must be established in another sentence, since sentences have only one time adverbial each); Flexible Anchoring adverbials, that anchor either to ST or to a time given explicitly. The chart in 15 shows this classification of temporal adverbials. Habitual adverbials have quite different properties, and are not included in the analysis given here; for some discussion see Smith 1978.

15. Types of adverbials (not an exhaustive list)

Deictics
← last week, a week ago, yesterday
= now, this moment
→ in three days, tomorrow

Dependents
← previously, earlier
= the same time
→ afterwards, later

Flexible Anchoring
← → on Tuesday, before John left

Adverbials with all relational values occur as Deictics and Dependents; Flexible Anchoring adverbials have the values of anteriority and posteriority.

These categories serve to organize the facts about participation in extended temporal structures. Sentences with no adverbials, or Dependents, demand capture; sentences with Flexible Anchoring adverbials are available for capture; sentences with Deictic adverbials are protected from capture.

Only adjacent sentences have been considered, so far, as possible captives. Sentences can also be captured from afar, although deciding what sentences go together may pose difficult problems. (See §3 for discussion, especially sentences 40-42). Consider for example sequence 16.

16. (a) We are deeply concerned with last Sunday's robbery of an important public figure. ... intervening sentences...
(b) Last night I talked to a reporter who had new information.
(c) The maid left before noon.

I am interested in the interpretation of (c). In sequence with (b), it may be interpreted as anchored to the time established in (b). If (a) is in the sequence, there is also a natural interpretation of (c) as anchored to (a); the shift of attention from reporter to robbery is made without an explicit transition, elliptically. If the robbery is in the forefront of consciousness no explicit reference to (a) is necessary.

There is also a vague interpretation of (c), in sequence with (b) only. On this interpretation (c) is taken to be anchored to a time not given explicitly, but anterior to that time - which may be known extralinguistically, or may be not known precisely. Vagueness like this is not at all unusual, especially in conversation. I shall return to this vague reading in section §2.

§1.3 Captors: Sentences that act as captors in extended temporal structures must establish a time, either independently or through sharing with another

sentence. Deictics, Flexible Anchoring, and Dependent adverbials can all appear in a capturing sentence.

Sentences with Deictics can capture other sentences, with the general limitations established above. Sentences with Flexible Anchoring adverbials can capture Dependents and sentences without adverbials. If two Flexible adverbials appear in temporally related sentences, they always have an orderly interpretation[11] because of the fact that clock and calendar time uses recurring sequences. Strictly speaking, such an orderly relation is not capturing, since neither of the adverbials is anchored to the other. Sentences with Dependent adverbials can also capture other sentences. When this happens the time of the captor sentence is part of the interpretation of the captive, even though the captor does not establish a time independently. As the examples show, there are lexical items that indicate chains of dependency with Dependent adverbials.

The sequences of 17, 18, and 19 show capturing by sentences with Deictics, Flexible Anchoring, and Dependent adverbials respectively. Habituals, which can capture other sentences, are not included here. Deictics do not appear in captive sentences, since they protect a sentence from capture.

17. (a) John arrived last Tuesday. (b) Mary was away in the country.
(c) Mary came on Sunday.
(d) Mary was leaving in 3 days.

18. (a) John arrived on Sunday. (b) Mary was away in the country.
(c) Mary came on Tuesday.
(d) Mary was leaving in 3 days.

19. (a) ..Sunday. Mary was leaving in 3 days. John was leaving later.
(b) ..noon. John called afterwards. He said he was coming in 5 minutes.
(c) ..10 o'clock. Oona bought her ticket earlier. John got his even earlier.

§1.4 General Principles: I can now state a generalization about extended temporal structures: A sentence can be captured to form part of such a structure, so long as it does not have a Deictic adverbial. A principle of this type might be used in interpreting texts; extended temporal structures would be sought according to the temporal expressions that occurred in sentences of the text.

A second principle is also necessary. This principle applies to tense, the other element needed to establish temporal reference. Principle 2:

---

11 Assuming that they refer to identical or compatible units - see section §3 for discussion.

A sentence can be captured only if it has the same tense as its captor. Sequence 20 shows the need for this second principle. The two sentences of 20 have different tenses, and cannot be interpreted as a temporal unit.

20. (a) Last summer Isobel went to Greece. (b) She will visit friends.

The principle involving Deictic adverbials has a certain logic, and one would espect it to hold generally among languages. If language has adverbials that anchor only to the time of speech, those adverbials would not be available for participation in extended temporal structures. The second principle, that participants in such structures have the same tense, depends on the way tenses function in a particular language. In fact both principles must be modified, as I now show.

§1.4.1 I consider now some examples which appear to cause difficulty for the principles given above. There are sentences that call into question the principle barring deictic adverbials from participation in extended temporal structures. Consider first sentences whose main verbs involve communication or consciousness, such as *say*, *announce*, *realize*. In sentences with verbs such as these, a deictic adverbial can have an anchor other than ST; in sequence with such sentences, a deictic can form an extended temporal structure. The examples illustrate.

21. (a) The office announced the news last night (b) The ambassador will resign in a month.

22. (a) John said something surprising at noon. (b) Mary had resigned an hour ago.

Sequence 21 exemplifies direct quotation, and 22 indirect. In both, the adverbials of the (b) sentences are taken as anchored to a time established in the (a) sentences. It should be noted that deictic adverbials vary as to whether they allow this type of re-anchoring. Some deictics are flexible, others are rigidly anchored to ST. The deictics in the examples (*in* - and - *ago*) are flexible for most speakers; *tomorrow* and *yesterday*, in contrast, are usually taken as anchored to ST and rarely appear in sentences like 21(b) or 22(b).

Verbs of communication and consciousness (CC verbs) allow the re-anchoring of deictics. One might attempt to save Principle 1 by treating these verbs as expressions governing the exceptions. Sentences with re-anchored deictics could be identified as exceptional because of their relationship to CC verbs. However, this approach is inadequate. It cannot deal with ellipsis, as sequences 23 and 24 show.

23 (a) We invited Tom to join us but he couldn't come.
   (b) He had eaten dinner an hour ago.

24 (a) Della left the party early.
   (b) She was flying to Istanbul in a few hours.

In both sequences, the (b) sentences can be taken as reported speech, or narrated monologue.[12] 23(b) presents the reason Tom gave for not coming to dinner; 24, on this interpretation, gives Della's explanation for her early departure. Although the (b) sentences present material that is communicated or in consciousness, they are not governed by CC verbs or explicitly related to such verbs.

The mode of reported speech exemplified in the (b) sentences above has a characteristic pattern of deixis which sets it off from direct and indirect speech. Reported speech presents material unrefracted through the consciousness of a speaker or narrator. It has neither the directness of quotation nor the ambiguity of indirect speech.[13] Characteristically, in reported speech pronouns are 3rd person, place and time adverbials are deictic (unshifted), and tense is past. The sentences of 25 illustrate.

25.

Direct:	I am ready now.	I will leave next week.
Indirect:	She is ready now.	She will leave next week.
Reported:	She was ready now.	She would leave next week.

The mode of reported speech has been discussed mainly in connection with literary texts.[14] However, reported speech also appears in other types of texts, especially historical narratives. I cite two examples from texts that are primarily historical rather than fictional.

26. An education at Oxford appealed to a new class of rich and well-to-do men who wished to use it to improve the prospects of their sons. The Colleges were now therefore able to charge fees proportionate to the social advantages likely to accrue.
Darlington, *Encounter,* 1967.

27. Having put his religious house in order, Mohammed now began to enjoy his power as the undisputed ruler of a large number of tribes.
van Loon, *The story of mankind,* 1921.

---

12 This mode is also known as "erlebte Rede", and the "style indirecte libre".

13 The essential ambiguity of indirect speech is emphasized by Lips (1927).

14 For instance by Banfield (1973), Cohn (1947), Friedman (1955), Hamburger (1957), Weinrich (1964).

From the point of view of this paper, reported speech must be distinguished not only because of its pattern of deixis, but also because its temporal forms cannot be fully interpreted alone. The sentences with *now* in the examples above, for instance, do not establish temporal reference. They lack a reference point that would give value to *now*. In the terms of this analysis, the sentences of reported speech are incomplete temporally and demand capture; to be fully interpreted, they require additional information.

Sentences with re-anchored deictics are in the mode of reported speech, and are dependent temporally as noted above. Their temporal dependency can be provided for in a revision of Principle 1, that accounts for the counterexamples and preserves the main point of the original principle. The revised principle: A sentence can be part of an extended temporal structure so long as it does not have an independent Deictic adverbial.

The phenomenon of reported speech shows that CC verbs need not appear in a text for consciousness to be conveyed. Reported speech involves ellipsis, which is comming in discourse of all kinds. Therefore an approach that depends on governing CC verbs is doomed to failure. Yet the connection between CC verbs and reported speech is undeniable. One interesting piece of evidence comes from the interpretation of complex sentences that have re-anchored deictics. Compare, for instance, the following sentences.

28. On Sunday Mary told me she had lost the watch a week ago.
29. On Sunday Mary lost the watch she had bought a week ago.

The interpretation of 28 is straightforward, with the deictic anchored to Mary's telling of Sunday. That of 30 is less so: the deictic must be anchored to *Sunday* because of the anterior *have*,[15] but there is no explicit consciousness to which the deictic can be attributed.

I suggest that we understand sentences like 30 in a conventionalized way, so that even without an explicit shift the narrative is taken to have assumed (if only momentarily) the point of view of Mary. Since language is often elliptical, semantic dependencies cannot always be associated with explicit governing forms.[16] One can explain certain consistent interpretations by recognizing the possibility of ellipsis. Thus I shall say that re-anchored deictics signal a consciousness at the time of re-anchoring, whether or not such a time is associated on the surface with a CC verb.

15 Auxiliary *have* indicates anteriority: see Smith (1976).

16 Smith & Whitaker (1979) investigate some different types of ellipsis.

Benveniste, in his important paper "Les relations de temps dans le verbe francais", proposes a fundamental distinction between texts, that of *histoire* and *discours*. *Histoire* is objective (science, history and the like) and *discours* subjective. In *histoire* no voice, no consciousness appears: "...A vrai dire, il n'y a même plus alors de narrateur. Les événements sont posés comme ils se sont produits à mesure qu'ils apparaissent a l'horizon de l'histoire. Personne ne parle ici: les événements semblent se raconter eux-mêmes." (Benveniste 1966:241).

Benveniste bases his distinction on the different tenses that appear in *histoire* and *discours* (he recognizes that the two often are mixed, as in a history with commentary). He notes that pronouns do not appear in *histoire*, but makes no other reference to deictics and none to temporal adverbials. Yet it seems evident that deictics of most if not all kinds must not occur in *histoire*, since they are evidence of consciousness.

In an analysis where both tense and adverbial combine to make temporal reference Benveniste's main insight about types of discourse can be easily stated. The essence of the distinction is that temporal reference involving subjectivity or consciousness appears in *discours* but not in *histoire*. I suggest that the distinction be recast in terms of deictic temporal expressions rather than with tense: deictics would occur only in *discours*. Languages differ in tense systems; the double system of French that Benveniste cites appears quite rarely (this is true whether or not one agrees with his particular claims about French). Therefore in generalizing these notions about discourse to other languages it would be necessary to refer to other indications of subjectivity and consciousness besides tense. The suggested recasting faces no such problem, since Deictics appear quite generally in language.

§1.4.2 The second principle for extended temporal structures also requires modification. According to this principle, sentences that form extended structures must have the same tense. But there are sequences that violate this principle, sequences with the Historical Present. Historical Present is the traditional name for the combination of tense and adverb exemplified in 2(c) above, present tense and past adverb, indicating Past time. The combination occurs frequently in informal discourse, such as joke or story-telling; it can also be found in more formal and sustained narratives.

Sentences with the Historical Present often occur in sequence with past tense sentences.[17] 30 and 31 illustrate.

30. Something funny happened last night. I'm sitting in my rocking chair as usual, when all of a sudden the door opens, apparently without human agency.

31. ...this American couple the other day. They walk in, total strangers, ask what kind of local ale is gook, I put in my bit of advice, and straightaway, we're carrying on talking and laughing like we'd all been chums for years. Then I taught him how to play carts. He won.
advertisement in *The New Yorker*, 1978.

The sentences of these examples have different tenses, yet they all indicate Past time. Thus the temporal interpretation is consistent although the tense forms are not.

These facts suggest a modification of Principle 2 that is more general than its first formulation above. The modified form is semantic rather than syntactic, referring to interpretation rather than particular tense forms. Revised Principle 2: The temporal forms in sentences of an extended temporal structure must establish the same time. The formulation allows for variation among languages, an important advantage over the earlier one. It is a peculiarity of English (although not exclusive to English) that present tense can indicate times other than present. This peculiarity means that extended temporal structures indicating Past may have past or present tense. In languages where tenses function differently, the application of Principle 2 would not allow the same sequences.

I have formulated two principles that account for important aspects of extended temporal structures, in English and other languages (but see section §3 for some caveats). These principles allow the recognition or formation of temporal structures, according to the forms that occur in individual sentences. The classification of adverbials according to their anchoring properties proved to be essential for understanding of temporal structures within and across sentences.

§2. I turn now to some different questions about temporal reference. The discussion above has prepared the ground, I hope, for a consideration of the general properties of system that make temporal reference.

---

17 Wolfson (1979) discusses the narrative function of switching from past to present tense in accounts of past events.

§2.1 The distinction between temporal expressions that anchor freely and those that anchor only to Speech Time (ST) has been emphasized above. This distinction is traditionally made, in slightly different form, for tenses. Tense is said to be "absolute" or "relative", depending on whether it anchors to ST or to other temporal expressions as well. The fact that the distinction applies to both tenses and adverbials is expected on the analysis given here, where temporal reference is based on both tense and adverbial.

In languages such as Japanese and Hindi tense is relative. This means that tense is interpreted either as anchored to ST or to the time established by another sentence: in embedded contexts, tense anchors to the matrix sentence and not to ST. For Japanese and Hindi the relational value of present tense is simultaneity, that of past tense is anteriority. These values are consistent although the anchors for the tenses vary. The way that temse in embedded sentences is interpreted, for these languages, is indicated in 32 and 33.[18]

32. John [say+past] that Mary [is+present] sick
    Interpretation: S_1:Past S_2: simultaneous with S_1
33. John [say+past] that Mary [is+past] sick
    Interpretation: S_1:Past S_2: anterior to S_1

It is interesting to compare the interpretations with those of the corresponding embedded sentences in English.

34. John said that Mary is sick.
    Interpretation: S_1 Past S_2: simultaneous with ST
35. John said that Mary was sick.
    Interpretation: S_1: Past S_2: simultaneous with S_1
36. John said that Mary had been sick.
    Interpretation: S_1 Past S_2: anterior to S_1

35 corresponds to 32 in Japanese or Hindi. The examples show that past tense is absolute in English: it is taken as anterior to ST in both embedded and non-embedded contexts. To indicate anteriority the anterior form *have* is used; thus 36 corresponds to 33 in Japanese or Hindi. There are dialects of English that do not use *have* as in 36; for such dialects, past tense is relative in embedded contexts.).

No counterpart to 34 exists in Japanese or Hindi. In 34, the embedded sentence is taken as simultaneous to ST rather than to the time of the matrix sentence; 34 would be appropriate only if Mary were sick at the moment of

18 I would like to thank Hermann van Olphen for information about Hindi. The Japanese information is given in Ota (1972).

speech, with the same sickness that afflicted her when John spoke of it. This interpretation of present tense as limited to speaker responsibility occurs only when it is embedded under a Past time. It is an absolute interpretation of present tense, with the relational value of simultaneity and the constant anchor ST. This exceptional fact corresponds to another: the Historical Present cannot be embedded. The speaker responsibility interpretation takes precedence, within sentences, in the system of English. Hindi does not have a use of present tense like the Historical Present, perhaps because tense is relative in that language to ST or to other sentences, rather than to adverbials.

The notion of relative and absolute anchors is illuminating for understanding of the English tense system. The English system is well-known for its asymmetry and irregularity, but just where the asymmetry lies is less well-known. Most analyses content themselves with a simple list of the interpretations of the tenses, as in 2 above. The facts of interpretation are these: past tense indicates a time anterior to ST except when embedded as a complement under a sentence establishing future time. In this exceptional case, it indicates anteriority to the matrix sentence.[19] Present tense indicates various times -see 2 above- except when embedded under a sentence establishing past time. In this exceptional case, it indicates a time simultaneous to ST. Embedded contexts are exceptional for both tenses.

It is clear that past and present tense have different anchoring properties in English, exceptions aside, and therein lies its irregularity. Past tense anchors consistently to ST: it is an absolute tense.[20] Present tense anchors to times established by adverbials or times in other sentences: it is a relative tense. The system has both relative and absolute tense. This is true for both tense and adverbials, of course. Deictic adverbials are absolute, Dependent adverbials are relative, and Flexible Anchoring adverbials may be either absolute or relative.

---

19 An example of this use of past tense:
The minister will announce at midnight that he burned the documents an hour ago.

On one reading the documents were burned an hour before ST, but on another they will be burned an hour before midnight, and the past tense indicates a time anterior to the RT of the main sentence.
Discussion of the limited syntactic contexts for this interpretation is presented in Smith, in press.

20 In the conditional and irrealis uses of past tense, it does not have the absolute interpretation.

§2.2 Temporal reference in language involves two kinds of anchoring, as we have seen. Absolute tense and adverbials are anchored to the time of speech; relative tense and adverbials are anchored to a time given in text or context. These facts lead to a new argument for how systems of temporal reference should be represented: I will argue that they support the approach of Hans Reichenbach to this question.

According to Reichenbach, temporal reference involves 3 times, Speech Time (ST), Reference Time (RT), and Event Time (ET). ST is the keystone of the system: other times are related to it and to each other by the relations of simultaneity or sequence. Examples that show the need for 3 times include (for English) the perfect, and embedded sentences that do not establish a time independently.

37. Michelle had already called last week.

38. They said on Sunday that they were leaving in 3 days.

In both of these sentences, 3 times must be distinguished for temporal interpretation: ST, a time anterior to ST (*last week*, *on Sunday*), and a third time, anterior or posterior (37 and 38, respectively) to the time established in the matrix sentence.

One might claim that a system of tense (or time) embedding would be adequate to account for sentences such as 35 and 36. In this view, there is no need for a temporal system such as Reichenbach's. The English perfect, for instance, could be analyzed as a past embedded under a past, without any special machinery. The analyses of tense logic exemplify this view. An embedding analysis can indeed account for sentences with 3 times, but whether it gives the best analysis of temporal reference is dubious. It would need special, ad hoc provisions to account for vague and discontinuous reference, and would need to be severely restrained in power since it predicts multiple embeddings of times that do not occur in natural language.[21]

A system of 3 times with ST at its center is predicted implicitly by the way times anchor in language. Absolute reference has ST (or another orientation time) as its anchor, and a related time, RT. Relative reference has a time given on the surface as its anchor, and a related time. It is this third time that corresponds to ET in Reichenbach's system.

---

21 I present a detailed discussion of this point in Smith, 1975, in which I argue against a proposal of McCawley's that English tenses be treated as higher verbs, and embedded as part of the general embedding system of English.

The close fit between the anchoring facts and a system of temporal reference such as Reichenbach's constitutes strong evidence in favor of that system.[22]

§2.3 A second argument for Reichenbach's system also emerges from the analysis of this paper. For this argument I return to example 16, repeated here.

16. (a) We are deeply concerned with last Sunday's robbery of an important public figure ...
    (b) Last night I talked to a reporter who had inside information.
    (c) The maid left before noon.

I noted in §1.3 above that sentence (c) anchors to (a); such a sentence might be uttered in sequence with (b), that is, without an explicit anchor. Its interpretation would then be vague, as noted in the discussion of this example. I have suggested[23] an analysis of this type of sentence, in which it specifies ET and shares RT specified in another sentence.

Cases like this are not unexpected in a system with 3 times. A sentence has only one time adverbial, specifying RT or ET. A related sentence may specify ET different from RT, or an RT that anchors ET. I see no natural way to account for discontinuous and vague interpretations in an embedding system. For discontinuous readings, a missing anchor or related time would have to be located and correctly interpolated. For vague readings the mediating anchor time would have to be specified as missing. But why, in such a system, should there be a missing time? And, if in principle one could allow for a missing time, why should not there be infinitely many such times? These and other questions seem to me extremely serious. No doubt an embedding system could handle them, but this is an indication of the general power of such systems and not of their appropriateness for dealing with temporal reference.

In this section I have developed some ramifications of the analysis of extended temporal structures. I have shown that tenses as well as time adverbials may anchor either to ST or a time given on the surface, and have explained the well-known asymmetry of English tenses in terms of different types of anchoring. The tense system of English is mixed, with past tense of the absolute type and present tense of the relative type; there is one

---

22 See McGilvray (1978) for an argument that Reichenbach's approach to temporal reference captures important points about consciousness and temporal referer.

23 See Smith (1978).

exception for each tense, that of an embedded complement context. I have also argued that a Reichenbach approach accords well with this analysis, which I take to have general scope well beyond the system of English. An embedding system, on the other hand, is ill-suited to predicting or explaining the way temporal reference works. I return in section §3 to extended temporal reference.

§3. Temporal reference is established by combinations of tense and time adverbials, and under certain conditions can have a domain beyond that of a single sentence. In §1 two principles were presented that account for the way temporal expressions form extended temporal structures. However, the principles do not, and cannot, tell us when such structures actually occur. There are two reasons for thsi limitation.

Semantic connections are essential to linguistic structures that have a domain larger than a sentence. But such connections are not within the scope of syntactically-based principles that deal with temporal expressions. Elliptical phenomena are also beyond the scope of such principles, which rely on explicit forms.

I discuss briefly some points of semantic interpretation that are not temporal, strictly speaking, but that affect extended temporal structures. My purpose here is to indicate the wide range of connections that are relevant rather than to give a complete account. To begin, note that temporal expressions themselves cannot be related without consideration of the type of unit to which they refer. For instance, in 39 (b) cannot be plausibly capturing according to the criteria developed in §1.

39. (a) Every weekday morning Jenkins did the same thing.
    (b) Last year he (Jenkins) went to Louisiana.

The units *morning* and *year* are not commensurate, and capture must therefore be blocked. Order is crucial in these sequences: if *year* preceded, a plausible extended interpretation could be made. Semantic facts about temporal expressions, then, are relevant to capturing.

A closer look at possible interpretations reveals a Pandora's box of difficulties. Temporal structures can be established only when certain connections obtain between sentences. These connections may have to do with subjects, objects, or other complements in the sentences; or with the relation between verbs in one sentence and nouns in another; etc. Thus in the pair of sequences below, 40 can be interpreted as an extended temporal structure and 41 cannot, yet the difference is minimal and not related to the temporal expressions.

40. (a) Last week Nick went to Oxford.
    (b) He saw Danforth on Tuesday.

41. (a) Last week Nick went to Oxford.
    (b) I saw Danforth on Tuesday.

Taking 40 as one temporal structure, *Tuesday* refers to a day of *last week;* and Danforth is probably someone whom Nick saw at Oxford. 41 has no such interpretation. Rather, *Tuesday* is deictic; Danforth is probably someone who has news of Nick's activities.

As a final example consider 42, followed by either 43 or 44:

42. (a) The Taylors went on a trip.
    (b) They sold their porcelain dog collection to raise the money.

43. The sale was last month.

44. The trip was last month.

For semantic reasons, 43 forms a temporal unit with 42(b), and 44 forms a unit with 42(a). *Sell* and *sale* connect the first pair, the two *trips* connect the second.

The immediate conclusion to be drawn from these examples is that informtion other than the temporal is needed to establish extended temporal structures. But there is a more general conclusion, which bears on the larger question of linguistic structures beyond the sentence. The conditions for temporal structures given here are sentence-based, depending on expressions that are associated with individual sentences. But it is clear that conditions of this type cannot, in principle, suffice for recognizing or forming structures beyond the sentence. Semantic connections are necessary, and semantic connections are not limited as to type of expression or syntactic context. Although this paper discusses only temporal structures, I suggest that the conclusion extends to other areas as well.

I would like, finally, to consider briefly the relation of extended temporal structures to structures in discourse. What in fact is the relation of syntactico-semantic units such as extended temporal units to discourse structure? Is there a correspondence between the two? Is it consistent, so that in locating the first, one has located the second? These are difficult questions. However, I would like to suggest a partial answer; somewhat negative, possibly useful for inquiry in these interesting and little-understood areas. There is no general relation, I would argue, between discourse structure and extended temporal units such as those I have investigated in this paper. In some discourse, temporal structures play a role that corresponds to important structures in discourse. In other discourse it does not.

Only when dealing with a limited genre, where it is known that temporal units have an organizing role, can a correlation between temporal units and discourse structure be assumed in advance. Even in such cases there are always exceptions.

In certain types of discourse, temporal and other deictic structures often have an organizing role: extended temporal units can be taken as signals of units of discourse. Narratives, especially informal and personal narratives, are of this type.[24] But in other discourse (interviews, didactic writing, thematically organized texts, for example) units of time or place may play a subsidiary role in the structure of the whole. To locate such units, then, does not necessarily mean that one has located the structures important to the general organization of a given text.[25] To decide on the organizing structures of a text, the nature of the particular language activity must be considered, and no correlation can be assumed in advance.

References

Stenning, K., 1978, Anaphora as an approach to pragmatics. In eds., Halle, M., Bresnan, J. & Miller, G., *Linguistic theory and psychological reality*. Cambridge, Mass: MIT Press.

Webber, B., 1978, A formal approach to discourse anaphora. *Technical Report 3761*. Cambridge, Mass: Bolt, Beranek and Newman.

Weinrich, H., 1964, Tempus. Stuttgart: Kohlhammer.

Wolfson, N., 1979, The conversational historical present alternation. *Language 55*.

---

24 Labov and Waletzky (1966) make a persuasive case that temporal and spatial structures are important in personal narratives. They are less persuasive in their argument that such structures are of general importance in discours.

25 See Levinson (1978) for an interesting discussion of differences in the structure of different types of linguistic activity.

FINITE AND NON-FINITE TEMPORAL CONSTRUCTIONS IN GERMAN

Rainer Bäuerle and Arnim von Stechow, Universität Konstanz

1. *Preliminary considerations*

Most textbooks on German grammar distinguish 'simple' and 'compound' tenses. This is a purely morphological classification: 'synthetic' verb forms are termed simple, 'analytic' forms, which involve the use of auxiliaries, are called complex.

Syntactically, this classification is not very revealing. In the indicative active, German has two infinitives, the infinitive present (*schlaf-en*) and the infinitive perfect (*geschlafen hab-en*). And it seems only natural to assume that the three tenses 'Present', 'Past', and 'Future' make finite forms out of these infinitives.

This leads to an intuitively plausible classification of the tense forms of German. We may apply perfectivization to form the second infinitive from the first one, and then one of the three tenses can be combined with one of the two infinitives to form the six tense forms of (the 1st person of) the German indicative active:

1.	schlaf(en)	+		+	Present	=	schlafe
2.	schlaf(en)	+		+	Past	=	schlief
3.	schlaf(en)	+		+	Future	=	schlafen werde
4.	schlaf(en)	+	Perfect	+	Present	=	geschlafen habe
5.	schlaf(en)	+	Perfect	+	Past	=	geschlafen hatte
6.	schlaf(en)	+	Perfect	+	Future	=	geschlafen haben werde

We leave it open how the morphological processes involved are to be described. The important points are that we take the infinitive to be the base for tense formation, and not the present tense, which is normally considered to be basic in 'tense logics'; that we assume the morphologically complex future forms to represent a 'simple' basic tense (which is justified by global syntactic and semantic considerations); and last, that we postulate a fixed order of application for these rules: first the perfect-rule works optionally on the present infinitive, and then one of the basic tenses 'Past', 'Present', or 'Future' makes a finite form out of this non-finite base. This is to say that there are no other forms than the ones arrived at by the application of these rules, in particular we do not allow for any such device as 'iteration of tenses'.

Iteration of tenses is the traditional tense logician's tool for explaining the perfect tenses. Nested operators such as P(Pp) or F(Pp) are said to account for pluperfect and future perfect, respectively. It is only a minor shortcoming that such an approach still leaves us with the present perfect as a primitive. TICHY (1978), e.g., defined a tense operator Perf (= Present Perfect), and BENNETT (1977) also discusses the semantics of such an operator. The major mistake of this approach is that it overlooks the fact that tense is a deictic phenomenon.

It has been stated most clearly in GABBAY (1976) that however deeply we nest or embed tenses, we always have to keep track of t_o, the time of utterance. Consider

(1) A child was born that will be king.

The 'be king' is not simply in the future of the date of birth, it is in future of t_o. We must therefore assume that the past, present and future operators are invariably evaluated at t_o, and that the time of evaluation cannot be shifted. But this would be a prerequisite for an explanation of P(Pp) as 'p is in the past of some past point t'.

Whereas the basic tenses are deictic expressions* that cannot shift their point of evaluation, 'perfect' certainly has a shiftable index, consider

(2) Es ist heute ein Makel,1933 in der Partei *gewesen zu sein*.
It is today a fault .... in the Party been to be .

and

(3) Es war damals ein Makel, 1933 in der Partei *gewesen zu sein*.
It was then a fault .... in the Party been to be .

In (2) the 'in der Partei sein' is before 'heute', in (3) it is before 'damals'.

Another point we argue for in this paper is that tenses must have a *definite* interpretation. The traditional tense logicians analysis, however, is an *indefinite* one. Under an indefinite interpretation, a sentence like

---

* Traditional grammars distinguish between *absolute* and *relative* tenses. The basic tenses are absolute, the perfect is relative. This terminology is also found in the article of C.Smith, contained in this volume.

(4) I turned off the stove.

is analysed as

(5) My turning-off of the stove occurred at *some* t before t_o.

HALL-PARTEE (1973) showed that the indefinite interpretation yields only unacceptable results for the negation of (4), namely either

(6) There is *no* t before t_o, at which I turned off the stove.

or

(7) There is some t before t_o, at which I didn't turn off the stove.

What is really meant, however, by an utterance of the negation of (4) is that the utterer did not turn off the stove at some *definite* (contextually specified) time-interval in the past. We therefore introduce two parameters for the interpretation of tenses, the time of utterance and the time-interval referred to. Basic tenses specify a subinterval of the time referred to, and this subinterval is the temporal frame for quantification over act-times by means of frequency adverbs like 'always, sometimes, twice' etc.; such a definite interpretation of tenses is worked out in in detail in BÄUERLE (1977, 1979).

A last remark concerns the syntactic status of tense operators. Morphologically, tense is a feature of the finite verb. But a sentence like

(8) Meine Freunde werden schlafen.
My friends will sleep .

is ambiguous, and can express either the proposition that my present friends will be asleep, or the proposition that my future friends will then be asleep. And in tense-logic this ambiguity is usually explained by the scope of the tense operator: in the first case the term 'meine Freunde' is not in the scope of the future operator, in the second case it is. Such a treatment assumes the tenses to be sentence-operators, because semantic considerations require (8) to have two representations:

(9) a) Meine Freunde $\hat{x}$ (F (x schlafen))
My friends sleep

b) F (Meine Freunde schafen)

This is not very natural from a syntactic point of view. One could try to better the situation by defining tenses as verb operators. But this has its own problems. First, one still has to account for the ambiguity of (8); and second, not all verbs are intransitive, so that the verb modifier would have to be defined not only for 1-place verbs, but also for verbs with a different number of places.

We will not adopt such an approach. Instead we will try to demonstrate, that it is possible to arrive at an intuitively satisfactory German syntax, if one introduces tenses transformationally. The perfect modifies an intransitive non-finite non-perfect VP (verb phrase). And the three tenses modify intransitive non-finite VP's, perfect or non-perfect.

2. *Syntax*

We can, of course, only describe a very small fragment of German. But the examples chosen are meant to illustrate certain peculiarities of German. There is, e.g., a characteristic difference in word order between main clauses and subordinate clauses:

(10) a) Ede *säuft* gerade ein Fass Bier *aus*.
Ede drinks at the moment a keg of beer out.

b) ... (daß) Ede gerade ein Fass Bier *aussäuft*.
Ede actually a keg beer outdrinks.

In (10 b) the finite verb occurs in final position, whereas in (10 a) it occupies the second position. Only the affix *aus-* is still in final position. These facts can be explained quite easily, if we take the word-order of the subordinate clause as the basic one and derive (10 a) transformationally from (10 b) by moving the finite part of the verb, but not the affix. This is quite an old observation (cf. e.g. BACH, BIERWISCH).

Infinitive clauses have a structure very similar to subordinate ones:

(11) *Ein Fass Bier auszusaufen* ist für Ede eine Kleinigkeit.
A keg beer out to drink is for Ede a bagatelle.

It seems only natural to assume that the infinitive clause in (11) is a tenseless intransitive verb phrase (IV). And that (11) and (10) have the same underlying IV. The respective structures of (10 b) and (11) are, roughly speaking, the following:

(10b') 1. $[[\textit{ein Faß Bier}]_{NP}\ [\textit{aus}]_{AFF}\ [\textit{sauf-}]_{V}]_{IV}$
[-finite]

2. $[[\textit{ein Faß Bier}]_{NP}\ [\textit{aus}]_{Aff}$

$[\textit{sauf + Pres + 3.Pers Sing}]_{FIN}]_{IV}$
[+finite]

= *säuft*

Introduction of Present

3. $[\textit{Ede}]_{NP}$

4. $[[\textit{Ede}]_{NP}\ [\textit{ein Faß Bier}]_{NP}\ [\textit{aus}]_{AFF}\ [\textit{säuft}]_{FIN}]_{S}$,

Subject introduction

(11') 1. $[[\textit{ist}]_{FIN}\ [\textit{für Ede}]_{PP}\ [\textit{eine Kleinigkeit}]_{NP}]_{IV}$
[+fin]

2. $[[\textit{ein Faß Bier}]_{NP}\ [\textit{aus}]_{Aff}\ [\textit{sauf-}]_{V}]_{IV}$
[-fin]

3. $[[\textit{ein Faß Bier}]_{NP}\ [\textit{aus}]_{Aff}\ [\textit{sauf + en}]_{V}$

$[\textit{ist}]_{Fin}\ [\textit{für Ede}]_{PP}\ [\textit{eine Kleinigkeit}]_{NP}$,

Subject introduction

There are two important differences between infinitive clauses and finite sentences: infinitive clauses have no basic tenses and they are also subjectless. The form of the finite verb in German is determined by a basic tense as well as by the person and number of the subject. Therefore it seems to be a good idea to introduce tense together with the subject, as MONTAGUE suggested in PTQ (S 17). A sentence like

(12) Ede soff ein Fass Bier aus.

will then be generated in the following way:

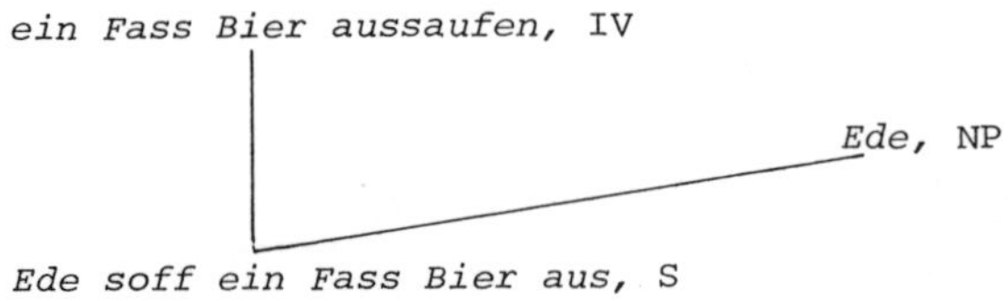

We do not follow exactly this approach, which will be justified later.

As we are not interested in the internal structure of noun phrases (NP), we assume the existence of rules that generate NPs like

$[\mathit{Ede}]_{[\mathrm{NP}}$ 3rd Person Sing Nominative]

or

$[\mathit{ein\ Fass\ Bier}]_{[\mathrm{NP}}$ 3rd Pers Sing Accusative]

The internal structure of NPs is neglected here, and the meaning of the features attached to the category label NP is obvious. Verbs have lexical entries of the following sort:

$[[\mathit{schlaf}\text{-}]_{\mathrm{V}}]_{\mathrm{VP}}$
sleep <NP nom>

or

$[[\mathit{aus}]_{\mathrm{Aff}}[\mathit{sauf}\text{-}]_{\mathrm{V}}]_{\mathrm{VP}}$
out drink <NP nom, NP acc>

or

$[[\mathit{soll}\text{-}]_{\mathrm{V}}\quad]_{\mathrm{VP}}$
[+mod] <NP nom, $\overline{\mathrm{VP}}$>

or

$[[\mathit{wünsch}\text{-}]_{\mathrm{V}}]_{\mathrm{VP}}$
want <δ>

where δ = <NP nom, $\overline{\mathrm{VP}}$> or <NP nom, $\overline{\mathrm{S}}$>

($\overline{\mathrm{S}}$ is short for [S. + *comp*]. This is the category of sentential complements. $\overline{\mathrm{VP}}$ stands for $[\mathrm{VP}, -\mathit{finite}]_{\langle A\delta\rangle}$. This is the category of infinitivals.
Instead of [V, + *modal*] we shall sometimes write *Mod.*)

Thus each verb entry is provided with one or more complement-frames, and the bracketings contain all the information needed. It does therefore not seem to be necessary for the purpose of this paper to be more precise and technical.

The rule for the generation of complex verb phrases is (S1):

(S1) *Verb phrases*

$$^{[\alpha]}[VP, <A_1\delta_1, \ldots A_n\ \delta_n\ >]$$

$$^{[\beta]}[A_n\delta_n]$$

$$n > 1 \text{ and } A_n = NP\delta,\ \overline{S},\ \overline{VP}$$

$$^{[\gamma\alpha]}[VP, <A_1\delta_1, \ldots, A_{n-1}\delta_{n-1}>], \text{ where}$$

$$\gamma = \begin{cases} ^{[\beta]}[A_n\delta_n], & \text{if } A_n = NP,\ \overline{S} \\ \beta_1[\beta_2 + Inf]_V, & \text{if } A_n = [\overline{VP} - zu] \text{ and } \beta = \beta_1[\beta_2]_V \\ \beta_1[zu\ \beta_2 + Inf]_V, & \text{if } A_n = [\overline{VP} + zu] \text{ and } \beta = \beta_1[\beta_2]_V \end{cases}$$

Before commenting on the rule, let us consider a sample derivation for the complex VP

(13) ... *ein Fass Bier aussaufen soll-*

In the derivation-tree (13a) we illustrate the strategy pursued in formulating our syntactic rules: a certain amount of structure is preserved throughout the derivation, but only as much as is needed for the formulation of word-order transformations. (13a/5.) does , e.g., not tell us that [*ein Fass Bier*] is the object of the transitive verb [*saufen*] - the information is not needed by the rules that apply later on. For other purposes, however, it may be necessary to preserve more of, or even all of the structural information. Such a situation may, for instance, arise in connection with the description of restrictions for possible anaphora (cf. BRESNAN, REINHART)

(13a) 1. $[[\textit{aus}]_{\mathrm{Aff}}[\textit{sauf-}]_{\mathrm{V}}]_{\mathrm{VP}}$, Lexicon
<NPnom, NP acc >

2. $[\text{ein Fass Bier}]_{\mathrm{NP}}$ $\begin{bmatrix}3.\text{Sing}\\ \text{acc}\end{bmatrix}$, (some appropriate rule)

3. $[[\text{ein Fass Bier}]_{\mathrm{NP}}[\textit{aus}]_{\mathrm{Aff}}[\textit{sauf-}]_{\mathrm{V}}]_{\mathrm{VP}}$, (S1 case 1)
$\begin{bmatrix}3.\text{Sing}\\ \text{acc}\end{bmatrix}$ <NP nom >

4. $[[\textit{soll-}]_{\mathrm{Mod}}]_{\mathrm{VP}}$, Lexicon
<NP nom,
$[\overline{\mathrm{VP}}$ - zu]>

5. $[[\textit{ein Fass Bier}]_{\mathrm{NP}}$ $[\textit{aus}]_{\mathrm{Aff}}[\textit{sauf+en}]_{\mathrm{V}}$
$\begin{bmatrix}3.\text{Sing}\\ \text{acc}\end{bmatrix}$

$[\textit{soll-}]_{\mathrm{Mod}}]_{\mathrm{VP}}$, (S1, second case)
<NP nom>

The rules used are a sort of 'generalized' transformations. Each rule consists of a structural description of the admissible input-structures plus a description of the output-structures. Where one and the same rule may yield different output-structures, as it is the case with (S1), it is of course always possible to split it up into several distinct rules. It should also be obvious that any transformation rule of the kind we use here can also be represented in the form of Montague-Syntax.

In the sample derivation (13a), the category symbol of the NP [*ein Fass Bier*] is further specified by the features [number], [person], and [case], i.e. 3rd person singular accusative, although these features are not mentioned at all in (S1). Our convention is this: features not mentioned explicitly in the syntactic rules carry over, with only features that get changed by the rule being listed.

Notice that rule (S1) determines the morphological realization of non-finite verbs. In German, there are two kinds of infinitives, the so called *0 - infinitive* and the *prepositional infinitive*. The 0 - infinitve of

$[\textit{ein Fass Bier}]_{\mathrm{NP}}[\textit{aus}]_{\mathrm{Aff}}[\textit{sauf-}]_{\mathrm{V}}$

is represented as

$[\textit{ein Fass Bier}]_{\mathrm{NP}}[\textit{aus}]_{\mathrm{Aff}}[\textit{sauf + Inf}]_{\mathrm{V}}$.

The prepositional infinitive, on the other hand, is represented as

$[ein\ Fass\ Bier]_{NP}[aus]_{Aff}[zu\ sauf + Inf]_V$

Inf is an abstract marker that triggers the appropriate morphological transformation. In the very simple example we have chosen this means only the addition of the infinitive suffix *-en:* [*saufen*]. German morphology is certainly more complicated than this, consider e.g. 'Umlaut'-phenomena as in [*muss* + *Inf*] → [*müssen*]. However, for a more sophisticated treatment we can only refer the reader to WURZEL.

The same holds true for the intricacies of perfect-formation. The perfect forms of non-modal verbs give us the basic principle:

```
(15)  a) Dornröschen hat      lange     geschlafen.
                    has a long time    slept.
                     |_________________|
                    Present Perfect

      b) ...   lange     geschlafen haben
           a long time    slept  |  have
                                 |
                      Infinitive Perfect
```

Here the verb is transformed into a past participle, and an auxiliary *(haben* or *sein)* is added which takes the inflexional endings. Again, the past participle occurs in a variety of forms, depending on whether the verb in question is a weak verb or a strong verb, and whether it is a prefixed verb or not.

But the behaviour of modal verbs is even more complicated:

```
(16)  Dornröschen hat     lange     schlafen müssen.
                 has a long time    sleep    must.
                  |_____________________|
                    Present Perfect
```

Instead of the past participle, these verbs show the infinitive. And yet we cannot wholly exclude the past participle, as in

```
(17)  Dornröschen soll       lange     schlafen gemußt haben.
                 is said a long time  sleep     must  |  have
                                                      |
                                          Infinitive Perfect
```

For all these reasons we will again postulate an abstract marker *Perf* and leave the actual surface realization to a more careful elaboration of morphological rules:

(S2) *Perfect*

$$\frac{[\alpha[\beta]_V]_{[VP_{\langle A\delta\rangle},\ -perf,\ -fin]}}{[\alpha[\beta{+}Perf]_V[\gamma]_V]_{[VP_{\langle A\delta\rangle},\ +\ perf]}}$$

$$\gamma = \begin{cases} hab\text{-}, & \text{in case the VP } \alpha[\beta]_V \text{ takes the auxiliary 'haben'} \\ sei\text{-}, & \text{in case the VP } \alpha[\beta]_V \text{ takes the auxiliary 'sein'} \end{cases}$$

Whether a VP takes 'haben' or 'sein' depends on the VP's main verb. An exact definition of the conditions used here is only possible via a recursive definition over the rules generating complex VPs. Cf. STECHOW (1978). A verb phrase is marked [+fin], if and only if it is specified as to tense, number, and person. The feature [+perf] is introduced in order to prevent an iteration of (S2). We will comment on the so-called 'double perfect' in German, on sentences like

(18) Ich habe ihn gesehen gehabt,

in the chapter on the perfect.

Let us now return to our example (13a) and apply the perfect rule:

(19) 1. $[[ein\ Fass\ Bier]_{NP}[aus]_{Aff}[sauf\text{-}]_V]_{VP_{\langle NP\ nom\rangle}}$, (= 13a/3.)

2. $[[ein\ Fass\ Bier]_{NP}[aus]_{Aff}[\underbrace{sauf + Perf}_{=\ \text{gesoffen}}]_V$
$[hab\text{-}]_V]_{VP\begin{bmatrix}\langle NP\ nom\rangle \\ +\ perf\end{bmatrix}}$, (S2)

3. $[[soll]_{Mod}]_{VP_{\langle NP\ nom,\ \overline{VP}\rangle}}$,Lexicon

4. $[[ein\ Fass\ Bier]_{NP}[aus]_{Aff}[gesoffen]_V$
$[haben]_V[soll]_{Mod}]_{VP_{\langle NP\ nom\rangle}}$, (S1)

We shall now introduce tenses. Contrary to Montague's idea, mentioned previously, tenses and subject are introduced separately. The reason for this will be explained in the comments on rule (S3):

(S3) *Basic Tenses*

$[\alpha[\beta]_V]_{VP}$, β is the one and only V in this construction which as
[-*fin*] yet has no suffix

---

1. $[\alpha[\beta+\gamma \text{ Pres}]_{V\gamma}]_{VP\gamma}$
2. $[\alpha[\beta+\gamma \text{ Past}]_{V\gamma}]_{VP\gamma}$
3. $[\alpha[\beta+ \text{ Inf}]_V[\text{werd} + \gamma]_{V\gamma}]_{VP\gamma}$, where γ = per(a) num(b)

The verbs which are not specified as to number and person are subclassified as -fin. The other verbs are +*fin* . Instead of V. +fin we shall write *Fin*.

As an example, let us consider how (S3) is applied to (19/2.):

(20) 1. $[[\textit{ein Fass Bier}]_{NP}[\textit{aus}]_{Aff}[\textit{gesoffen}]_V$

$[\textit{hab-}]_V]_{VP}$, (cf. (19)/2.)

$\begin{bmatrix}\langle NP\ nom\rangle \\ +\ perf\end{bmatrix}$

2. $[[\textit{ein Fass Bier}]_{NP}[\textit{aus}]_{Aff}[\textit{gesoffen}]_V$

$[\underbrace{\textit{hab + Inf}}_{=\text{ haben}}]_V[\underbrace{\textit{werd-} + \text{3rd pers + sing}}_{=\text{ wird}}]_V$ [+fin]

$]_{VP}$

$\begin{bmatrix}\text{3rd pers sing} \\ \langle NP\ nom\rangle\end{bmatrix}$

In (20/1.) the only verb without suffix is the auxiliary 'hab-' (by "suffix" we mean the markers Inf, Perf, Pres, Past etc., which transform "stems" from the lexicon into word-forms), and the application of alternative 3 of (S3) yields

(21a) ein Fass Bier ausgesoffen haben wird
a keg beer out drink have will

as in

(21b) ..., daß Ede morgen *ein Fass Bier ausgesoffen haben wird.*
that Ede tomorrow a keg beer out drunk have will.

The reason for not introducing tenses and subject at the same time is the existence of co-ordinate verbs with the same subject, but different tenses:

(22a) Ede *hat* das Problem *gelöst* und
Ede ha|s the problem sol|ved and
Present Perfect

wird morgen darüber *berichten.*
will tomorrow thereabout report
Future

Such a complex predicate cannot be generated by (S17) in Montague's PTQ. There is, however, a possibility to obtain the structure underlying (22a) by using the technique of 'quantifying in':

(22b) 1. er$_{17}$ hat das Problem gelöst, S
he$_{17}$ has the problem solved

2. er$_{17}$ wird morgen darüber berichten
he$_{17}$ will tomorrow thereabout report

3. er$_{17}$ hat das Problem gelöst und er$_{17}$ wird morgen darüber berichten, S (Conjunction)

4. Ede, NP

5. Ede hat das Problem gelöst und er wird morgen darüber berichten, S (Quantifying in)

6. Ede hat das Problem gelöst und wird
Ede has the problem solved and will
morgen darüber berichten , S (Pronoun-deletion)
tomorrow thereabout report

Our rule for tenses could easily be reformulated in such a way that it conforms with the idea developed above. But this approach is not much in the spirit of a surface oriented syntax rule like (S17) in PTQ, which allows for a direct coordination of intransitive VPs.

It would also certainly be more economical to introduce person and number together with the subject, as it is done in CHOMSKY's 'Aspects'. But the result would be an intransitive verb phrase that is specified as to a basic tense but not specified as to person and number. And such an IV never occurs in surface structures. In accordance with Barbara PARTEE's "well-formedness constraint", however, we try to generate only expressions that actually occur in the surface structure.

The next step in the derivation is now the introduction of the subject:

(S4) *Subject Introduction*

$$\frac{\begin{array}{l}[\alpha]_{VP\ \langle NP_{cas(a)}\rangle\ num(b),\ pers(c)} \\ [\beta]_{NP\ cas(a),\ num(b),\ pers(c)}\end{array}}{[[\beta]_{NP}\ \alpha]_S}$$

Notice that rule (S4) wipes out the VP-label.
Applied to (20/2.) this yields

(23) 1. [[*ein Fass Bier*]$_{NP}$[*aus*]$_{Aff}$[*gesoffen*]$_{V}$[*haben*]$_{V}$
[*wird*]$_{V}$]$_{VP}$, (20/2.)
[+fin] [3rd pers sing / <NP nom>]

2. [*niemand*]$_{NP}$, Lexicon
[3rd pers / sing / nom]

3. [[*niemand*]$_{NP}$ [*ein Fass Bier*]$_{NP}$[*aus*]$_{Aff}$
no one [3rd sing / nom] a keg beer out

[*gesoffen*]$_{V}$[*haben*]$_{V}$[*wird*]$_{V}$]$_{S}$, (S4)
drunk have will [+fin]

The structure thus obtained exhibits the word order of subordinate clauses. We will now have to provide rules in the form of 'movement transformations' that take account of the word-order in other sentence types.

The structure of the main clause is arrived at by fronting the finite verb of a VP:

(S5) *Fronting of finite verb*

$$\frac{[\alpha[w]_{FIN}\ \beta]_{VP\,[+fin]} \quad , w \neq e}{[[w]_{FIN}\ \alpha[e]_{FIN}\ \beta]_{VP}}$$

FIN stands for either Modal$_{[+fin]}$ or V$_{[+fin]}$.
The element moved leaves a "trace". Although this is obviously motivated by the so-called 'trace theory', the traces in our system have a purely technical status. Their purpose is to serve as a "memory" for movement transformations. Without some such device, movement rules could be iterated and iterated, yielding a lot of rubbish. For an example, we go back to (23/1.). (23/1.) is transformed in (24) into what one might call the 'normal' or 'unmarked' word order of the German declarative main clause. There are, of course, other possible structures. To exemplify this, we will also introduce rules for interrogative word-order (as in the so-called yes/no-interrogatives without question word) and for topicalization.

(24) 1. $[[\textit{ein Fass Bier}]_{NP}[\textit{aus}]_{Aff}[\textit{gesoffen}]_{V}[\textit{haben}]_{V}$

$[\textit{wird}]_{FIN}]_{VP}$, (23/1.)

[3rd pers sing
<NP nom>]

2. $[[\textit{wird}]_{FIN}[\textit{ein Fass Bier}]_{NP}[\textit{aus}]_{Aff}[\textit{gesoffen}]_{V}$

$[\textit{haben}]_{V}\ [e]_{FIN}]_{VP}$, (S5)

[3rd pers sing
<[nom]>]

3. $[\textit{niemand}]_{NP}$, Lexicon

[3rd pers sing
<[nom]>]

4. $[[\textit{niemand}]_{NP}[\textit{wird}]_{FIN}[\textit{ein Fass Bier}]_{NP}$
no one will a keg beer

$[\textit{aus}]_{Aff}\ [\textit{gesoffen}]_{V}[\textit{haben}]_{V}[e]_{FIN}]_{S}$, (S4)
out drunk have

(S6) *Interrogatives*

$$\frac{[[\alpha]_{NP}\beta[w]_{FIN}\gamma]_{S}}{[[w]_{FIN}[\alpha]_{NP}\beta[e]_{FIN}\gamma]_{S}}$$, and w is the only FIN in this construction

The condition that w be the only FIN is imposed to guarantee, among other things, that w has not yet been moved. Any movement would have left a trace $[e]_{FIN}$, and with two FINs in the structure, the rule cannot be applied. (S6) allows for the derivation of

(25) $[[\textit{wird}]_{FIN}[\textit{niemand}]_{NP}[\textit{ein Fass Bier}]_{NP}[\textit{aus}]_{Aff}$
$[\textit{gesoffen}]_{V}[\textit{haben}]_{V}[e]_{FIN}]_{S}$

from (23/3.), but not from (24/4.)! The uniqueness condition for FIN would inevitably be violated, were (S6) applied after (S5). This condition also prevents the derivation of an undesirable (26b) from (26a):

(26a) $[[\textit{Irene}]_{NP}[\textit{Ede}]_{NP}[\textit{liebt}]_{FIN}[\textit{und}]_{conj}[\textit{Hubert}]_{NP}[\textit{haßt}]_{FIN}]_{S}$
loves and hates

(26b) *$[[\textit{liebt}]_{FIN}[\textit{Irene}]_{NP}[\textit{Ede}]_{NP}[e]_{FIN}[\textit{und}]_{conj}[\textit{Hubert}]_{NP}[\textit{haßt}]_{FIN}]_{S}$
loves and hates

We take the interrogative word order to be the basis for topicalization, a process that can be treated only rather superficially in this paper.

(S7) *Topicalization*

$[[w]_{FIN}\alpha[\beta]_{A}\gamma]_{S}$, where

β is an NP other than the subject, or else

A = PP, VP, S, Adv

$[[\beta]_{A}[w]_{FIN}\alpha[e]_{A}\gamma]_{S}$

We may call (S7) "Drach's rule" (cf. DRACH). With (25) as input, we can now derive (27), for instance:

(27) [[*ein Fass Bier*]$_{NP}$[*wird*]$_{FIN}$[*niemand*]$_{NP}$[e]$_{NP}$
a keg beer will nobody

[*aus*]$_{Aff}$[*gesoffen*]$_{V}$[*haben*]$_{V}$[e]$_{FIN}$]$_{S}$
out drunk have

The rule is meant to illustrate the idea rather than to give an account of the actual facts of German syntax. Although we can get (28b) from (28a) by (S7), (S7) does not give us the more complicated (28c):

(28a) [[*wird*]$_{FIN}$[*niemand*]$_{NP}$[*wünschen*]$_{V}$[e]$_{FIN}$
will no one want

[*daß Ede ein Fass Bier aussäuft*]$_{S}$]$_{S}$
that Ede a keg beer out drinks

(28b) [[*daß Ede ein Fass Bier aussäuft*]$_{S}$[*wird*]$_{FIN}$
[*niemand*]$_{NP}$[*wünschen*]$_{V}$[e]$_{FIN}$[e]$_{S}$]$_{S}$

(28c) [[*wünschen*]$_{V}$[*daß Ede ein Fass Bier aussäuft*]$_{S}$
[*wird*]$_{FIN}$[*niemand*]$_{NP}$]$_{S}$

For a more careful elaboration of the rules and a more general description of topicalization, see STECHOW (1978).

Here a comment should be inserted on the work recently published by C.L. THIERSCH. Thiersch tries to account for German word order by essentially two ordered rules (we neglect some further details). The final position of the finite verb is taken to be basic, and there is a rule fronting the finite verb and a topicalization rule. A sentence like

(29a) Jeder kennt Schwarz
(Everybody knows Schwarz)

is derived in roughly the following way:

(29b) 1. $[[jeder]_{NP}[[Schwarz]_{NP}[kennt]_{FIN}]_{VP}]_S$

Fronting of finite verb

2. $[[kennt]_{FIN}[jeder]_{NP}[[Schwarz]_{NP}[e]_{FIN}]_{VP}]_S$

Topicalization of subject

3. $[[jeder]_{NP}[kennt]_{FIN}[e]_{NP}[[Schwarz]_{NP}[e]_{FIN}]_{VP}]_S$

THIERSCH could also derive

(29c) $[[Schwarz]_{NP}[kennt]_{FIN}[jeder]_{NP}[[e]_{NP}[e]_{FIN}]_{VP}]_S$

from (29b/2.) by object-topicalization.

The idea is intuitively appealing, and also the movement rules are a little simpler than the ones we use. But remember our discussion of (22a/b): we wanted to account for predicate conjunction directly, which is impossible in THIERSCH's system. What THIERSCH's proposal and our own have in common, however, is that 'topicalized' sentences are derived via underlying 'interrogative' structures.

The rule for the co-ordination of predicates concludes the chapter on syntax.

(S8) *Co-ordination of finite VPs*

$$[\alpha_1]_{VP\begin{bmatrix}pers(a)\\num(b)\end{bmatrix}} \quad \dots \quad [\alpha_n]_{VP\begin{bmatrix}pers(a)\\num(b)\end{bmatrix}}$$

where the α_i must not be analyzable as $\beta_i[w]_{conj}\gamma_i$, and where, furthermore, the non-empty finite verbs have to occupy the same positions in all α_is, i.e. either front position or final position in every α_i (i = 1...n).

---

1. $[\alpha_1 \ [und]_{conj} \ \dots \ [und]_{conj}\alpha_n]_{VP\begin{bmatrix}pers(a)\\num(b)\end{bmatrix}}$
2. $[\alpha_1 \ [oder]_{conj} \ \dots \ [oder]_{conj}\alpha_n]_{VP\begin{bmatrix}pers(a)\\num(b)\end{bmatrix}}$

where all conjunctions except the last one may afterwards be optionally deleted.

(S8) correctly generates sentences like

(30a) Niemand *dient* *dem Teufel* und *gefällt* *dem Herrgott*.
no one serves the devil and pleases the Lord God.

It does not allow, however, for the generation of

(30b) Niemand *kann* ((dem Teufel dienen) und (dem Herrgott gefallen)).
no one can the devil serve and the Lord God please

The derivation of the latter sentence will be roughly as indicated in (30c).

(30c) 1. [*dem Teufel dienen*]$_{VP}$
[-fin]

2. [*dem Herrgott gefallen*]$_{VP}$
[-fin]

3. [dem Teufel dienen UND dem Herrgott gefallen]$_{VP}$
[-fin]

4. [[*könn-*]$_{M}$]$_{VP}$

5. [[*dem Teufel dienen und dem Herrgott gefallen*]$_{VP}$
the devil serve and the Lord God please
[*könn-*]$_{Mod}$]$_{VP}$
can

etc.

The example shows that infinitives have to be built up separately, not dependent on the verb that governs them, as it was done in (S1). Thus infinitive formation should be formulated as a separate rule in a full-fledged syntax. But for our present purposes, these rudimentary remarks on German syntax may suffice. It should be clear by now that the 'surface' is obtained from the structures generated by simply deleting the labelled bracketings and the 'traces'.

3. *Semantics*

As is usually done, we assume that there is a set W of possible worlds, a set T of time-intervals, and a set A of possible individuals. Thus we can establish the following terminology (for similar concepts see CRESSWELL or LEWIS):

3.1. (a) A *proposition* is a subset of W x T.

(b) The proposition p *is true* in world w at time t iff $\langle w,t\rangle \in p$.

(c) The set of propositions M *entails* the proposition q iff the intersection of M ($\bigcap M$) is a subset of q.

(d) The set M of propositions is *consistent* iff $\bigcap M \neq \emptyset$

Furthermore, some other entities will also be of interest to us, namely *characters* and *properties*.

3.2. A character (the term is taken from LEWIS) is a function from W x T into properties of individuals.

3.2. (a) An *n-place property (of individuals)* is a function from $\underbrace{A\times\ldots\times A}_{n \text{ times}}$ into the set of propositions.

(b) If X is the set of all characters, then an *n-place property (of characters)* is a function from $\underbrace{X\times\ldots\times X}_{n \text{ times}}$ into the set of propositions.

There are 'mixed' properties as well. The German verb *wünschen* ("Ich wünsche mir, daß dieser Aufsatz bald fertig ist"), for instance, expresses a property of characters and propositions. In essence, an n-place property is taken to be a function from products of sets of the appropriate kind into propositions. Therefore:

(c) If ω is an n-place property and $\zeta_1, \ldots, \zeta_n$ are entities of the appropriate kind, then ω is *true* of $\zeta_1, \ldots, \zeta_n$ in w at t iff $\langle w, t\rangle \in \omega(\zeta_1, \ldots \zeta_n)$, *false* iff $\langle w, t\rangle \notin \omega(\zeta_1, \ldots \zeta_n)$.

The set T of time intervals may be thought of as constructed on the basis of some set of linearly ordered time points.*

---

* This is the orthodox view. It is probably better to think of times as primitives meeting certain conditions. Vide the appealing paper of Ruth Manor contained in this volume.

We shall not go any further in the justification and explanation of the semantic vocabulary. In what follows, the context will make our assumptions understandable.

As is very well known, natural language expressions are context-dependent. A sentence like

(31) Er schläft gerade.
He sleeps just

may express the proposition that Ede is asleep at 1 p.m. on 1/1/79, or the proposition that Klaus-Jürgen is asleep at ..., and so on. The proposition expressed by (31) depends on the occasion of utterance. And given a context of utterance, it is the meaning of (31) that tells us which proposition is expressed. Therefore, we have to distinguish between the meaning of an expression and what is expressed by it at a particular occasion. The following terminological distinctions seem useful:

Semantic entities such as propositions, properties, characters, and even possible individuals are referred to as *contents*. And functions from utterance contexts to contents are called *meanings* (thus meanings here are what KAPLAN calls characters). Sentence meanings are thus functions from contexts to propositions, NP-meanings are functions from contexts to characters, and so on. For the sake of simplicity, we identify contexts with pairs of a world $\underline{w}$ and a time $\underline{t}$. We may then represent a sentence meaning as a function,

$$f: W \times T \rightarrow \text{the set of propositions}$$

Since each proposition may be identified with its characteristic function, this amounts to *double indexing*: a sentence meaning could as well be represented as a function (see LEWIS)

$$f: (W \times T) \times (W \times T) \rightarrow \text{the set of truth values.}$$

In order to interpret our syntax, we assign meanings to the lexical items. Furthermore, the syntactic rules are paired with 'semantic operations' that tell us how the meanings of the input structures are combined to yield the meaning of the output expression. More accurately:

To a syntactic rule r

$$\frac{\begin{array}{l}\alpha_1\\ \vdots\\ \alpha_n\end{array}}{\begin{array}{l}\text{output}_1^r\ (\alpha_1 \ldots \alpha_n)\\ \vdots\\ \text{output}_k^r\ (\alpha_1 \ldots \alpha_n)\end{array}}$$

we assign k semantic operations V_i^r (i = 1...k).
V_i^r $(\zeta_1 \ldots, \zeta_n)$ is the meaning of the ith output of r, i.e. of $\text{output}_i^r(X_1 \ldots X_n)$.

In the system presented here, semantic operations take a more special form: they are functions from contents to contents. How this is to be understood can be seen from the following recursive definition of the meaning assignment to the expressions of our language. We start from a given context $\langle w_o, t_o \rangle$ and a given function V^L that tells us the meaning of every lexical element.

3.4. (a) If α is a lexical element, then

$$V_{w_o t_o}(\alpha) := V^L_{w_o t_o}(\alpha)$$

(b) If β is a complex expression obtained as the ith output from input $\alpha_1 \ldots \alpha_n$ by syntactic rule r, then

$$V_{w_o t_o}(\beta) := V_r^i\,(V_{w_o t_o}(\alpha_1), \ldots, V_{w_o t_o}(\alpha_n))$$

The definition presupposes that a complex expression is built up in exactly one way. It is, by the way, not the most general recursive meaning assignment possible. A more general approach would have

$$V(\alpha) := V^L(\alpha)$$

$$V(\beta) := V_r^i\,(V(\alpha_1), \ldots, V(\alpha_n)).$$

Definitions of this kind can be found in CRESSWELL, reasons for choosing 3.4. instead are given in KAPLAN and will not be repeated.

$V_{w_o t_o}(\beta)$ may be read as "the content of expression β with respect to the context $\langle w_o, t_o \rangle$". When formulating our meaning rules, we will often proceed in an informal manner and omit the functional notation.

Let us first introduce the interpretation of the lexicon:

3.5. Let ω_{sleeps} be that property of individuals which is true of an $a \varepsilon A$ in a world w at time t iff a is asleep in w at t.

Let ω^+_{sleep} be the corresponding higher-order property of characters which is true of a character χ in w at t iff $\omega_{sleep} \varepsilon\ \chi(w, t)$.

Then $V^L_{w_o t_o}([[\textit{schlaf-}]_V]_{VP\,(NP\ nom)}) := \omega^+_{sleep}$.

Notice that an untensed verb like *schlaf-* expresses the same property at every context of utterance. (KAPLAN would call V(*schlaf-*$_{VP}$) a 'stable character'. Since we reserve the term 'character' for NP-contents, we will instead speak here of a '*stable meaning*'.) In such a case we may identify the meaning of a word with its content. Therefore, we simply say that the VP *schlaf-* expresses the property ω^{+}_{sleep} . Tensed verbs, however, do not have stable meanings.

3.6. (a) $V^{L}_{W_O T_O}$ (*Ede*$_{NP}$) (w, t) = {ω:ω is true of the individual Ede in w at t} : = χ_{Ede}(w, t)

(b) $V^{L}_{w_o t_o}$ (*ein Fass Bier*$_{NP}$) (w, t) =

{ω: there is an a∈A, a is a keg of beer in w at t, and ω(a) is true in w at t = : $\chi_{a\ keg\ of\ beer}$(w, t)

(c) Let ω_{drink} be that two-place property of individuals which is true of (a, b)∈A^2 in w at t iff a drinks b in w at t. Then ω^{+}_{drink} is that two-place property of characters which is true of (χ_1, χ_2) in w at t iff $\lambda^{*}a[\lambda^{*}$ (w, t)$[\lambda^{*}$ b ω_{drink} (a, b)$\in\chi_2$(w, t)]] $\in\chi_1$(w, t)

(λ^{*} denotes the metalinguistic λ-operator).

Now, $V^{L}_{w_o t_o}$ (*aussauf*$_{VP \left[\begin{smallmatrix} <NP\ nom \\ NP\ acc> \end{smallmatrix} \right]}$) = ω^{+}_{drink}

The next thing to do is to interpret the syntactic rules. (S5), (S6), and (S7) are movement rules that do not change the content of the input expression. We therefore stipulate

3.7. The semantic operation assigned to (S5), (S6), and (S7) is identity mapping.

The interpretation of (S1) reduces n-place verb phrase meanings (i.e. n-place properties of characters) to (n-1)-place verb phrase meanings (i.e. (n-1)-place properties of characters).

3.8. *Interpretation of (S1)*

Let α be the first premiss of (S1), i.e.: α is an n-place infinitive VP. Suppose further that β is the other input structure of (S1), i.e. an appropriate NP, PP, S or infinitive phrase.

The semantic operation associated with (S1) can now be described by the follwoing equation:

$V^{(S1)}(V_{w_o t_o}(\alpha), V_{w_o t_o}(\beta))$

is that (n-1)-place property

ω_{n-1} which is true of the appropriate entities $\zeta_1, \ldots, \zeta_{n-1}$ in w at t *iff* $\langle w, t\rangle \in V_{w_o t_o}(\alpha)(\zeta_1, \ldots, \zeta_{n-1}, V_{w_o t_o}(\beta))$.

To give an example:

$V^{(S1)}(\omega^{+}_{drink}, \chi_{a\ keg\ of\ beer}) = \omega^{+}_{drink\ a\ keg\ of\ beer}$,

which is true in w at t of χ_{Ede}, iff

ω^{+}_{drink} is true of $(\chi_{Ede}, \chi_{a\ keg\ of\ beer})$ in w at t.

According to 3.6. c) this is the case iff

$\lambda^{*} a[\lambda^{*}(w, t)[\lambda^{*} b\ \omega_{drink}(a, b) \varepsilon \chi_{a\ keg\ of\ beer}(w, t)]] \varepsilon \chi_{Ede}(w, t)$,

(which by 3.6. a) and λ^{*}-conversion gives)

$\Leftrightarrow [\lambda^{*} b\ \omega_{drink}(Ede, b)] \varepsilon \chi_{a\ keg\ of\ beer}(w, t)$

(and by 3.6. b) and λ^{*}-conversion)

$\Leftrightarrow$ There is an $a \varepsilon A$: a is a keg of beer in w at t, and $\omega_{drink}(Ede, a)$ is true in w at t

(and by the definition of ω_{drink} in 3.6. c) we get)

$\Leftrightarrow$ There is an a : a is a keg of beer in w at t and Ede drinks a in w at t.

So $\omega_{drink\ a\ keg\ of\ beer}(\chi_{Ede})$ is the proposition that Ede drinks a keg of beer.

The interpretation of the rule of subject introduction (S4) consists of an appropriate functional application:

3.9. *Interpretation of (S4)*

Let α and β be input expressions for (S4). Then

$V^{(S4)}(V_{w_o t_o}(\alpha), V_{w_o t_o}(\beta)) = V_{w_o t_o}(\alpha)(V_{w_o t_o}(\beta))$ for any $\langle w_o, t_o\rangle$.

However, our most immediate concern in the present paper is the interpretation of the rules for the tenses and the perfect, (S3) and (S2), and accordingly these will be considered in more detail. We shall start with the semantics of tense, because this only involves a re-formulation of ideas already stated in BÄUERLE (1977), SCHPAK - DOLT (1977), and TICHY. The formulation of these ideas in the present paper is taken from KRATZER (1978).

4. *Tense*

It is quite an old observation that tensed sentences may be true at one time, and false at another. So it is true today that Arnim cut his finger, although this was not true a week ago. Any sentence of the form "past α", that is to say, depends on a context of utterance that specifies a time t_o, the 'now' of the utterance. (That what counts as the utterance-time t_o in a context c may not exactly coincide with the time interval needed for the production of the relevant sound-waves is discussed in KRATZER; we cannot go into details here).

But as has become obvious from the discussion of PARTEE's example (cf. (4)-(7)), this is not enough. Arnim may have cut his finger quite a number of times in the last thirty years, but this doesn't make (32) an eternal truth:

(32) Arnim cut his finger.

The reference of (32) obviously is to some definite finger-cutting at some definite time. Therefore we also have to evaluate sentences of the form "past α" at a time t, the time referred to. And it is also obvious that the tenses establish a temporal relationship between t_o and t.

And this is why we do not believe in 'iterated basic tenses': however deeply basic tenses are embedded, they refer back to t_o. And it is this deictic character of basic tenses that distinguishes them from the perfect. There is, by the way, a similar situation in the class of temporal adverbs: adverbials like 'heute, letztes Jahr, ...' refer back to t_o, regardless of their position in the sentence structure, whereas adverbials like 'am Tag zuvor, im Jahr zuvor, ...' readily shift their index.

Now it is often assumed that 'past' establishes a relation of absolute precedence between t_o and t: $t < t_o$. But this is a rather simplistic view, and t_o and t may certainly overlap, as in

(33) *Heute* schnitt sich Arnim in den Finger.
Today cut himself in the finger

where $t_o \subset t$.

Therefore we stipulate that the tense operator '*past*' selects that part of t that is before t_o:

4.1. "past α", uttered at time t_o in world w_o, is true in world w at time t iff there is a subinterval t^* of t which is before t_o, and α is true in w at t^*, where t^* is the *maximal* subinterval of t which is before t_o.

Thus, "past α" may be used appropriately in situations a) and b), but not in situation c), because the existential claim 'there is a t^* (subinterval of t) before t_o' is not fulfilled:

(a)

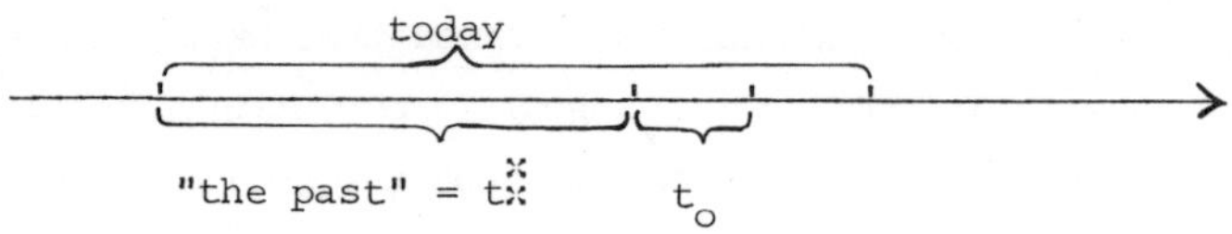

(b)

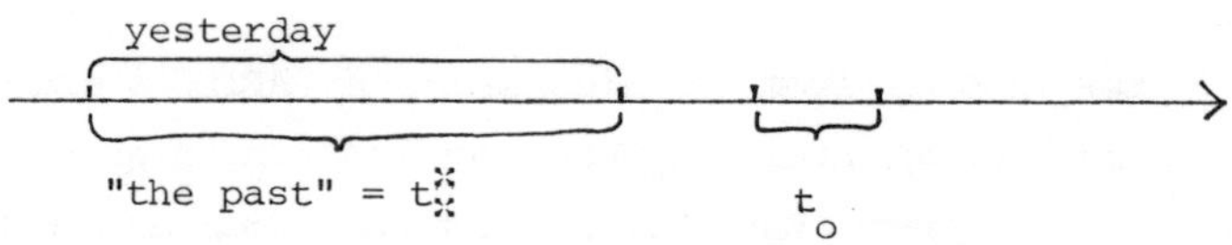

(c)

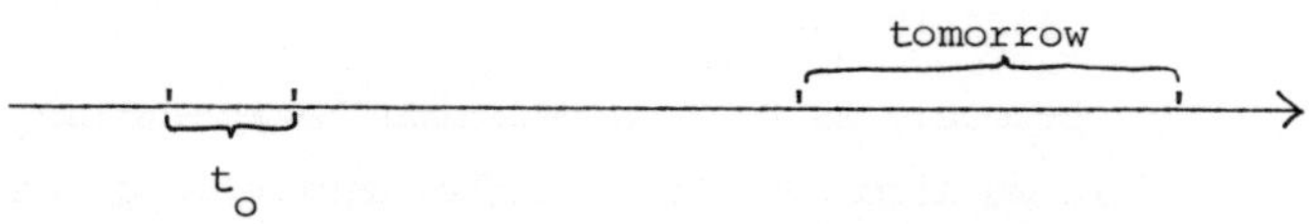

Given situation (a) or (b), 'past α' is true iff α is true in t^*. But in (c) 'past α' will always be false. (Or without truth-value, if you like - but this would also require some changes in our semantic framework).

The future tense can, of course, be described analogously:

4.2. "fut α", uttered at time t_o in world w_o, is true in world w at time t iff there is a subinterval t^* of t which is after t_o, and α is true in w at t^*, where t^* is the maximal subinterval of t which is after t_o.

Consider an example: according to 4.1., the sentence (33) (= Arnim cut his finger today), uttered by Rainer at 4 p.m. (= t_o), is true *today*) = t) iff Arnim was cutting his finger during the whole part of today that is before 4 p.m. (= t^*) - which is nonsense. What Rainer was really trying to say is that Arnim cut his finger *at some time during* t^*.

But the nonsense is only apparent. We consider it a mistake to burden tense semantics with the *at some time during*-restriction. Such an approach has, e.g. been chosen by GUENTHNER and KUHN. GUENTHNER gives the following semantics for the past-operator:

(34) $[P\emptyset]^{I}_{t,g} = 1$ iff $[\emptyset]^{I}_{t'} = 1$ for some $t' \varepsilon I$ and $t' < t$.

Note that we have to do with three temporal indices:

t: the 'now' of utterance (= t_o)

I: the time referred to (= t)

t': the (actual) act-time

But binding the act-time-variable by the existential quantifier should not be built into tense semantics, because quantification over act-times is the job of frequency adverbs:

(35) Josef ging heute / went today {mindestens einmal (at least once) / zweimal (twice) / oft (often) / manchmal (sometimes) / überhaupt nicht (not at all)} in die Cafeteria / to the Cafeteria

It should be obvious from (35) that tenses really specify a time t^{*} as defined in *4.1* and *4.2*, and that within t^{*} things are said to occur once, twice, often, or not at all. The relation between the 'now' of the utterance and the time referred to is taken care of by the tenses, and the relation between the time referred to and the (actual) act-time(s) is explained by adverbs of frequency. So we will be able to account for the seeming nonsense of the above interpretation of (33) when we - later on - introduce frequency adverbs.

But we still have to account for the German present tense.

4.3. "pres α", uttered at time t_o in world w_o, is true in world w at time t iff
there is a subinterval t^{*} of t which is *not before* t_o (= either t^{*} and t_o overlap, or else $t_o < t^{*}$) and
α is true at t^{*} in w, where t^{*} is the maximal subinterval of t which is *not before* t_o.

This definition reflects the fact that the German present is a non-past, i.e. it can always replace the future, and it is also used when t^{*} overlaps t_o, whereas the past has $t^{*} < t_o$, consider

(36) a) Nächstes Jahr *bin* ich in Neuseeland.
next year am I in New Zealand

b) Heute *arbeite* ich an einem Aufsatz.
   Today work I at an article

We are now in a position to give a semantic interpretation for (S3). A slight complication is due to the fact that tenses are treated as predicate modifiers*, but the basic idea is exactly analogous to the explications developed above.

4.4. *Interpretation of (S3)*

Suppose we are given an input expression α for rule (S3). Then let $\langle w_o, t_o \rangle$ be an arbitrary context, and $\langle w, t \rangle$ any world-time-pair, and χ any character.

a) The Present Tense

$\langle w, t \rangle \in V_1^{(S3)} (V_{w_o t_o}(\alpha))(\chi)$ iff

$(\exists t^* \subseteq t)$ (t^* is not before t_o)

and

$V_{w_o t_o}(\alpha)(\chi)$ is true in w at t_*^*, where t_*^* is the maximal subinterval of t which is not before t_o.

b) The Past Tense

$\langle w, t \rangle \in V_2^{(S3)} (V_{w_o t_o}(\alpha))(\chi)$ iff

$(\exists t^* \subseteq t)$ (t^* is before t_o)

and

$\langle w, t_*^* \rangle \in V_{w_o t_o}(\alpha)(\chi)$, where t_*^* is the maximal subinterval of t which is before t_o.

c) The Future Tense

$\langle w, t \rangle \in V_3^{(S3)} (V_{w_o t_o}(\alpha))(\chi)$ iff

$(\exists t^* \subseteq t)$ (t^* is after t_o) and

$\langle w, t_*^* \rangle \in V_{w_o t_o}(\alpha)(\chi)$, where t_*^* is the maximal subinterval of t which is after t_o.

---

* For syntactic reasons, of course. The same syntactic motivations are found in the article of E.Bach, contained in this volume.

Now we are able to evaluate sentence

(12) Ede soff ein Fass Bier aus.
Ede drunk a keg beer out

Remember that the sentence is generated in the following way:

(12a) 1. $[[aus]_{Aff}[sauf\text{-}]_V]_{VP}$, (Lexicon)
<NP nom, NP acc>

2. $[ein\ Fass\ Bier]_{NP}$, (appropriate rules)
[acc]

3. $[[ein\ Fass\ Bier]_{NP}[aus]_{Aff}\ [sauf\text{-}]_V]_{VP}$, (S1, case 1)
[acc] <NP nom>

4. $[[ein\ Fass\ Bier]_{NP}[aus]_{Aff}\ [\underbrace{sauf\text{-} + \text{3rd sing} + \text{past}}_{= \text{soff}}]_{FIN}]_{VP}$,
<NP nom>

(S3, case 2)

5. $[[soff]_{FIN}[ein\ Fass\ Bier]_{NP}[aus]_{Aff}[e]_{FIN}]_{VP}$, (S5)
<NP nom>
[3rd sing]

6. $[Ede]_{NP}$, (Lexicon)
[nom
3rd sing]

7. $[[Ede]_{NP}\ [soff]_{FIN}[ein\ Fass\ Bier]_{NP}[aus]_{Aff}\ [e]_{FIN}]_S$, (S4)
[nom]

Now let $\langle w_o, t_o\rangle$ be an arbitrary context at which (12) is uttered. It has already been demonstrated in the preceding chapter that

$$V_{w_o t_o}\ (12a/3.) = \omega^{+}_{\text{drink a keg of beer}}\ .$$

The content of the tensed VP (12a/4.) is arrived at by applying the semantic operation $V_2^{(S3)}$ to this property:

$$V_{w_o t_o}\ (12a/4.) = V_2^{(S3)}\ (\omega^{+}_{\text{drink a keg of beer}})\ .$$

By *3.7.*, (12a/5.) has the same content. And by *3.9.* and *3.6(a)* we get

$$V_{w_o t_o}\ (12a/7.) = V_{w_o t_o}\ (12a/4.)(\chi_{Ede}).$$

For any element $\langle w, t\rangle$ of $W \times T$,

$$\langle w, t\rangle \in V_{w_o t_o}\ (12a/7.)$$

iff

$(\exists t^{*} \subseteq t)(t^{*}$ is before $t_o)$ and
$\langle w, t^{*}_{*} \rangle \in \omega_{\text{drink a keg of beer}}(\chi_{\text{Ede}})$, where
t^{*}_{*} is the maximal part of t which is before t_o

Obviously this is fulfilled only if

a) $(\exists t^{*} \subseteq t)(t^{*}$ is before $t_o)$, and

b) Ede drinks a keg of beer in w at t^{*}_{*}, where t^{*}_{*} is the maximal part of t which is before t_o.

And this is indeed the proposition we were looking for.

Our treatment of tenses as predicate modifiers gives rise to a rather special problem. Predicates are the contents of one-place verb phrases. And there are such awkward verbs as e.g., *regnen*, which are intransitive verbs syntactically, but nevertheless express propositions:

(37) Es regnet.
It rains

Here we adopt a trick and stipulate the following meaning rule:

4.5. $V^{L}_{w_o t_o}$ $([[regn\text{-}]_V]_{VP\,\langle NP_{es}\rangle}$ (χ)

is that proposition which is true in a world w at a time t *iff* it is raining in w at t.

The content of *regnen* is thus a one-place property of characters, but it is a constant property. The proposition expressed does not depend on our choice of character. It could, for instance, also be the "deictic character" *es* (= *it*). Accordingly, the analysis of (37) is

(37a) 1. $[[regn\text{-}]_V]_{VP\,\langle NP_{es}\rangle}$, Lexicon

2. $[[regn\text{-} + \text{3rd sing} + \text{pres}]_{V\,[+fin]}\ ]_{VP\,\langle NP_{es}\rangle}$, (S3 , case 1)

3. $[es]_{NP\,(es)}$, Lexicon

4. $[[es]_{NP}[regnet]_{FIN}]_S$, (S4)

It seems clear that this analysis also gives us the correct meaning.

In the preliminary considerations, we also discussed the scope-ambiguity which occurs in (8). In the present account, the nominal *meine Freunde*$_{NP}$ is always in the scope of the tenses. In order to account for the *de re*-reading, one has to work with a quantifying-in-rule that generates (8) in essentially the following way:

(8a) 1. [*sie*$_{17}$ *werden schlafen*]$_S$
  they$_{17}$ will sleep

  2. [*meine Freunde*]$_{NP}$
  my friends

3. [*meine Freunde werden schlafen*]$_S$, (quantifying in)

'*Sie*$_{17}$' functions as variable which is bound by the nominal *meine Freunde*. The techniques necessary for such a treatment are exactly the same as those developed in MONTAGUE's PTQ.

Another, more attractive, technique has been developed by COOPER/PARSONS. One could generate (8) by means of the syntactic rules we introduced, and add a scope indication. (8) would then be given the analysis

(8b) 1. [*werden schlafen*]$_{VP}$

  2. [*meine Freunde*]$_{NP}$

3. [[*meine Freunde*]$_{NP}^{Q,i}$ *werden schlafen*]$_S^{S,i}$

The superscript Q,i indicates that the NP is a quantifier whose scope is the sentence with the superscript S,i. The techniques for the interpretation of structures of this kind are, however, somewhat tedious, and we shall not pursue the matter any further in this paper.

A further possibility is discussed in KRATZER, where it is claimed that the nominal is ambiguous between an attributive and a referential reading.

5. *Adverbs of frequency*

In the preceding section, we gave some arguments as to why tenses should specify the whole past, present, or future part of the time referred to. But it was also shown that this can only be maintained, if adverbs of frequency are introduced at the same time. We argued that, in analogy to sentences like

He is $\left\{\begin{matrix}\text{never}\\ \text{always}\end{matrix}\right\}$ asleep,

what is expressed by an utterance of

Angelika sneezed yesterday

is something like

Angelika sneezed (*at least*) *once* yesterday.

Since we have sentences like

(38) Siebenmal zu niesen bringt Unglück ,
seven times to sneeze brings misfortune

frequency adverbs should be treated as modifiers of non-finite verbs. The syntax of such a "quantified" VP is roughly as in (39).

(39) 1. $[[\textit{nies-}]_V]_{VP}$ $\langle$NP nom$\rangle$ $\left[\begin{matrix}-\text{fin}\\ -\text{freq}\end{matrix}\right]$, Lexicon

2. $[\textit{immer}]_{Adv}$ [freq] , Lexicon

3. $[[\textit{immer}]_{Adv}$ $[\textit{nies-}]_V]_{VP}$, (S9, see below)
always [freq] sneez- $\langle$NP nom$\rangle$ [+freq]

(S9) *Adverbs of frequency*

Syntax:

$$[\alpha_1\alpha_2]_{VP\langle\gamma\rangle[-\text{freq}\ -\text{fin}]}$$

where α_1 is a, possibly empty, sequence of NPs, and the first member of α_2 is not a NP.

$$[\beta]_{Adv\ [+\text{freq}]}$$

$$[\alpha_1[\beta]_{Adv}\alpha_2]_{VP\ \gamma\ [+\text{freq}]}$$

Semantics:

$$V^{(S9)}\ (V_{w_o t_o}\ ([\alpha_1\alpha_2]_{VP}),\ V_{w_o t_o}\ ([\beta]_{Adv}))$$
$$= V_{w_o t_o}\ ([\beta]_{Adv})\ (V_{w_o t_o}\ ([\alpha_1\alpha_2]_{VP}\ ))$$

As an example for the interpretation of frequency adverbs, consider

5.1 $V^{L}_{w_o t_o}$ ([*einmal*]$_{\text{Adv} \atop [\text{freq}]}$) (ω) (χ)

is true in w at t iff

$(\exists t' \subseteq t)$ (ω(χ) is true in w at t')

5.2 $V^{L}_{w_o t_o}$ ([*immer*]$_{\text{Adv} \atop [\text{freq}]}$) (ω) (χ)

is true in w at t iff there is a set T of subintervals of t such that $t = \bigcup T$, *and* ω(χ) is true in w at t' for every $t' \varepsilon T$ (see KRATZER, p.66).

Given these rules, we can now interpret

(40) Angelika nieste **immer.**
 sneezed always.

The derivation of the sentence is given in (40a)

(40a)

1. [[*immer*]$_{\text{Adv}}$[*nies-*]$_{\text{V}}$]$_{\text{VP}}$ <NP nom> , (39/3.)

2. [[*immer*]$_{\text{Adv}}$[*nieste*]$_{\text{FIN}}$]$_{\text{VP}}$ <NP nom> [3rd sing] , (S3, case 2)

3. [[*nieste*]$_{\text{FIN}}$[*immer*]$_{\text{Adv}}$[e]$_{\text{FIN}}$]$_{\text{VP}}$ <NP nom> [3rd sing] , (S5)

4. [*Angelika*]$_{\text{NP} \atop \text{nom}}$, (Lexicon)

5. [[*Angelika*]$_{\text{NP}}$ [*nieste*]$_{\text{FIN}}$[*immer*]$_{\text{Adv}}$[e]$_{\text{FIN}}$]$_{\text{S}}$, (S4)

Now, given a context $<w_o, t_o>$ and some $<w, t>$,

$V_{w_o t_o}$ (40a5.) is true in w at t iff

$V_{w_o t_o}$ (40a/3.) is true of χ_{Angelika} in w at t.

This, in turn, is true iff there is a subinterval

$t^{*} \subseteq t$ which is before t_o, such that

$V_{w_o t_o}$ ([[*immer*][*nies-*]]$_{\text{VP}}$) is true of χ_{Angelika} in w at t_{*}^{*}, where t_{*}^{*} is the maximal subinterval of t which is before t_o.

The latter is true iff there is a subinterval

> t^{*} of t which is before t_o, and there is a set T of subintervals of t_{*}^{*} (where t_{*}^{*} is the largest subinterval of t which is before t_o), such that $t_{*}^{*} = \bigcup T$ and
>
> $V_{w_o t_o}$ ([*nies-*]$_{VP}$) is true of $\chi_{Angelika}$ in w at t', for every t' ε T.

Notice that the semantics of *immer* automatically yields the right subdivision of the time referred to into a set of act-time intervals. In the case of 'sneezing', two intervals of T may not overlap, whereas in the case of 'sleeping' they may very well do so. A more accurate description of the 'interval properties' of verbs would reveal this consequence.*

6. *Temporal modifiers*

Words like *heute* (today), *gestern* (yesterday), and *morgen* (tomorrow) are used in a number of quite different functions, as can be seen from the following examples:

(41) a) Es regnet *heute* (It is raining today)

b) Er versprach *heute* zu kommen. (He promised to come today)

c) *Heute* ist Freitag, der 13. März 1979. (Today is Friday, 13.3.79)

d) *Heute* ist ein schöner Tag. (It's a beautiful day today)

e) *Heute morgen* ist eine Sitzung. (There is a meeting this morning)

f) Die Versammlung *heute* war langweilig. (Today's meeting was boring).

In a) and b), *heute* is an adverb modifying a tensed verb and an untensed verb, respectively. In c) and d), however, *heute* seems to function as a name.** In e), *morgen* is an attributive modifier of *heute*, which in turn modifies the tensed verb. And in f) we certainly have to do with an attributive *heute*, that modifies the common noun *Versammlung*.

---

* Compare the articles of Gabbay & Moravcsik and of Manor, contained in this volume.

** Manfred Bierwisch doubts that *heute* may function as a name. He thinks that d) has the structure [[*heute*]$_{Adv}$[*ist*]$_{FIN}$]]$_{Predicate}$[*ein schöner* Tag]$_{Subject}$]$_S$ (Personal communication). This might be right, but it does not touch the general point discussed here.

These examples make plausible that one should not assign exactly one syntactic category to the adverbs mentioned. The adverbs are polyfunctional, and our syntax should be able to explain how this is possible. Let us first consider the adverbial use in (41a):

(41g) 1. $[[\textit{regnet}]_{FIN}]_{VP}$ $\langle NP_{es}\rangle$, (cf. 37a/2.)

2. $[\textit{heute}]_{N}$ $[+time]$, Lexicon

3. $[\textit{heute}]_{Adv}$ $[+time]$, (S10, see below)

4. $[[\textit{heute}]_{Adv}[\textit{regnet}]_{FIN}]_{VP}$ $\langle NP_{es}\rangle$, (S11, see below)

5. $[[\textit{regnet}]_{FIN}[\textit{heute}]_{Adv}[e]_{FIN}]_{VP}$ $\langle NP_{es}\rangle$, (S5)

6. $[\textit{es}]_{NP_{es}}$, Lexicon

7. $[[\textit{es}]_{NP}[\textit{regnet}]_{FIN}[\textit{heute}]_{Adv}[e]_{FIN}]_{S}$, (S4)

The N *heute* has a very trivial interpretation: it denotes 'the present day'.

6.1 $V^{L}_{w_o t_o}([\textit{heute}]_{N\,[+time]})$ = the day which counts as the present day in w_o at t_o. ("today (w_o, t_o)" for short)

The adverbial use of *heute* is accounted for by rule (S10):

(S10) *Formation of temporal adverbs*

Syntax :

$$\frac{[\alpha]_{N\,[+time]}}{[\alpha]_{Adv\,[+time]}}$$

Semantics:

$$<w, t> \varepsilon\ V^{(S10)}(V_{w_o t_o}(\alpha))(\omega)(\chi)$$

$$\Leftrightarrow\ <w, V_{w_o t_o}(\alpha)> \varepsilon\ \omega\ (\chi),$$

where ω is a one-place property and χ a character.

It is easily seen that by this interpretation we ascribe a frame-setting effect to adverbs like *heute*. The neutral proposition $\omega(\chi)$ is changed into a proposition which is *specific* with respect to the time *today* (w_o, t_o). (A proposition p is specific as to time t, if the following holds:

$$<w, t^{*}> \varepsilon\ p \text{ iff } <w, t> \varepsilon\ p, \text{ for any } <w, t^{*}>.)$$

(S11) *Insertion of temporal adverb*

Syntax:

$$[\alpha_1 \alpha_2]_{VP<\gamma>}$$

where α_1 is a, possibly empty, sequence of NPs, and the first member of α_2 is not a NP.

$$[\beta]_{\substack{Adv \\ [+time]}}$$

$$[\alpha_1 [\beta]_{Adv} \alpha_2]_{VP<\gamma>}$$

Semantics:

$$V^{(S11)}\ (V_{w_o t_o}([\alpha_1 \alpha_2]_{VP<\gamma>}),\ V_{w_o t_o}([\beta]_{\substack{Adv \\ [+time]}})$$

$$= V_{w_o t_o}([\beta]_{Adv})(V_{w_o t_o}([\alpha_1 \alpha_2]_{VP}))$$

Rule (S11) only requires that $[\alpha_1 \alpha_2]$ is a one-place VP; it is not important whether the VP is tensed or not.

We are now in a position to evaluate (41g). As can easily be seen,

$V_{w_o t_o}$ (41g/1.) is true of a character χ in w at t iff

$(\exists t^{*} \subseteq t)(t^{*}$ is not before $t_o)$, and it is raining in w at t^{*}_{*}, where t^{*}_{*} is the maximal part of t which is not before t_o (for any <w, t>).

Now, by (S11),

$V_{w_o t_o}$ (41g/4.) is true of a character χ iff

$V_{w_o t_o}$ ([*heute*]$_{Adv}$) ($V_{w_o t_o}$ (41g/1.)) (χ) is true in w at t.

And this holds good, by (S10), iff

<w, $V_{w_o t_o}$ ([*heute*]$_{Adv}$)> ε $V_{w_o t_o}$ (41g/1.) (χ),

which is the case exactly if

($\exists t^* \subseteq$ today $(w_o t_o)$) (t^* is not before t_o), and it is raining in w at t^*_*, where t^*_* is the maximal subinterval of today $(w_o t_o)$ which is not before t_o.

As $V_{w_o t_o}$ (41g/4.) expresses a constant property not affected by subject introduction, this is already the truth condition for the whole sentence.

The above example also clarifies the remark on the frame-setting effect of *heute*: the evaluation time t is shifted to the evaluation time *today* $(w_o t_o)$. The treatment of sentences like (41b), where *heute* modifies an untensed verb, required us to allow a temporal adverb like *heute* to occur in the scope of the tenses. But we cannot allow for this in the generation of (41a), because the resulting interpretation is grossly counter-intuitive. To illustrate this, let us modify (41g) accordingly:

(41h) 1. [[*regn-*]$_V$]$_{VP}$ ⟨NP_{es}⟩ , (Lexicon)

2. [[*heute*]$_{Adv}$ [+time] , (S11)

3. [[*heute*]$_{Adv}$[*regn-*]$_V$]$_{VP}$ ⟨NP_{es}⟩ , (S11)

4. [[*heute*]$_{Adv}$[*regnete*]$_{FIN}$]$_{VP}$ ⟨NP_{es}⟩ , (S3, case 2)

The sentence thus generated is

Heute regnete es,

and the proposition expressed should be that it *was* raining today. But according to our rules of interpretation, $V_{w_o t_o}$ (41h/4.) is true in w at t iff

$(\exists t^{*} \subseteq t)(t^{*}$ is before $t_o)$, and $V_{w_o t_o}$ (41h/3.) is true in w at t^{**}, where t^{**} is the maximal part of t which is before t_o. And $V_{w_o t_o}$ (41h/3.) is true in w at t^{**}, iff $V_{w_o t_o}$ (41h/1.) is true in w at today$(w_o t_o)$.

And this is roughly the proposition that it *is* raining today.
One should be able to avoid constructions like (41h) by imposing certain restrictions on the syntax. In the present paper, we do not pursue this any further, but go on instead to show how (41b) can be treated.

For an interpretation of (41b), we need an appropriate lexicon entry for *versprechen*. This entry should be like the following rule:

6.2. $V_{w_o t_o}$ ([*versprech-*]$_{VP}$ [NP nom, NP dat, Zu-Inf])
$= \omega^{+}_{promise}$, where $\omega^{+}_{promise}$ is true in w at t of χ_1, χ_2, ω *iff*
$\lambda^{*}a[\lambda^{*}(w, t)[\lambda^{*} b\ \omega_{promise}(a, b, \omega) \varepsilon\ \chi_2(w, t)]]\ \varepsilon\ \chi_1(w, t)$.
$\omega_{promise}$ is the property which is true of *a*, *b*, and ω, iff *a* promises *b* ω

Let us now expand (41b) a little and generate the sentence:

(41i) *Er versprach mir gestern, heute vorbeizukommen.*
He promised me yesterday today to come

(41j)
1. [*versprech-*]$_{VP}$ <NP nom, NP dat, [$\overline{VP}$+zu]> , (Lexicon)
2. [[*heute*]$_{Adv}$[*vorbei*]$_{Aff}$[*komm-*]$_{V}$]$_{VP}$ <NP nom> , (S11)
3. [[[*heute*]$_{Adv}$[*vorbei*]$_{Aff}$[*zu kommen*]$_{V}$]$_{\overline{VP}}$[*versprech-*]$_{V}$]$_{VP}$ <NP nom, NP dat> (S1)
4. [*mir*]$_{NP\ dat}$, Lexicon
5. [[*mir*]$_{NP}$[*heute vorbei zu kommen*]$_{\overline{VP}}$[*versprech-*]$_{V}$]$_{VP}$ <NP nom> , (S1)
6. [[*mir*]$_{NP}$[*heute vorbei zu kommen*]$_{\overline{VP}}$[versprach]$_{FIN}$]$_{VP}$ <NP nom> , (S3)

7. [*gestern*]$_{Adv}$, (S10)
[+time]

8. [[*mir*]$_{NP}$[*gestern*]$_{Adv}$[*heute vorbei zu kommen*]$_{\overline{VP}}$

[*versprach*]$_{FIN}$]$_{VP}$, (S11)
<NP nom>

9. [[*versprach*]$_{FIN}$[*mir*]$_{NP}$[*gestern*]$_{Adv}$

[*heute vorbeizukommen*]$_{VP}$[e]$_{FIN}$, (S5)
<NP nom>

10. [*er*$_1$]$_{NP}$, (Lexicon)

11. [[*er*$_1$]$_{NP}$[*versprach*]$_{FIN}$[*mir*]$_{NP}$[*gestern*]$_{Adv}$

[*heute vorbeizukommen*]$_{VP}$[e]$_{FIN}$]$_S$, (S4)

Suppose now, that at a given context of utterance $\langle w_o, t_o \rangle$, *mir* denotes Ede and *er*$_1$ denotes Arnim. Then $V_{w_o t_o}$ (41j/2.) is the property of coming today ($w_o t_o$), and $V_{w_o t_o}$ (41j/5.) is the property of making Ede a promise to come today($w_o t_o$).

$V_{w_o t_o}$ (41j/6.) is then true of an *a* in w at t iff there is a t^* in t which is before t_o, and *a* promises Ede at t^* that *a* comes today($w_o t_o$) (where t^* is the maximal subinterval of t which is before t_o). By (41j/8.) we are told that t^* is in fact yesterday($w_o t_o$). So that $V_{w_o t_o}$ (41j/11.) is true in w at t iff Arnim promises Ede at *yesterday*($w_o t_o$) that he comes *today*($w_o t_o$).

This account of the truth-conditions of (41i) is, of course, very insufficient, because we have entirely neglected the fact that there are higher-order properties involved. But here we are only interested in illustrating the argumentation with regard to tenses and adverbs.

The 'name use' of *heute*, as it is found in (41c, d), poses no problems. It could easily be integrated into our present fragment of German grammar by the addition of an extra rule for the copula verb *sein*. And the treatment of the attributive use in (41f) is very similar to that of the adverbial use. What is needed is a rule that allows for the derivation of (41k):

(41k) 1. $[\textit{Versammlung}]_{CN}$

2. $[\textit{heute}]_{N}$

3. $[\textit{heute}]_{\substack{AdC \\ [+time]}}$

4. $[[\textit{Versammlung}]_{CN}[\textit{heute}]_{\substack{AdC \\ [+time]}}]_{CN}$
meeting today

As we have not gone into the internal structure of nominals in this paper, we shall not give any further details. The general strategy of the approach adopted here should by now have become clear enough:

A distinction has been made between word meaning and grammatical function. The former is taken from the lexicon, and the latter explained by means of syntactic rules. (In a categorial syntax, these two aspects are intertwined which makes categorial syntax rather clumsy at times. For an explanation of the different uses of *heute*, categorial grammar would either have to set up a corresponding number of lexical categories, or else make extensive use of λ-abstraction.)

7. *The Perfect*

For two reasons, the interpretation of (S2), which introduces the perfect, has been deferred until this last chapter of our paper.

First, what we presented in the preceding pages is a well-established account of the non-perfect tenses, which has been developed over the past few years in the SFB 99 at Konstanz university. Our interpretation of the perfect, however, is rather novel, and there are as yet no attempts that we know of in truth-conditional semantics to keep basic tenses and the perfect entirely separate.

Second, our interpretation of basic tenses and the perfect as separate phenomena necessitates the introduction of an additional temporal index. Although there is no difficulty with this as far as logic is concerned, we did not wish to complicate all the rules presented so far by specifying this index. And we also feel that one should be on one's guard against additional indices, unless they can be motivated intuitively.

So our policy was to present chapters 1-6 in a self-contained manner, and whoever does not object to our additional index can easily change the rules presented so far in such a way that they agree with what is proposed here.

Furthermore, an entirely satisfactory account of the German perfect seems to depend on an adequate semantics for the intransitive VP, which cannot be given here. Just consider the difference between

(42a) Seit einer Stunde habe ich ihn beobachtet.
For one hour have I him watched

(42b) Seit einer Stunde hat er seine Jacke ausgezogen.
For one hour has he his jacket off taken

In (42a), we talk about the duration of *watching him*, whereas in (42b) we refer to the duration of the state resulting from *taking off his jackett*. The failure to be explicit about perfect meaning vs. verb phrase meaning seems to be the source for some theories on the perfect (*resultative tense* etc.), where some examples support this theory, and other examples quite another. For a survey of traditional theories, see McCOARD.

As has become obvious from our preliminary considerations, the 'Present Perfect' is treated as a perfect in the scope of the present tense. This may be perfectly acceptable in English, where, to quote BENNETT, whenever we use the present perfect

"...there is always a reference to an interval of time that includes the present moment, ..., as well as past moments."

(For a more detailed elaboration see TICHY). In German, however, a present perfect form is also used together with adverbs that exclude any reference to the present moment:

(43) Gestern *bin* ich im Kino *gewesen*.
yesterday am I in the cinema been

According to our definition of the tenses, this cannot be explained as a 'Present + Perfect' construction. We have to assume, therefore, that the Past Tense in German has, morphologically speaking, an analytic and a synthetic form (which is one of the reasons why we had to be careful about the morphological realization of tenses):

(44) Past (schlafen) → ich schlief
Past (schlafen) → ich habe geschlafen
Pres Perf (schlafen) → ich habe geschlafen

And this is also the solution for the so-called "double perfect" mentioned earlier, namely that a sentence like

(45a) Ich habe ihn gesehen gehabt.
I have him seen had

should be analysed as

(45b) Ich (Past$_{\text{analytic form}}$(Perf (ihn seh-))).

And this is equivalent to

(45c) Ich hatte ihn gesehen.
I had him seen

with the past tense in its synthetic form.

The need for an additional index becomes obvious from the following example:

(46a) Er gab gestern zu, 1933 in der Partei gewesen zu sein.
He admitted yesterday, in the party been to be

(46a) is true if he admitted yesterday the truth of the embedded proposition, which in turn is true if he was a party member during that part of 1933 which is in the past of the time of admitting the membership. The same holds true for

(46b) Er gab gestern zu, in der Partei zu sein.
in the party to be

The embedded proposition is true if he is a party member at the time of admitting it. So we do not only have to keep track of t_o, which may also be needed in the evaluation of an embedded sentence, e.g. in (46c)

(46c) Er gab zu, gestern in der Partei gewesen zu sein,
yesterday in the party been to be

because *gestern* always denotes the day before t_o. We also have to keep on record another contextually specified time interval, in our examples (46a - c) the time of admitting.

Note, by the way, the difference between these examples and

(46d) Er gab zu, daß er 1933 in der Partei war.
that he in the party was

where the embedded proposition has a tensed verb. The only information we get here is that the admitting is in the past of t_o, and that being a party member is also in the past of t_o.

No temporal relationship between *being a party member* and *admitting it* is explicitly established, except that the semantics of *to admit* in itself may require that doing something precedes admitting it. But the temporal relationship can easily be made explicit by adding the perfect:

(46e) Er gab zu, daß er 1933 in der Partei gewesen war.
that he in the party been was.

By the basic tense, the embedded sentence of (46e) is true if his being a party member is before t_o, and by the perfect, it must also be before the time of admitting.

That is to say, we have to do with two points of view. There always is an absolute point of view, the time of utterance; and in addition a relative, and that is shiftable, point of view may become relevant.

In (46a - e) it was the matrix sentence that provided this second point of view, the relative index. Now we only have to apply the meaning-preserving movement rule (S5) to the structure underlying the embedded sentence in (46e), and we get the main clause

Er war 1933 in der Partei gewesen.
He was in the party been

In that case the relative point of view is to be provided from the context, as in

(46f) Der Staatssekretär wurde entlassen. Er war 1933 in der Partei gewesen.
The secretary of state was sacked. He was in the party been

In (46f) the relative point of view is the time of the firing of the secretary of state.

Before stating more precise rules, let us consider a last example. If we were to leave Ede a note saying

(47) Wir sind ins Kino gegangen.
We are to the cinema gone.

then this can by no means be interpreted as a past tense in its analytic form. We could not have written

(48) Wir gingen ins Kino.
We went to the cinema

because this is an event of 'current relevance'. It is not a thing past, but an indication of where we are to be found at present. The time of reading the note is an obvious candidate for t_o, but what is our contextually specified relative point of view? Seen from t_o, the present tense gives us a non-past interval (e.g. one overlapping t_o), which captures the 'current relevance'. And the *going to the cinema* is said to have happened at the past part of this non-past interval. In other words: the relative point of view is also t_o.

Which is not too astonishing, because at the beginning of a conversation, both the absolute and the relative index start at t_o. And it is only during the conversation that we shift the relative index, in that we introduce, via the past, present, or future times we talk about, other indices.

In (46f), for instance, we start with both indices at t_o and evaluate

$$\langle w,t\rangle \varepsilon \left|\text{Der Staatssekretär wurde entlassen}\right|^{t_o}_{t_o}$$

For the second sentence, we keep track of t_o as the absolute index, but the relative index is now shifted to the time of sacking the secretary of state:

$$\langle w,t\rangle \varepsilon \left|\text{Er war 1933 in der Partei gewesen}\right|^{t'_o}_{t_o}$$

(assume t' is the time of firing)

Whereas the basic tenses invariably refer to t_o, the absolute index, German has only one form, one 'relative tense', so to say, that refers to the shiftable index: the perfect. This may be different in languages that also have a future infinitive. But the relative index is also important for a subclass of temporal adverbs. We know that *gestern* always refers to the absolute index, but consider

(49) Im Januar wurde der Staatssekretär entlassen.
In January got the secretary of state sacked

Seit einem Jahr hatte man das geahnt.
For one year had one this sensed.

Here the phrase *seit einem Jahr* denotes the interval from one year before January up till and including January: January, the time of firing, is the relative index for an evaluation of *seit einem Jahr*.

It is perhaps best to first state the perfect rule and illustrate it by interpreting (47), where both indices are at t_o, and then to go on to discuss the shifting of the index.

7.1 *The Perfect*

$$<w,t> \varepsilon V^{(S2)}(V^{t'}_{w_o t_o}(\alpha))(\chi) \Longleftrightarrow (\exists t_i \subseteq t)(t_i \text{ is before } t') \text{ and}$$

$$<w,t_j> \varepsilon V^{t'}_{w_o t_o}(\alpha)(\chi), \text{ where } t_j \text{ is the maximal part of } t \text{ which is before } t'.$$

The rule is exactly analogous to the interpretation of the past tense, except that in this case we work with the relative index instead of the absolute one.

Consider now (47), which is

Wir sind ins Kino gegangen.

This we analyse as

Wir sind (mindestens einmal) ins Kino gegangen.
at least once
Adv_f

Now

$$<w,t> \varepsilon \,|\text{Pres}(\text{Perf}(\text{Adv}_f(\text{ins Kino gehen})))|_{w_o t_o}^{t_o} \text{ (we)}$$

is true iff (by the tense rule)

$$(\exists t^{*} \subseteq t)(t^{*} \text{ is not before } t_o) \text{ and}$$
$$<w,t_{*}^{*}> \varepsilon \,|\text{Perf}(\text{Adv}_f(\text{ins Kino gehen}))|_{w_o t_o}^{t_o} \text{ (we)}$$
$$(\text{where } t_{*}^{*} \text{ is the maximal part of } t \text{ which is not before } t_o).$$

This holds iff, by the perfect rule, the following is true:

$$(\exists t_i \subseteq t_{*}^{*})(t_i \text{ is before } t_o) \text{ and}$$
$$<w,t_j> \varepsilon \,|\text{Adv}_f(\text{ins Kino gehen})|_{w_o t_o}^{t_o} \text{ (we)}$$
$$\text{where } t_j \text{ is the maximal part of } t_{*}^{*} \text{ which is before } t_o.$$

Let us stop here, because it is now clear that what is expressed by the sentence can be depicted as follows

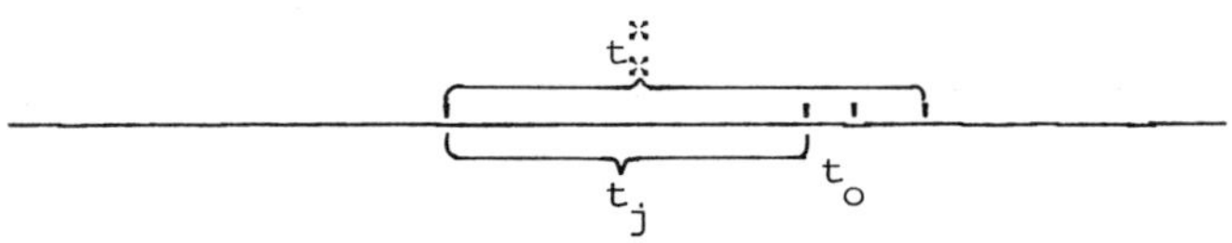

where t_j is the time interval at which *(mindestens einmal) ins Kino gehen* is true of a.

A straightforward example for the shifting of the index is given by (46a):

$\langle w,t\rangle \varepsilon\,|$gab gestern zu, 1933 in der Partei gewesen zu sein$|^{t_o}_{w_o t_o}$ (he)

is true iff (cf. tense rules)

$\langle w,t_*\rangle \varepsilon\,|$zugeben, 1933 in der Partei gewesen zu sein$|^{t_o}_{w_o t_o}$ (he)

(where t_* is the maximal subinterval of *yesterday*(t_o) which is before t_o).

Now this holds iff he admits at t_* the proposition

$|$1933 in der Partei gewesen zu sein$|^{t_*}_{w_o t_o}$(he)

Thus the relative index for the embedded proposition is shifted to t_*: the relative index of the embedded proposition is the actual act-time arrived at in the matrix sentence. So we can now go on with

$\langle w,1933\rangle \varepsilon\,|$in der Partei gewesen sein$|^{t_*}_{w_o t_o}$ (he)

which is true iff

$(\exists t_i \subseteq 1933)$ (t_i is before t_*) and

$\langle w,t_j\rangle \varepsilon\,|$in der Partei sein$|^{t_*}_{w_o t_o}$ (he),

where t_j is the maximal part of 1933 which is before t_*.

So, if he admitted the truth, the situation is as follows:

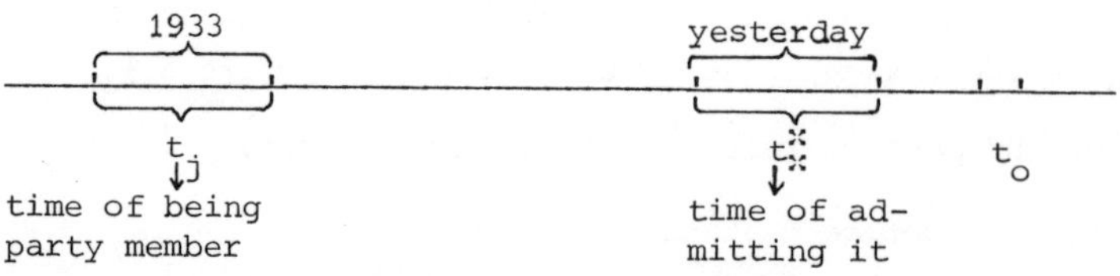

If the relative index of the embedded sentence is shifted to the actual act-time arrived at in the matrix-sentence, it seems obvious that for (46f)

(46f) Der Staatssekretär wurde entlassen.
The secretary of state got sacked .

Er war 1933 in der Partei gewesen.
He was in the party been

the relative index of the second sentence is shifted to the act-time arrived at in the first sentence, which is indeed the case:

Applying the appropriate rules, we first get some $\langle w,t_*\rangle$, such that

$$\langle w, t_{*}\rangle \varepsilon\, |\text{entlassen werden}|_{w_o t_o}^{t_o} \text{ (der Staatssekretär),}$$

and now the second sentence is true iff

$$\langle w, t\rangle \varepsilon\, |\text{war 1933 in der Partei gewesen}|_{w_o t_o}^{t_{*}} \text{ (he)}$$

which is true iff

$$\langle w, 1933\rangle \varepsilon\, |\text{Past(Perf(in der Partei sein))}|_{w_o t_o}^{t_{*}} \text{ (he).}$$

And this holds (see the tense rule) iff

$$\langle w, t'\rangle \varepsilon\, |\text{Perf(in der Partei sein)}|_{w_o t_o}^{t_{*}} \text{ (he)}$$

(where t' is the maximal part of 1933 which is before t_o).

And as a last step we get (perfect rule):

$$\langle w, t_j\rangle \varepsilon\, |\text{in der Partei sein}|_{w_o t_o}^{t_{*}} \text{ (he)}$$

(where t_j is the maximal part of t' which is before t_{*}).

Notice that our interpretation of (47) captures nicely the difference between

Seit 1955 beobachte ich ihn.
Since watch I him

and

Seit 1955 habe ich ihn beobachtet.
Since have I him watched

The first sentence is true at a non-past interval that began in 1955 and includes t_o, the second sentence is true at the past part of such an interval - t_o is now excluded.

And notice also that we can deal with sentences like

Gestern gab er zu, *gestern* in Konstanz gewesen zu sein,

because the matrix sentence is analysed as

$$\langle w, \text{yesterday}(t_o)\rangle \varepsilon\, |\text{Past}(\underbrace{\text{mindestens einmal}}_{\text{Adv}_f}(\text{zugeben} \ldots)|_{w_o t_o}^{t_o} \text{ (he)}$$

and therefore the *in Konstanz sein* is true of him at the maximal subinterval of *yesterday* which is before t_o (by the tense rule) and before that time during *yesterday* at which he admits the embedded proposition.

Although the system explained our examples satisfactorily, we are aware that it is not invariably the act-time of the immediately preceding sentence that the relative index is shifted to. Consider

(50) Der Sekretär wurde entlassen. Er war 1933 in der Partei gewesen.
The secretary got sacked . He was in the party been

Außerdem hatte er gelogen.
Furthermore had he lied

Clearly the relative index of the second and third sentence is the act-time of the first sentence. The reason seems to be that both sentences give independent reasons for the sacking of the secretary that is reported in the first sentence. Which is to say that for a solution of this problem one has to venture into the field of the structure of texts, which, obviously, is a matter of further research.

Bibliography

Bach, E., 1962, *The order of elements in a transformational grammar of German.* Language 38, p.263-69.

Bäuerle, R., 1977, *Tempus, Temporaladverb und die temporale Frage,* Papiere des SFB 99, Nr.13 and 16, Konstanz.

Bäuerle, R., *Temporale Deixis, temporale Frage,*Tübingen, Narr, to appear.

Bennett, M., 1977, *A guide to the logic of tense and aspect in English,* photocopied manuscript.

Bierwisch, M., 1973, *Grammatik des deutschen Verbs,* Studia Grammatica II, 8. Aufl., Berlin.

Bresnan, J., 1978, *A realistic transformational grammar,* in: Halle/Bresnan/ Miller (eds.), *Linguistic theory and psychological reality,* MIT Press.

Chomsky, N., 1965, *Aspects of the theory of syntax,* MIT Press.

Cooper, R., and Parsons, T., 1976, *Montague grammar, generative semantics, and interpretive semantics,* in: Partee, B. (ed), *Montague grammar,* New York-San Francisco-London, Academic Press.

Cresswell, M., 1973, *Logics and languages,* London, Methuen.

Drach, E., 1940, *Grundgedanken der deutschen Satzlehre,* 3.Auflage, Frankfurt.

Gabbay, D.M., 1976, *Investigations in modal and tense logics,* Dordrecht,Reidel.

Guenthner, F., 1979, *Time schemes, tense logic, and the analysis of English tenses,* in: Guenthner/Schmidt, S.J. (eds), *Formal semantics and pragmatics for natural languages,* Dordrecht, Reidel.

Hall-Partee, B., 1973, *Some structural analogies between tenses and pronouns in English,* The Journal of Philosophy 70, p.601-609.

Hall-Partee, B., 1977, *Montague grammar and the well-formedness constraint,* L.A.U.T. Nr.45.

Kaplan, D., 1977, *Demonstratives,* UCLA manuscript.

Kratzer, A., 1978, *Semantik der Rede,* Königstein, Scriptor.

Kuhn, S., 1979, *The pragmatics of tense,* Synthese 40, p.231-263.

Lewis, D., 1970, *General semantics,* Synthese 22, p.18-67.

Mc Coard, R.W., 1978, *The English perfect: tense-choice and pragmatic inferences,* Amsterdam, North Holland.

Montague, R., 1974, *The proper treatment of quantification in ordinary English,* in: R.H.Thomason (ed), *Formal Philosophy, Selected papers of Richard Montague,* Yale University Press.

Reinhart, T., 1976, *The syntactic domain of anaphora, doctoral thesis,* MIT.
Schpak-Dolt, N., 1977, *Zur Semantik von Tempus und Aspect des Russischen im Rahmen einer λ-kategorialen Sprache,* Diss., Konstanz.
Stechow, A. von, 1978, *Deutsche Wortstellung und Montague-Grammatik,* Papiere des SFB 99, Nr.25.
Thiersch, C.L., 1978, *Topics in German Syntax,* Diss., MIT.
Tichy, P., 1978, *The logic of temporal discourse,* photocopied manuscript.
Wurzel, W.U., 1970, *Studien zur deutschen Lautstruktur,* Studia Grammatica VIII, Berlin.